ファイバーシティ
縮小の時代の都市像

大野秀敏＋MPF
渡辺洋・波形理世（英訳）

Fibercity
A Vision for Cities in the Age of Shrinkage

OHNO Hidetoshi + MPF

English translation by WATANABE Hiroshi and
NAMIGATA Riyo

東京大学出版会
University of Tokyo Press, Komaba, Tokyo, Japan

Fibercity: A Vision for Cities in the Age of Shrinkage
Text and Illustrations by OHNO Hidetoshi + MPF
Translation by WATANABE Hiroshi and NAMIGATA Riyo

Copyright © 2016 OHNO Hidetoshi

All rights reserved. No part of this book may be reproduced in any form by any electronic or mechanical means, including photocopying, recording, electronical retrieval, without written permission.

First published August 2016
by University of Tokyo Press
4-5-29 Komaba, Meguro-ku, Tokyo 153-0041, Japan
Tel: +81 3 6407 1069 Fax: +81 3 6407 1991
E-mail: info@utp.or.jp   URL: http://www.utp.or.jp/

Printed in Japan
ISBN 978-4-13-066855-2

Cover design by YAHAGI Kijuro

謝辞

この研究が下記の諸賢のご示唆とご教示と支援なくしては生まれなかったことを記し、深く感謝したい。

本書は東京大学大野研究室の研究活動の成果である。研究室の助手（2007年以降は助教）を務めていただいた小嶋一浩、中村研一、本江正茂、鵜飼哲矢、岡部明子、日高仁、大島耕平の各氏の鋭い洞察力はこの研究の最大のエネルギー源であった。個々の研究については、AUSMIP制度で来日した欧州の学生諸氏を含めて大野研究室に在籍した歴代の学生諸氏の調査研究に多くを負っている。それゆえ、大野研究室のOB会であるMPF（Metropolis Forumの略記）を本書の共著者とした。

東京大学の故北沢猛、及川清昭、加藤孝明、横浜国立大学の佐土原聡の各氏には専門的見地から直接間接にご教示を得た。

デルフト工科大学のフランシーヌ・フーベン氏からは、2003年の第1回ロッテルダム建築ビエンナーレ展へ出展の招待を受け、本研究の進展の直接のきっかけとなった。2007年には、フィリップ・オズワルド氏と縮小をテーマとする展覧会を共同で企画した。2008年から3年間は、中村勉氏のお誘いで、環境省地球環境総合推進費による委託研究に参加でき、長岡市長の森民夫氏の支援によって長岡のケーススタディを進めることができた[1]。続く2012年から3年間は、藤井俊二氏の協力を得て都市の縮小問題を議論する場をもつことができた[2]。参加いただいた多くの方々から貴重な示唆を得た。

本研究には、地図情報が欠かせない。東京大学空間情報科学研究センターには4年のあいだ、貴重な情報利用の便宜を得た。

本書の編集から資料整理まで、東京大学大学院新領域創成科学研究科社会文化環境学専攻の大島耕平氏と松宮綾子氏、岩井茂氏に格別な支援をいただいた。工学系研究科の山崎由美子氏には研究から出版に至るプロジェクトの管理で尽力いただいた。また、建築計画学の大月敏雄氏には草稿段階で目を通していただき貴重な助言をいただいた。

図版の表現の検討と作図には、MPFの梅岡恒治氏と久保秀朗氏に全面的な協力を得た。また、表紙と頁の基本構成はグラフィックデザイナーの矢萩喜從郎氏のデザインである。英訳は、建築分野の翻訳家として高い評価を得ている渡辺洋氏に本文をお願いした。註釈と図版説明は、MPFの波形理世氏に担当していただいた。

研究の遂行から発表のための展覧会には大学の通常の研究室予算では賄いきれない費用が必要である。下記の方々から経済的支援をいただいた。年代順に記して謝意を表したい。国際交流基金、ユニオン造形文化財団、東日本旅客鉄道株式会社、東京ガス株式会社、NTT都市開発株式会社、ダイビル株式会社、鹿島建設株式会社、秋葉原クロスフィールド、株式会社YAMAGIWA、東京大学21世紀COE「都市空間の持続再生学の創出」、東京大学大学院新領域創成科学研究科環境学研究

Acknowledgments

I would like to express my deep appreciation to everyone listed below whose guidance and support made the study possible.

This book is the fruit of research undertaken by the Ohno Laboratory, the University of Tokyo. The keen insights of successive laboratory assistants (who have been referred to in English since 2007 as "assistant professors"), Kojima Kazuhiro, Nakamura Ken'ichi, Motoe Masashige, Ukai Tetsuya, Okabe Akiko, Hidaka Jin and Ōshima Kōhei, sustained that research. With respect to individual works of research, I am greatly indebted to the surveys and studies of students who have been enrolled over the years in the Ohno Laboratory, the University of Tokyo, including students from the EU who came to Japan under the AUSMIP program. For that reason, Metropolis Forum (MPF), the organization of former members of the Ohno Laboratory, is cited as a co-author of this book. I benefited from guidance provided, both directly and indirectly, by the late Kitazawa Takeru, Oikawa Kiyoaki and Katō Takaaki of the University of Tokyo, and Sadohara Satoru of Yokohama National University.

In 2007, an exhibition on the theme of shrinkage was planned in collaboration with Philipp Oswalt. The discussions I have had with Mr. Oswalt, who was alert from an early stage to the problem of shrinkage caused by the rapid decline in population in the former East Germany after reunification, have been for me extremely rewarding intellectually. In 2008-2011, I was able to participate in a commissioned study for the Environment Research and Technology Development Fund, Ministry of the Environment, at the invitation of Nakamura Ben, and to undertake a case study of Nagaoka with the support of Mori Tamio, mayor of Nagaoka City.[1] In 2012-15, I had a venue for discussing urban issues of shrinkage,[2] thanks to the cooperation of Fujii Shunji. I gained valuable suggestions from the many individuals who participated in those discussions.

Detailed maps and information were indispensable in carrying out this research. Since 2011, the Center for Spatial Information Science, the University of Tokyo, has given us access to valuable information. Sekisui House, Ltd., provided support including financial support for research on the nature of the suburbs. Ōshima Kōhei, Matsumiya Ayako, Wada Natsuko, Iwai Shigeru and Moriya Kayo of the Department of Socio-Cultural Environmental Studies, Graduate School of Frontier Sciences, the University of Tokyo, were especially supportive in matters ranging from the editing of the text to the sorting of data. The project, from research to publication, was successfully managed thanks to the kind efforts of Yamazaki Yumiko of the School of Engineering, the University of Tokyo. Ōtsuki Toshio of the Department of Architecture, the University of Tokyo, read my manuscript and gave me invaluable advice.

I received full cooperation from Umeoka Kōji and Kubo Hideaki of MPF in considering terminology for illustrations and preparing drawings. The cover and basic composition

系、一般財団法人住総研、公益財団法人LIXIL住生活財団、積水ハウス株式会社、株式会社総合資格である。

　2006年に発行された「ファイバーシティー／東京2050」が海外の専門家からも大きな関心をいただいたので、本書も日英併記の希望を伝えたところ、東京大学出版会には英断をいただいた。発刊にあたっては、東京大学出版会の岸純青氏に多大な尽力をいただいた。

　著者は、槇文彦氏が東京大学で教授であった時代に助手を務めた。槇氏退職直前の研究プロジェクトのなかで、東京の空間特性を表現するキーワードとして発見された概念が「線分性」であった。西欧や中国の都市には都市空間全体を支配する無限に伸びる直線が認められるのに対して、東京では断片的で短い線が卓越し、それらが東京の場所の性格を支配しているという発見である。この用語が一つの啓示として著者の心に宿った。ファイバーシティは、この分析概念を計画概念に発展させたとも言える。本書が、槇氏の滞米時代の名著『グループ・フォーム』の思想をいささかでも引き継ぎ、それを発展させると同時に、建築家の社会に対する責務を果たす論考になればと大それたことを願っている。

　本書を槇文彦氏に献ずる。

大野秀敏　2016年　初夏

of the pages were designed by the graphic designer Yahagi Kijūrō. Watanabe Hiroshi, whose translations in the field of architecture are highly regarded, translated the main text into English. Namigata Riyo of MPF was in charge of translating the notes and the captions for illustrations.

　Carrying out studies and organizing exhibits for presenting the results of those studies require funding that is beyond the means of an ordinary university research budget. I received financial support from the following organizations, listed in chronological order, and to them I would like to express my gratitude: Japan Foundation; Union Foundation for Ergodesign Culture; East Japan Railway Company; Tokyo Gas Co., Ltd.; NTT Urban Development Builservice Co.; DAIBIRU Corporation; Kajima Corporation; Akihabara Crossfield; Tokyo Gas Co., Ltd.; YAMAGIWA Corporation; Center for Sustainable Urban Regeneration, 21COE, The University of Tokyo; Institute of Environmental Studies, Graduate School of Frontier Sciences, The University of Tokyo; Housing Research Foundation JUSOKEN; LIXIL JS Foundation; Sekisui House, Ltd.; and SOGO SHIKAKU Co., Ltd.

　Experts overseas showed great interest in *Fibercity/Tokyo 2050*, published in 2006. That led me to hope that this book too might be published in a bilingual (Japanese and English) edition. University of Tokyo Press generously acceded to my request. I am grateful to Kishi Junsei of University of Tokyo Press for his exceptional efforts in connection with the publication of this book.

　The author was an assistant to Maki Fumihiko when he was professor at the University of Tokyo, and "segmentedness," a concept used to express a spatial characteristic of Tokyo, was discovered in a research project in the Maki Laboratory just before his retirement. Whereas straight, infinitely extended lines that exercise control over the entire urban space are found in the cities of the West and China, short, fragmentary lines are conspicuous in Tokyo and determine the character of place in the Japanese capital. This concept was to me a revelation and has stayed with me since. Fibercity can be said to be a development of that analytical concept into a planning concept. It is my hope that this book will carry on and elaborate, however modestly, the philosophy of "group form" developed in Maki's famous essay written during his period in the United States and enable me to fulfill my responsibility to society as an architect.

　I dedicate this book to Maki Fumihiko.

Ohno Hidetoshi　Summer 2016

目次		Table of Contents
謝辞		Acknowledgments
序	1	Preface
第一部:観察と分析	7	Part 1. Observation and Analysis
第一章:深い危機	8	Chapter 1. A Profound Crisis
1. 長く続く縮小		1. Long-Term Shrinkage
2. 縮小の時代の都市		2. Cities in an Age of Shrinkage
第二章:モダンの都市、ポストモダンの都市	17	Chapter 2. Modern Cities, Postmodern Cities
1. モダンの都市		1. Modern Cities
2. ポストモダン都市		2. Postmodern Cities
第三章:21世紀の都市のための10箇条	35	Chapter 3. Ten Articles for the Development of Twenty-First Century Cities
第四章:日本の都市の診断書	41	Chapter 4. A Diagnosis of Japanese Cities
1. 隙間に息づく自然		1. Nature in Gaps
2. 線形性好み		2. A Predilection for Linearity
3. 人工物と自然の混成系		3. A Composite System of Man-Made Objects and Nature
4. 芝生・スポーツ・ショッピング・アメリカ		4. Lawns, Sports, Shopping, the United States
5. 土地神話の崩壊		5. The Collapse of the Myth of Land
6. 新築依存症		6. A Dependence on New Construction
7. 日本橋と二条城		7. Nihonbashi and Nijōjō
8. ハコモノ		8. White Elephants
9. 孤立と互恵性		9. Isolation and Reciprocity
10. お一人様支援技術		10. Party-of-One Support Technologies
11. 遠く・速く・大量に		11. Further, Faster and on a Larger Scale
12. 自家用車過依存		12. An Overdependence on Automobiles
13. 狭い道路		13. Narrow Streets
14. 立派な交通基盤と貧弱な連携		14. Excellent Transportation Infrastructure and Poor Coordination
15. 水上交通は都市の宝石		15. Water Transportation Ought to Be Prized by Cities
16. 地方都市のダイナミズム		16. The Dynamism of Local Cities
17. 都市の住宅と家族		17. Houses and Families in Cities
18. 死者の眠る場所		18. Where the Dead Sleep
19. 多島海化する日本		19. An Archipelago-like Structure
20. コンパクトシティは目標足りうるか		20. Can Compact City Achieve Its Objectives?

第二部：理論とデザイン	77	Part 2. Theory and Design
第一章：流れと場所の計画論	78	Chapter 1. The Planning of Flow and Place
1. 流れも場所も		1. Both Flow and Place
2. 見取り図		2. A Sketch
3. ファイバーシティは何をめざすのか		3. What Are the Objectives of Fibercity?
第二章：デザインプロジェクト	104	Chapter 2. Design Projects
汀		Shore
壁をなす／リアス（新宿御苑）		Folding / Rias (Shinjuku Gyoen)
めぐらせる／暖かい巡回		Circulating / Orange Rounds
川		River
置き換える／緑の網		Replacing / Green Web
数珠つなぎにする／緑の間仕切り		Stringing Together / Green Partitions
組み合わせる／暖かい網		Combining / Orange Web
運河		Canal
結びつける／青い首飾り-品川		Connecting / Blue Necklace—Shinagawa
結びつける／青い首飾り-秋葉原		Connecting / Blue Necklace—Akihabara
乱流		Turbulence
縁飾りを付ける／緑の花輪		Edging / Green Garland
囲う／生命の回廊		Enclosing / Cloister of Life
庭		Garden
灌漑／暖かい食卓		Irrigation / Orange Tables
小さい庭／緑の指+パッチワーク		Small Gardens / Green Fingers + Patchwork
結語	152	Conclusion:
「重建設主義」と「大きい流れ」に打ち勝つために		Overcoming "Constructionalism" and "Big Flow"
註	155	Notes
図版出典	171	Illustration Credits
ファイバーシティに関する著者による公表物	177	Author's Publications on Fibercity.

ファイバーシティはわれわれの造語であり、それは具体的な都市ではない。ファイバーシティは、成長の時代の後に続く縮小の時代を都市が乗り切り、それを実り豊かな時代とするための都市計画理論であり同時に具体的な提案（プロジェクト）で示す都市戦略群である。ファイバーシティの形態的特徴は、都市の線状要素に注目し、それを操作することで都市の流れと場所を制御しようとすることであり、理論的特徴は、従来の都市デザインが基礎をおいてきた場所論に加えて流れの視点を導入することにある。

本書では、私たちは、東京首都圏と日本海に近い長岡市を対象とする。規模が違う二つの都市を取り上げることで、未来の都市のあり方に対して広く示唆ができればと考えている。

本書で論じることは、建築設計分野と都市計画分野の両方に関わる。この二つの専門分野は本来連続的でなければならないのだが、外部の人々が想像する以上に、現代の建築家と都市計画家のあいだで関心事と方法論は離れている。この専門分野の乖離が、現代都市の問題が適切に扱われていない原因

序

Preface

Fibercity is a neologism and does not refer to any specific city. Fibercity is both a city planning theory and a collection of urban strategies illustrated by actual projects that are intended to help cities not only survive the age of shrinkage that has followed the age of growth but thrive. Fibercity is distinguished morphologically by a focus on linear elements in cities and an attempt to control flow and place in cities through the manipulation of those elements and theoretically by the introduction of the idea of flow into urban design, which has hitherto been based only on the idea of place.

In this book we target the Tokyo Capital Region and Nagaoka, a city close to the Sea of Japan. By focusing on these two cities that are quite different in size, we hope to suggest the way future cities in not only Japan but developed countries in general might be organized.

The subject of this book is related to both the field of architectural design and the field of city planning. These two specialized fields ought to be in essence continuous, but contemporary architects and city planners have very different concerns and methodologies. Those differences

になっている。われわれは、建築家の感性と能力を最大限活かして、縮小時の時代における都市空間の再組織化の方法を探求してみようと思う。一方、都市計画理論における「流れ」と「場所」の乖離もまた大きな問題である。建築にも都市にも、交通や水やエネルギーや情報などさまざまな「流れ」があり、重要な計画対象なのだが、都市の骨格を決め構築物を計画する専門家は「場所」に関心を示し、「流れ」は多くの場合、技術的分野の対象に閉じ込められてきた。ところが、21世紀になり、人々の住み方や働き方にこれまで以上に「流れ」の影響力が強くなっている。都市空間の再組織化の方法論のなかで、「流れ」と「場所」の統合を試みてみようと思う。

丹下健三氏の「東京計画1960」(→ 001)[3]を含めて、1960年代の前半には世界中で、多くの建築家や都市計画家が競って未来都市の姿を描いた。1960年代の後半になると、都市の未来に対する悲観が大きくなり異議申し立ての政治的熱風が巻き起こったが、それもオイルショックによって沈静化し、その後は、日本全体が経済に狂奔し、建築家たちも遠い将来より明日の仕事にかまけていた。しかし、世紀が変わるころから雲行きが変わり始めた。再び都市のビジョンが求められていると著者は感じた。ただし、1960年代前半に比べると、状況ははるかに複雑かつ悲観的になっている。かつては、誰もが、成長と発展という目標を共通にかつ楽観的に信じることができたのだが、環境、資源、人口、消費いずれにおいても、少なくとも先進諸国においては、縮小が不可避な状況にあることがいよいよ誰の目にも明らかになりつつあるからである(→ 002)。福島の原子力発電所事故は、このような観測を決定的にした。先進諸国では綱渡り的な経済運営にすがり、結果として社会の歪みも拡大している。多くの政策は明らかに未来の先食いである。日本では、人口でも経済でも大都市だけが強大になり、多くの地方中小都市の衰退が止まらない。その打開策として唱えられているコンパクトシティ政策は実現性に展望がなく、これによって地方中小都市の活力が増し、住みやすくなるとは到底期待できない。いま、はっきりしていることは、楽天的な60年代の夢を再現することではなく、新たな思想と方法論で都市を計画、運営しなければならないということである。

001 丹下健三の「東京計画1960」(1961)
Tange Kenzō's "Tokyo Plan 1960" (1961)

丹下健三は、時代や人々の夢を建築化するうえで特別の才能を発揮した建築家である。東京湾の洋上に交通路を延ばし、その沿道に建築施設を配する、この提案も高度経済成長期の日本の活力を讃えている。

Tange Kenzō was an architect skilled at capturing and expressing the age and people's dreams in architecture. This proposal, in which a traffic route with buildings arranged along it stretches across Tokyo Bay, is a paean to the vitality of Japan during the period of high economic growth.

are much greater than outsiders imagine. That estrangement is the reason contemporary urban problems are not dealt with properly. We intend to use to the fullest extent our sensibilities and abilities as architects in order to search for ways to reorganize urban spaces for the age of shrinkage. Meanwhile, the split between *flow* and *place* in city planning theory is also a major problem. Diverse flows such as those of transportation, water, energy and information exist in both buildings and cities and are important subjects of planning, but experts who determine the framework of cities and plan structures are concerned with place and consign flow in many instances to fields of engineering. However, the influence of flow on the way people live and work has become much greater in the twenty-first century. We wish to attempt to integrate flow and place in the methodology for the reorganization of urban spaces.

In the 1960s many architects and city planners, including Tange Kenzō with his "Tokyo Plan 1960,"[3] (→ 001) vied with one another to depict the city of the future. In the late 1960s, pessimism about the future of cities grew and firestorms of political protest occurred. However, those too were extinguished by the oil crisis of 1973. Japan as a whole was caught up in efforts to bolster the economy, and architects became busy with actual practice—that is, work having to do more with tomorrow than with a distant future. However, the author feels that the situation has begun to change since the start of the new century and that people are once more in search of an urban vision. However, conditions today are far more complex and the outlook far more pessimistic than they were in the 1960s. In the past people had a shared, optimistic belief in growth and development, but as will be explained in detail in the following pages, it is at last becoming clear to everyone that shrinkage, as far as the environment, resources, population and consumption are concerned, is becoming unavoidable, at least in the developed countries of the world (→ 002). This pessimistic view was borne out by the Fukushima nuclear power plant disaster. The developed countries cling to risky economic policies, and as a result, social distortions are becoming greater. Many policies clearly involve the deferment of payment for what we spend now. In Japan, the population and the economy are

新しい都市計画の理論と方法を求める私たちの研究は、1999年に大野研究室が、東京大学工学系研究科建築学専攻から新設された新領域創成科学研究科社会文化環境学専攻に移籍したときに開始された。研究の成果の最初の公表は、2005年に開かれた「サステーナブルビルディング国際会議・東京」である（fc 25JE：巻末ファイバーシティに関する著者による公表物一覧参照）。翌年の2006年には、これを発展させて「Japan Architect 63号」に発表した（fc 25JE）（→ 003）。翌年には、「Shrinking City & Fibercity @ Akihabara」と銘打つ展覧会とシンポジウムを東京で開催し（fc 48JE）、同時に『シュリンキング・ニッポン』（fc 16J）として単行本を刊行した。2011年には、ファイバーシティの一つの戦略である「緑の網」を補強して『東京2050//12の都市ビジョン展』で展示した（fc 42JE）。2008年度から2010年度までの3年間は、地方中都市の代表として長岡市を対象として取り上げ、調査と提案を行った（fc 41J, fc 104-97JE）。これらの一連の都市空間の再組織化に関わる提案（プロジェクト）を洗練させ、統合して本書に示す。

2005年の会議での発表をファイバーシティVersion 1.0とすると、本書はVersion 3.0となる。一方、縮小する時代には、拡大成長する時代の計画理論とは異なる都市計画理論が求められる。その核となるのが「流れ」と「場所」の統合だろうと考えている[4]。

ところで、計画の目標年次として2050年を設定しているのだが、その理由は著者が責任を取れる一番遠い時期という意図からである。もし、これを2100年とすれば、われわれ提案者も、そしてこの本の読者のほとんども立ち会うことができない。それゆえ自由に発想できるが、無責任にもなる。一方、2020年となると近すぎて、大胆な提案ができない。2050年とはそういう時間である。それゆえ、本書で提案した解決策は、社会の根本的な変革を構想する人からすれば、妥協的な産物に見えるかもしれない。しかし、現在の日本の都市政策は「都市再生」を叫んで、相変わらず建設投資によって経済成長を促すという、日本がまだ発展途上の時代にあった頃に抱いた成長モデルに囚われ、時間を巻き戻そうとしている。まずはそこから抜け出し、前

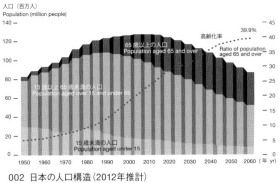

002 日本の人口構造（2012年推計）
Demographic structure of Japan (estimate in 2012)

growing only in metropolises; decline continues in many local cities. The compact city policy that is being advocated as a solution has little chance of realization; it cannot be expected to invigorate small and medium-sized local cities and make them more livable. What we need to do today is not to recreate the optimistic visions of the 1960s but to employ new ideas and methodologies in the planning and administering of cities.

Our study of new city planning theories and methods began when the Ohno Laboratory moved from the Department of Architecture in the School of Engineering to Environmental Studies in the newly-created Graduate School of Frontier Sciences in 1999. The first presentation of the fruits of this study was "Sustainable Building International Conference: Tokyo" held in 2005. In 2006, a developed form of that report was published in Volume 63 of *Japan Architect* (fc 25JE: See publications by the author on Fibercity listed at the end of this book) (→ 003). The following year, an exhibition and symposium entitled "Shrinking City & Fibercity@Akihabara" were held in Tokyo, and subsequently published in book form as *Shrinking Nippon* (2008). In 2011, the idea of "Green Web" was further elaborated and presented at "Tokyo 2050//Twelve Urban Visions Exhibition"[7] (fc 48JE). In a three-year period from 2008 to 2010, we targeted Nagaoka, a representative local city of medium size, and undertook a survey and a proposal (fc 40J, fc 104-97JE). These projects involving the actual reorganization of urban spaces have been refined and integrated in this book. If the presentation made at the 2005 conference is Version 1.0, this is Version 3.0 (00013). Meanwhile, an age of shrinkage demands a city planning theory different from that of the age of expansion and growth. We believe that the integration of *flow* and *place* will be the nucleus of such a theory.[4]

2050 was made the target year for the plan, not for technical reasons, but because if it were set any further into the future we would no longer be held accountable for our ideas. If it were set in, say, 2100, we who are making these proposals and most of the readers of this book would no longer be around to see what had become of the world. We would be able to give our imaginations free rein but no longer be held responsible. Meanwhile, 2020 would be too

向きになることが何より急務である。私たちが主張するように21世紀を「縮小の時代」と規定することが妥当ならば、日本は間違いなく課題先進国である。先鞭を切って、この困難な課題に取り組むべきである。解決を見いだした地域は21世紀文明の先頭に立つことになるだろう。19世紀の半ば、明治維新の時代に日本人が示した叡智と勇気を再度奮う時期だと著者は思っている。

本書は二部で構成した。第一部では、ファイバーシティの理論と具体的提案に関わる社会的背景を説明する。第一章から第三章までは、都市を巡る状況を分析し、続いて20世紀の都市を歴史的に概観し、最後に現代の先進諸国の都市が向かうべき方向を提案する。第四章は、ファバーシティの提案の対象として選んだ東京首都圏と越後長岡の都市、社会、文化の分析である。これは二つの都市の体質と病歴の分析とたとえられる。第二部の第一章では、ファイバーシティの理論的枠組みを「流れと場所」という観点から示す。第二部第二章では東京あるいは首都圏と長岡を事例とした提案（プロジェクト）である。第二部から見ると、第一部の第四章は、これらの提案の註釈ということでもある。

なお、念のために断れば、これらの提案はいずれも自治体などからの依頼によるものではなく、われわれの自主的な提案である。

本書の読者は、都市計画や建築設計を学ぶ学生や実務者が多いと思われるが、自分の町の行く末や、この国の将来の姿に関心をもつ人々にも広く読んでいただきたいと考え、門外の方々にも親しめるようにつとめた。本書は、また、日本の現代建築や現代都市の形成に社会や経済、そして歴史がどのような役割をしたかという基礎的な知識を、日本人の若い読者のみならず、英語で理解できる読者にも提供するという役割も果たしたいと考えた。

Version 1.0 (2005)	『サステーナブルビルディング国際会議・東京』	
		"The 2005 World Sustainable Building Conference"
Version 1.1 (2006)	『Japan Architect 63 号』	"The Japan Architect 63"
Version 1.2 (2011)	『東京2050//12 の都市ビジョン展』	
		"Tokyo 2050//Twelve Urban Visions Exhibition"
Version 2.0 (2011)	『地方都市の素晴らしさを実感できる魅力的な住環境の提案』	
		"Proposals for attractive living environments designed to experience the splendor of the regional city"
Version 3.0 (2016)	『ファイバーシティ ------- 縮小の時代の都市像』	
		"Fibercity: A Vision for Cities in the Age of Shrinkage"

003 ファイバーシティの発展
Development of Fibercity

soon to allow us to make bold proposals. Hence the selection of 2050. For that reason, the solutions proposed in this book may appear to those visualizing a fundamental reform of society products of compromise. However, urban policy in contemporary Japan, in calling for "urban regeneration," still remains captive to the model of growth embraced in the period of development—the notion of promoting economic growth through investment in construction—and seeks to turn back the clock. Freeing ourselves from that belief and taking a forward-looking approach is our first priority. If the twenty-first century is indeed an age of shrinkage, as we assert, then Japan is without question a developed country in the sense that the issues that will need to be addressed are at an advanced stage here. We ought to get a head start and come to grips with these difficult issues. The region that discovers solutions will no doubt be on the leading edge of twenty-first century civilization. The author is convinced that the time has come for us to summon the wisdom and courage the Japanese displayed in the mid-nineteenth century at the time of the Meiji Restoration.

This book is divided into two parts. Part 1 explains the social context for the theory of Fibercity and the specific projects suggested in Part 2. In Chapters 1 through 3, conditions in cities are analyzed, a historical survey of the twentieth-century city is provided, and finally the direction that cities in developed countries today need to take is proposed. Chapter 4 is an analysis of the cities selected for Fibercity projects—the Tokyo Capital Region and Nagaoka—and their society and culture. It can be said to be an analysis of the nature and clinical history of those two cities. Chapter 1 of Part 2 indicates the theoretical framework of Fibercity from the perspective of *flow* and *place*. Chapter 2 is given over to projects that are case studies of Tokyo or the Capital Region and Nagaoka. Part 1, Chapter 4 can be regarded as annotations for Part 2.

It should be kept in mind that these projects were not commissioned by local governments but are our own independent proposals.

Many of the readers of this book are likely to be students or practitioners of city planning or architectural design, but, wishing to have it read by all those concerned with the future of their own communities and the future appearance of this country, I have tried to make it accessible to non-professionals as well. I also wanted to provide, not only young Japanese readers but readers with an understanding of English, basic knowledge on the role played by society, the economy and history on the formation of contemporary architecture and cities of Japan.

Note to readers: In this volume the names of Japanese individuals are all given in Japanese order: family name first, then given name.

第一部：観察と分析

Part 1. Observation and Analysis

第一章：深い危機

1. 長く続く縮小

人類がアフリカ大陸の東側で誕生して以来、一時的には、疫病や長期の戦乱で人口が減ることも文明の後退や長い停滞もあったが、大局的に見れば、量的にも質的にも概ね成長、拡大してきた。特に近代になると、穀物生産高、工業生産高、移動する速さや物を運べる量、通信の速度と情報量、使用エネルギー、人間の寿命、人口など何をとっても成長・拡大基調であった（→101）（→102）。人類は、技術を磨いて未開地を切り拓き活動空間を広げ続けてきた。

世界の安定と繁栄が期待される裏では、深い危機が静かに進行している。それは、環境問題と人口構造の変化と経済システムの行き詰まりなどが主因で相互に関連している。それらの問題群の先にあるのが長期的な縮小である。これによって、産業革命が近代都市を準備したと同じように、都市と建築は再び大きい衝撃を受けるだろう。そして、これらの問題群を解決しようとしても、拡大と成長を前提条件に組み立てられた近代都市計画理論では到底対応できない。今後成長が望めないので

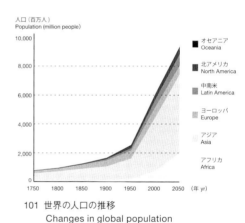

101 世界の人口の推移
Changes in global population

Chapter 1. A Profound Crisis

1. Long-Term Shrinkage

Humans have generally thrived since their evolution in East Africa although epidemics and extended periods of warfare have at times caused the population to decline and civilization to falter and stagnate. In the modern era in particular, everything basically increased, from grain yield, industrial output, speed of travel, capacity of transport, speed of communication, volume of information and energy utilization to longevity and size of population. (→101) (→102) Humans have continued to refine technology, open up undeveloped land and expand their sphere of activity.

However, even as hopes for the stability and prosperity of the world soared, a profound crisis has been quietly coming to a head. A combination of factors including environmental problems, changes in the population structure and the dead end at which the present economic system finds itself makes long-term shrinkage inevitable. The Industrial Revolution had an enormous impact on the world and paved the way for the modern city; the effect of shrinkage will be no less profound. Premised on expansion and growth, modern city planning theory is simply not up to

あれば、縮小、そしてその先の定常状態[5]を前提に再度都市計画の理論を組み立て直さなければならない。

既に、多くの議論が積み重ねられていることなので、読者には耳新しいことではないかもしれないが、縮小の理由をまとめてみよう。

環境的制約

環境問題は、地球上での人間の活動量には限界があるということを教えてくれた。人口が増え、一人一人が恵まれた生活を求めるようになると、いろいろな限界が露呈してくる。一つは、資源の有限性である。多くの資源は、再生するのに長い時間がかかる。森林などは、比較的早く再生するが、石油、天然ガス、石炭など植物枯死体からできる化石燃料の再生には気の遠くなる年月が必要である。食物生産と生物の生存に不可欠な水の利用可能量にも限界がある。もう一つの有限性は、人間活動に由来する汚染を浄化する能力の有限性である。たとえば、微生物による有機物の分解や植物の窒素固定などがある。

2015年現在の世界人口は73億人であり、今後さらに増え続け、2030年に84億人から86億人、2100年には95億人から133億人に達すると予想され[6]、事態はますます厳しくなる。人間活動を地球の環境容量以下にしない限りはこの世界は持続可能にはならないのだが、現実には、既に人類の活動量は、地球の環境容量を50%弱超過しているという[7]。これは、現在地球上に生きる人間が、将来の世代が使うべき資源を「収奪」しているということである。

国連の環境計画によれば、産業革命前からの気温上昇を摂氏2度以内とするためには、二酸化炭素の排出を地球全体で2030年までに2010年比で10〜39%の削減、2050年までに49〜63%の削減が必要である（→103）。省エネ技術、低炭素技術は日進月歩であるが、現在のように経済成長を最優先にする社会が続く限り、目標数値の実現すら楽観を許さない。生産と消費の抑制がどうしても必要であり、そのためには価値観そのものを改めなければならない。

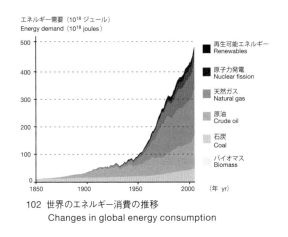

102 世界のエネルギー消費の推移
Changes in global energy consumption

the task of solving the problems we face. If it is unrealistic to hope for growth in the future, then a city planning theory premised on shrinkage followed by a "steady state"[5] must be formulated.

Although the reader may be familiar with the reasons for shrinkage since they have been the subject of much public discussion, I would like to summarize them here.

Environmental Constraints

Environmental problems have taught us there are limits to the level of human activity on earth. The increase in population and the growing demand for a better life for every individual have exposed diverse limits. One is the finite amount of available resources. Many resources require a long period of regeneration. Forests are restored relatively quickly, but fossil fuels created from the remains of dead vegetation such as petroleum, natural gas and coal require eons for regeneration. The amount of water, a resource indispensable to food production and life, that is available for use is also limited. Still another limit is the capacity to clean up the pollution originating in human activity, for example, the capacity of microorganisms to break down organic matter or plant life to contribute to nitrogen fixation. The world's population is 7.3 billion in 2015 and expected to continue to grow, reaching 8.4 billion in 2030, and between 9.5 billion and 13.3 billion in 2100.[6] The situation will become even more severe. This world is unsustainable unless human activity is reduced below the environmental capacity of the earth, but in fact the volume of human activity already exceeds the earth's environmental capacity by nearly 50 percent.[7] This means that humans living today are "exploiting" future generations, that is, dispossessing them of resources they might use.

According to the United Nations Environment Program, limiting the rise in earth's temperature to two degrees C. or less relative to the pre-industrial era will require a reduction in CO_2 emission of 10~39 percent by 2030 relative to 2010 by the entire world, and a reduction of 49~63 percent by 2050 (→103). Advances are constantly being made in energy conservation and low-carbon technologies, but as long as society places priority on economic growth, the achievement of any target figures set is uncertain.

低い出生率と延びる寿命

20世紀には環境問題が進行すると同時に人口動態も大きく変化した。世界的にみれば、人口減少する豊かな地域と人口爆発を続ける貧しい地域が同時に存在しているが、全体としてみると、増加地域の勢いが勝っている。ただし、もう一つの傾向で言えば、経済的に豊かになると出生率は下がるので、先進諸国は一般的に出生率が低く、現在多産な地域も豊かになれば少子化すると考えられる。出生率が人口置換水準を割り込めば、その地域の人口は減り始める。しかし、先進国のなかでもアメリカと北欧諸国の出生率は高く人口置換水準に近い。アメリカは労働市場の流動性が高く産後の社会復帰がしやすいことが挙げられ、北欧諸国は育児に対する公的補助が手厚い。一方出生率の低い国は、男は外、女は家庭といった因習的な観念が残っている地域や労働市場の流動性が低い地域である（→104）。

日本は出生率の低いグループに属する。2005年に合計特殊出生率1.26の最低値を記録し、その後2013年には1.4程度まで盛り返したとはいうものの、既に母親になる世代の人口が少ないので、長期的な人口減少が始まっている。今後の日本の総人口に関しては、日本政府の機関である国立社会保障・人口問題研究所は、出生率が1.12から1.60の仮定幅で、2050年に9200万人から1億300万人、2100年には、3100万人から5900万人まで減ると予測している。

平均寿命の方は、20世紀、特に第二次世界大戦後に保健思想の普及、栄養の改善、医療技術の進歩と医療体制の整備、労働環境の改善などにより、世界的に延びた（→105）。なかでも、日本は現在世界一の長寿国（2014年で男女平均83.1歳）である。長寿は人間の夢の一つであるが、それを受け入れる社会制度や物的環境が追いついていない。日本の団塊世代の高齢化は大規模な超高齢社会の運営実験を最初に提供することになるだろう。われわれ居住空間の専門家にとっての最大の課題は、高齢者が多数派になる社会の都市や建築の姿を思い描くことである。それに失敗すると人類は真に幸せになれない。

飽和する欲望

高度経済成長の時代の消費は、日本の家庭生活を一変させ

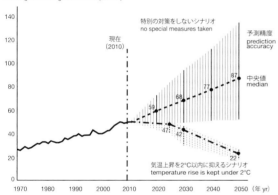

103 二酸化炭素排出量と必要な排出削減
CO₂ emissions and necessary reductions

産業革命前より、2100年の気温上昇を摂氏2度以内に抑えるためには、2010年以降に排出できる二酸化炭素量は1000Gtである。そのためには、2030年までに2010年比で10〜39％の削減、その後2050年までは更に49〜63％の削減が必要である。

In order to keep the rise in temperature from the pre-industrial era to the year 2100 below 2 ℃, the amount of carbon dioxide that can be emitted after 2010 is 1,000Gt. For this, a reduction of 10-39% relative to 2010 is necessary by 2030, and a further 49-63% reduction is necessary by 2050.

Production and consumption must be controlled, and for that to happen, our value system must change.

Low Birthrate and Increasing Longevity

As environmental problems worsened in the twentieth century, major changes took place in demographics as well. There was a population explosion in the poorer regions of the world; meanwhile, affluent regions experienced a decline in population but not such a decline as to offset the increase elsewhere. However, inasmuch as the trend has been for the birthrate to decline as a society becomes more prosperous—birthrates in developed countries are generally low—it is believed that the high current birthrates in poorer regions will decline once those regions become richer. Once the birthrate dips below the level necessary to maintain the population in a region, that population begins to decline. However, among developed countries, birthrates in the United States and Scandinavian nations are high and approach the levels required to maintain their populations. The reasons cited for this are, in the case of the United States, the mobility of the labor market and the ease with which women can reenter the workforce after giving birth, and in Scandinavia, the generous public assistance provided for child care. Meanwhile, low birthrates are found in countries where conservative notions concerning the participation of women in the workforce survive or where mobility in the labor market is limited (→104).

Japan belongs to the group of countries with low birthrates. In 2005, it recorded its lowest total fertility rate of 1.26. Although this rebounded somewhat to approximately 1.4 in 2013, the current generation of potential child-bearers is already small in size; a long-term decline in population has begun. The National Institute of Population and Social Security Research (IPSS), an agency of the Japanese Government, projects that the total population will decline to between 92 million and 103 million by 2050 and to somewhere between 31 million and 59 million by 2100, assuming the birthrate is between 1.12 and 1.60.

Average longevity increased in the twentieth century especially after World War II as the result of factors such as improvements in public health, nutrition and the work environment, advances in medical technology and

た。洗濯機・冷蔵庫は家事労働を大きく変え[8]、主婦に自由時間をもたらした。自動車は人々の行動範囲を広げ郊外開発を推しすすめ、山間の集落にも文明の恵みを届けるのに大きな力となった。やがて必要なものが一通り揃ってしまうと消費意欲は減退しはじめるが、企業としては、それでは困る。本来はさほど必要ではないものを欲しがらせなければならない。それを担うのが宣伝広告である。商品の優れた点を訴えるのは、もはや古くさい。人気のスポーツ選手と契約して自社商品を独占的に使用させる、商品とは無縁な環境活動に資金提供するなど、硬軟とりまぜあらゆる手段を駆使して、消費者に向かってわれわれの商品を使えば、もっと素晴らしい人生が待っていますよと説得し続ける[9]。相当用心深い消費者でも、こうした環境に曝され続けていると値段が張るブランド品に手が伸びてしまう。宣伝費とは、消費者が欲しくないものを欲しがらせるための費用であり、それを当の消費者が自分で支払っているのが現代社会の姿である。

欲望は社会が作りだすものであり、満ち足りた地域では欲望を作りだすコストが高くなっている。しかも、消費が幸福に結びつかなくなっている。もっともわかりやすい例が肥満である。貧しい時代には、栄養価の高い食品の摂取によって病気に対する耐性がつき長寿に繋がった。ところが、豊かな時代になると、自制しなければ栄養摂取は過度になり肥満に繋がる。見た目だけではなく健康を害して生活の質を落とす。

消費に飽和感が忍び寄っているとしても、企業は作り続けなければならない。経済成長を至上目的とするシステムとは、そういうものである。さらに、当然のことなのだが、過剰消費が地球の余命を短くしている。現在食料の1/3が人々の胃に入ることなく廃棄されている[10]。発展途上国では、コールドチェーンの未整備などで生産者から店先に至るまでに痛んでしまうためであるが、そのようなことが克服された豊かな地域では、「盛りだくさん」という蕩尽を演出する宴会料理や、苦情を恐れて余裕をもって設定された賞味期限での廃棄や、家庭の冷蔵庫のなかでの買いだめなどが原因となっている。

収奪の限界
先進諸国の経済成長の基礎には技術革新があったからなの

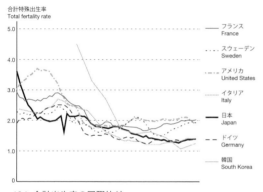

104 合計出生率の国際比較
International comparison of total fertility rates

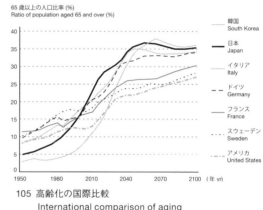

105 高齢化の国際比較
International comparison of aging

development of a system of medical care (→ 105). Japan is now the country with the longest average life expectancy in the world. (The combined figure for men and women is 83.1 years.) Long life has always been a dream of humankind, but the social system and the material environment have not been able to keep up with the increase in longevity. The graying of Japan's baby boomers will provide the first large-scale experiment in the world in administering a society that is highly advanced in age. The biggest task for us architects is to envision the cities and buildings of a society in which seniors constitute the majority of the population. True happiness cannot be achieved if we fail.

Satiation of Desire
Consumption in the age of intensive economic growth revolutionized life in Japanese households. Washing machines and refrigerators radically altered the nature of household labor[8] and gave housewives more free time. Automobiles enabled people to travel further afield, promoted suburban development and helped to deliver the blessings of civilization even to remote mountain communities. Once necessities have been obtained, enthusiasm for consumption begins to wane, but for corporations that presents a problem. The desire for things that are not absolutely required must be stimulated, which is where advertisements come in. Drawing consumers' attention to the excellence of a product is an out-dated stratagem. Using every available means, including signing popular athletes to endorse their products or funding environmental activities that have nothing to do with what they are selling, corporations attempt to convince consumers that using those products will make life more delightful.[9] Continually exposed to such an environment, even relatively cautious consumers will purchase brand products at inflated prices. The cost of advertising is the cost of making consumers crave something they do not really want, and in contemporary society consumers themselves end up paying that expense.

Desire is something created by society, and the cost of creating desire is higher in those communities where people are already content. Moreover, consumption is no longer directly linked to happiness in such communities. The most obvious example of an unhappy consequence is obesity. In an age of poverty, ingesting nourishing foods

だが、技術革新の成果が先進諸国の人々の生活を大々的に変えることができたのは三種類の「収奪」があったからである。一つは鉱物資源の地球環境からの「収奪」、もう一つは他地域(民族)からの「収奪」、最後は、既に述べた未来の世代からの「収奪」である。地球に物理的限界がある以上、資源はいつかは使い尽くされる。鉱物資源を他の天体に求めることも可能であるが、経済的に見合うかは疑問である。

一方、地域(民族)的な「収奪」は15世紀の「地理上の発見」以降今日まで続く。民族間の収奪は、「遅れた」周縁的な地域から安い労働力と鉱物資源を調達することである。さらに、周縁的な地域は中心的な地域の市場として収奪され、これが中心的な地域の産業を支える。資本主義の繁栄のために周縁の存在が必須なのである。中心と周縁の水位差こそ発展の原動力である。そして20世紀中頃からは植民地主義は、いまではグローバリズムに姿を変えて周縁的な地域からの収奪が今も続いている。同様な「収奪」は一つの国のなかでも行なわれてきた。たとえば、1950年代半ばに始まる集団就職は日本における中心/周縁性の存在を示す象徴的なことであった。

しかし、この収奪は永遠には続かない。地球からの収奪には資源の有限性があり、周縁の生活水準が上がり購買力が増すことは市場としての魅力を高めるのだが、安い労働力と資源の調達場所ではなくなる。日本での集団就職は1960年代一杯で終わった。その後の日本にとっては、中国から東南アジアが新たな周縁として位置づけられていたが、韓国、中国(中国国内では、内陸部が沿岸部に対して周縁化されている)が周縁性を薄め、今やベトナムやラオス、ミャンマーなど東南アジアに移動している。これらの地域もやがて豊かになるだろう。次はインド、さらにその先になるとアフリカに移って行くが、その先はもうない。

このような経済の行き詰まりや人口構造の変化は行政コストの上昇に繋がる。日本の公共財政は年金と医療費の膨張と、たびたび行なわれた景気浮揚策によって膨大な債務を抱えている。政府ならびに地方政府の債務は2014年現在、1202兆円であり、同年の名目GDP 487兆円の2.3倍を超えている(→106)。経済学者水野和夫は、日本の国債(長期)の利回りは1997年以降2%を切っており、続いて英米独の国債利率も低下している。これほど低い利率は、まさに中世から近代に移行する時期

built up resistance to disease and led to increased longevity. However, in an age of affluence, unless one exercises self-control, the intake of nourishment becomes excessive and leads to obesity, which adversely affects not only appearance but health and lowers quality of life.

Corporations must continue to make things even when consumption approaches the saturation point. That is in the nature of a system in which economic growth is the supreme objective. Furthermore, excessive consumption shortens earth's own lifespan. At present, one-third of food is unconsumed and wasted.[10] In developing countries, that is because cold chains are undeveloped and food is spoiled in transit from producers to stores, but in affluent regions where such problems have been overcome, the reasons for waste include banquets where a surfeit of food is offered for the sake of ostentation, sell-by dates that are set much earlier than necessary by stores out of fear of complaints, and the overstocking of refrigerators.

Limits of Exploitation

The economic growth of developed countries was based on technological innovation, but three types of "exploitation" enabled the fruits of technological innovation to change people's lives. First, there was the exploitation of mineral resources and the global environment; second, there was the exploitation of other regions (and peoples); and finally, there was the aforementioned exploitation of future generations. Resources will eventually be exhausted because there are material limits to the earth. Mineral resources may be obtained on other celestial bodies, but the economics of such schemes is questionable.

Meanwhile, the exploitation of other regions (and peoples) has continued since they were "discovered" in the fifteenth century. Exploitation between peoples takes place when cheap labor and mineral resources are procured from "backward" peripheral regions. Peripheral regions are also exploited to the extent that they become markets for central regions, and this exploitation helps support industries in central regions. A periphery must exist for capitalism to prosper. The gap that exists between center and periphery is the driving force of development. Since the middle of the twentieth century, colonialism has been superseded by globalism, but the exploitation of peripheral regions continues to this day. A similar exploitation can take place in one country. For example, factories in need of an unskilled workforce drew trainloads of youths seeking employment from farming areas to cities beginning in the mid-1950s in Japan, and this phenomenon known as "group employment" (*shūdan shūshoku*) was symbolic of the gap that existed and still exists between center and periphery in this country.

However, this exploitation cannot go on forever. A rise in the standard of living and increased purchasing power on the periphery enhance its appeal as a market, but the periphery no longer serves as a place providing labor and resources. By the end of the 1960s, "group employment" had ended. For a time, China and South Korea were regarded by Japan as the new periphery, but today those countries have in large measure outgrown that status and been replaced by Southeast Asian nations such as Vietnam, Laos and Myanmar. (China, for its part, has made its interior regions the periphery for its coastal areas.) Those countries too will eventually become affluent. India and Africa may take their place for a time, but after them there will be no more regions left to exploit.

This economic dead end and changes in demographics are leading to a rise in administrative costs. Japan is burdened with an enormous public debt as the result of increasing pensions and medical costs and frequent measures taken to stimulate the economy. The total debt of Japan's central government and local governments was 1,202 trillion yen in 2014 or more than 2.3 times Japan's

の17世紀にジェノバで起こって以来のことであり、今回は近代資本主義の終焉を予告しているという[11]。実体経済がだめなら、金融経済でバブル経済を作り出すしかなくなる。日本の土地バブルもアメリカのサブプライムローンも、それだというのである。しかし、バブル経済はマネーゲームなので、金融市場に対するカンフル剤的効果はあっても持続できず、効果が切れた後の副作用が激烈であり体力を消耗してしまう。

いずれにしろ、経済成長の基礎をなす二つの形の「収奪」は、いずれも早晩困難になることは明白である。残るは、未来の世代が使うべき資源を現在使う第三の収奪である。しかし、これが無責任な態度であることは言うまでもない。真の解決策は、それぞれの地域が空間的にも時間的にも自足してやってゆくこととしかない。この結論には筆者の終局回避の期待が混じっているが、それが唯一の希望の方向なのである。

消費と生産の平衡を目標にすることは地球社会の存続のために必須である。そのような平衡に達するためには、未だに貧困のなかにある発展途上国の国民の収入や教育水準や栄養水準の改善が先になされる必要があるから、先に発展した地域は、いままで以上に省資源の工夫と消費の削減をして、地球全体で過剰消費にならないように自己管理をするのが道義であろう。このように考えていくと、21世紀の世界では、少なくとも先進地域においては縮小は不可避であると結論せざるをえない。そして、日本はその先端を走るのだから、都市計画学もここから思考を開始すべきである。縮小は決して一時的で短期的なものではなく、われわれは、まさに文明の転換点に差し掛かっていることを理解する必要があるだろう。

2. 縮小の時代の都市

人口が減り、経済が成長しなくなり、環境問題が厳しくなると都市はどのような影響を受けるのだろうか。

まず、人口減はストックと需要のあいだのアンバランスをもたらす。そこらじゅうに空き家と空き地が増える。土地需要は大きく後退するので特別な場所を除いて地価は下がり続けるはずである。地方中小都市では土地を売却できないところが増えるだろう。住民の減少で自治体の税収が減り、都市の拡大に対応

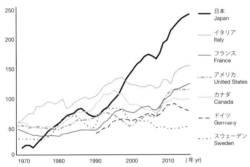

106 公的債務の国際比較
International comparison of public debt

nominal GDP that year of 487 trillion yen (→ 106). Economist Kazuo Mizuno points out that interest on Japan's (long-term) government bonds has been less than two percent since 1997 and that interest rates on the government bonds of England, the United States and Germany are also falling. He states that seventeenth-century Genova at a time of transition from the Middle Ages to the modern era was the last time and place the rate of yield was this low, and that this presages the end of modern capitalism.[11] If the real economy no longer works, the only alternative is to create a bubble economy through monetary measures. Japan's real estate bubble and America's house price bubble fueled by subprime mortgages are examples of this stratagem. However, a bubble economy is speculative; though it may provide the financial market with a shot in the arm, it cannot be sustained. Once the effect wears off, there are severe reactions that exhaust the strength of the economy.

In any case, it is clear that the two forms of exploitation that have been fundamental to economic growth will sooner or later become difficult to practice. That will leave only the third form of exploitation, the use of resources that ought to be left to future generations. Needless to say, that is an irresponsible attitude. The only solution is for each region to become self-sufficient in both a spatial and a temporal sense. The conclusion that this course is still open to us may be overly optimistic, but no other alternative exists.

Balance must be achieved between consumption and production if global society is to continue to exist. To achieve such a balance, first, incomes, levels of education and standards of nutrition of people still in poverty in developing countries must be improved. For already developed regions, the morally correct course is to manage themselves—by doubling efforts to conserve resources and reduce consumption—so as to prevent excessive consumption over the earth as a whole. If that is the case, we can only conclude that in the twenty-first century, shrinkage is unavoidable, at least in developed regions. Since Japan is on the leading edge in this regard, it is incumbent on us to begin a reconsideration of city planning. By no means will shrinkage be temporary or short-term. We need to understand that we are at a turning point of civilization.

して伸びきった基盤施設網（インフラストラクチャー）はメンテナンスをする余裕が無くなり放置される。各地で廃村だけでなく廃市も続出するだろう。かつて地元の商店街を一掃した大型ショッピングセンターは店を引き払ってしまうので、日々の生活に困る地域が広がる。公共交通は採算割れして運行は最小限か廃線になる。そうなると、経済的に余裕のある層は不便な地区から逃げ出し、郊外には車を利用できない高齢低収入層が取り残されるであろう。不在地主が多数になり空き地と空き家の管理が行き届かず犯罪の温床になるかもしれない。

日本では、2053年には高齢者の人口が40％になると予測されている[12]。高齢者は社会の多数派を形成する。高齢者の運動能力、知覚能力が都市設計や商品や施設設計の基準となり、若者向けの産業は小さい市場になるだろう。政治的にも、有権者の過半数が高齢者になり、高齢者に有利な立法・政策を唱える政治家が支持を集め、若者は冷遇される。そうなると将来に対する投資、たとえば教育投資などが疎かになる[13]。

少子化は働き手の不足に繋がる。たとえ資金力と技術力があっても、海外に需要があっても、働き手が不足すれば経済成長の足を引っ張る。女性の労働市場への参加も遅れ、移民政策にも慎重な日本では、労働力不足の問題は今後の経済運営、都市運営における大きな制約条件になる。

こうして簡単にみただけでも、縮小がもたらす問題の深さに思わず目がくらむが、敢えて言えば、縮小は決して不幸な状況ではない。一番身近な高齢社会を考えてみよう。日本が世界で一番長生きできる国だということは、われわれが理想社会に一歩近づいたということであり、祝福すべきことではないだろうか。ただ、われわれは高齢社会に慣れていないだけである。約1世紀前の1921年には日本人は42〜43歳で亡くなり、1965年でも男で67.7歳、女は72.9歳までしか生きていない[14]。今より10年以上短命だったのである（→107）。それから8年後の1972年に高齢者の痴呆を題材にした有吉佐和子の『恍惚の人』が出版され[15]、話題になった。高齢社会は介護や医療で費用がかかり、高齢者の増加は生産性を落とすが、これは長寿という理想社会を維持するための、相応の費用だと理解しなければならない。そして、それに対応できるように都市や社会の仕組みを手直ししなければならない時期にいるのだという理解が必要な

2. Cities in an Age of Shrinkage

What sort of effect will the decline in population, end of economic growth and severe environmental problems have on cities?

First, the decline in population will upset the balance between stock and demand. Unoccupied houses and vacant lots will increase everywhere. Since demand for land will shrink substantially, land prices will continue to fall except in certain special places. In local cities of small and medium size, places where land is unsellable will likely increase. The decline in population will result in a decrease in tax revenue, and municipalities will no longer have the wherewithal to maintain infrastructures developed in times of urban expansion. There will be deserted cities as well as deserted villages. The large shopping centers that once drove local shopping streets out of business will themselves close, making it difficult for people in those areas to get by. Public transport will no longer pay for itself; as a result, service will be reduced to a minimum or routes discontinued. Those better off economically will flee from inconvenient districts, and low-income people of advanced age who cannot use automobiles will be left in the suburbs. Absentee landlords will become the norm, and vacant lots and unoccupied houses, inadequately managed, will become hotbeds of crime.

By 2053, seniors are expected to constitute 40 percent of Japan's total population.[12] The majority of society will be seniors. The capacity of the elderly for locomotion and perception will become the standards for urban design and the design of products and facilities; there will be only a small market for products targeting youths. The fact that seniors are in the majority will have an impact on politics as well. Politicians advocating legislation and policies advantageous to the elderly will gain support, and less heed will be paid to youths. As a consequence, investment in the future, for example, investment in education, will be neglected.[13]

A declining birthrate will lead to a shortage of workers. Even if there is financial and technological capability, and even if there is overseas demand, economic growth will still be hindered if workers are in short supply. In Japan, where the rate of participation by women in the labor market is still low compared to rates in many other countries and where government policy with respect to immigration remains highly cautious, the problem of a limited workforce will be a major constraint on the management of the economy and cities in the future.

The seriousness of the problems caused by shrinkage, as revealed by even a cursory examination such as the above, is enough to make one's head spin. Nonetheless, shrinkage is not necessarily a bad thing. Let us consider, for example, the aspect of it with which we are most familiar: Japan's aging population. The fact that the Japanese have the longest life expectancy in the world means we are one step closer to the achievement of an ideal society. It is something to be celebrated. The problems associated with it only mean we are still unaccustomed to this development. In 1921, approximately a century ago, the average Japanese lived to only 42 to 43 years; even as recently as 1965, the life expectancy of men was 67.7 years and that of women 72.9 years[14], or more than ten years less than what they are now (→ 107). In 1972, Sawako Ariyoshi published her much-discussed novel *The Twilight Years (Kōkotsu no hito)* dealing with senile dementia.[15] The aging of the population means additional costs for care and medical treatment, and an increase in the number of seniors leads to a drop in productivity. However, these must be understood to be the reasonable cost of maintaining what is by the measure of longevity an ideal society.

To consider shrinkage is to anticipate the future and to be proactive, steering society in a direction appropriate to the

のである。
　縮小を考えるということは、新たな千年紀の未来を先取りし積極的に新しい時代に適した社会に向けて舵を切ることである。

　よく知られているように、日本の経済成長が鈍化しはじめた頃の1972年に、ローマクラブは『成長の限界』[16]で、現在のままで人口増加や環境破壊が続けば、資源の枯渇や環境の悪化によって100年以内に人類の成長は限界に達すると警鐘を鳴らし、まさに縮小の時代を予告したのである。ところが、オイルショックを乗り切った後に経済的繁栄が再びやってきたので、人々はこの託宣を意識の隅に追いやってしまった。しかし、構造的な衰退はいつまでも覆い隠せるものではなかった。21世紀の最初の10年間に起こったアメリカ合衆国に対する同時多発テロ攻撃（2001）、リーマンショック（2008）、津波による福島第一原子力発電所の事故（2011）などは、経済、文化における欧米および日本が保って来た力の衰退と多極化、そして資本主義経済の構造的限界などを示し、ポストモダンは近代の次などではなく、近代という大舞台の最終幕であることを明らかにしたのである。

　ローマクラブの報告書の理論的支柱の一人に、アメリカの経済学者で、持続可能な社会を目指す世界に大きい影響を与えたハーマン・E・デイリーがいる。デイリーによれば、かつては「空いている世界」であったのが、20世紀の最後になって「いっぱいの世界」に変わったとして、「いっぱいの世界」では経済成長は問題解決に繋がらないという。そして、「まず、総量を減らし、（次に）効率改善で対応する（括弧内は著者）」[17]ことが必要だと説く。ここでも基本は、人間の活動の総量が地球の環境収容力を超えないことにある。そのためには定常経済を目指さなければならないという。

　また、デイリーより原理的な思考を展開するフランスの経済哲学者セルジュ・ラトゥーシュは「経済成長」、「発展」は宗教のようなものであり、その呪縛から逃れ〈脱成長（デクロワサンス）〉へと価値転換を起こさなければ人類に未来はないと説く[18]。

　縮小が永遠に続けばやがて人間という種の存在が危うくなるから、いずれは平衡状態に達しなければならない。先進諸国

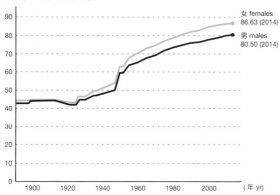

107 日本人の平均寿命の推移
Changes in average life expectancy for the Japanese population

平均寿命とは各年におけるゼロ歳児の平均余命である。

Average life expectancy describes the average remaining life span for zero-year-olds for each year.

age to come.

　In 1972, just as Japan's economic growth was beginning to slow, the Club of Rome in effect predicted the age of shrinkage, issuing a warning in its report *The Limits to Growth*[16] that if population growth and environmental destruction continued unchecked, development would reach its limits within a hundred years because of depletion of resources and deterioration of the environment. However, economic prosperity returned once the world weathered the energy crises; as a result, people turned a deaf ear to this prophecy. Nevertheless, structural decline could not be concealed forever. Events in the first ten years of the twenty-first century—9.11 (2001) and the financial crisis (2008) in the United States and the Fukushima Nuclear Power Plant accident (2011) caused by a tsunami in Japan—revealed the decline and multipolarization of the economic and cultural power once possessed by the West and Japan and the structural limits of a capitalist economy. It became clear that postmodernism was not the next thing after modernism but the final act of the grand performance called modernism.

　One of the key theorists behind the report by the Club of Rome was Herman E. Daly, an American economist who contributed to the worldwide acceptance of the idea of a sustainable society as an objective. According to Daly, the "empty world" of the past had, by the end of the twentieth century, changed into a "full world." In the latter, economic growth does not lead to the solution of problems. He argues that first, the total volume of human activity must be reduced, and next, improvements in efficiency must be made.[17] Here too, the basic idea is that the total volume of human activity must not exceed the earth's environmental capacity to accommodate such activity. To achieve that goal, we must aim for a steady-state economy.

　The French economist Serge Latouche, whose work is of a more fundamental character than Daly's, has written that "economic growth" or "development" is like a religion and that unless we break its spell and adopt the idea of *décroissance* (degrowth), humankind will have no future.[18]

　We must eventually achieve a balanced, steady state because endless shrinkage will endanger the survival of the human species. Developed countries have a responsibility to break free of the idea of economic growth for the sake of

は早急に経済成長のための経済成長という発想から脱して、縮小過程を巧みに管理し、定常経済を基礎とした社会に向けて舵を切る義務がある。

economic growth, skillfully manage the process of shrinkage and transform themselves into societies based on a steady-state economy.

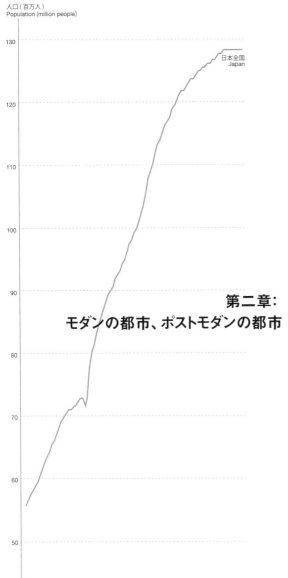

われわれの目標は、縮小の時代の都市のデザインの方法を考えることなのだが、そのためにも、まずは、なぜわれわれがここにいるのかを知らなければならない。手始めに、現在と地続きの20世紀の都市の歴史を概観すると、20世紀は、都市空間の再組織化[19]を誰が推進したかで大きく二期に分けることができる。1970年代までは政府が都市の再組織化の主役だった。1970年代以降は民間企業が主役になるという区分である。様式的にいえば前者はモダン、後者はポストモダンということになる。ポストモダンという用語は建築美学を超えて使われている。これはモダンより近代の「後＝ポスト」の意味であり、モダンの対概念である。

なお、本書では、第二部で都市の未来を考えるうえで、巨大都市の代表として東京、そして地方都市の代表として長岡を取り上げる、(→108)(→109)。本章でも、この二都市が実例としてよく登場するのはこの理由である。

第二章:
モダンの都市、ポストモダンの都市

109 東京都と長岡市の位置
Locations of Tokyo and Nagaoka City

Chapter 2.
Modern Cities, Postmodern Cities

Our objective is to consider the way to design cities in an age of shrinkage, but to do that, we must first consider why we are at this point now. A review of the history of cities of the immediate past shows that the twentieth century can be divided into two periods by the agents that promoted the reorganization[19] of urban spaces. Until the 1970s, the government was the main agent of urban reorganization. From the 1970s, private enterprises assumed that role. From a stylistic perspective, the former period was modern and the latter postmodern.

In considering the future of cities in Part 2 of this book, I will examine Tokyo as an example of a metropolis and Nagaoka, Niigata Prefecture, as an example of a local city. That is the reason these two cities will frequently appear as actual examples in this chapter (→108)(→109).

1. Modern Cities

A few key phrases summarize the salient qualities of the modern city: industrialization; city of equality; healthy city; transportation city; department stores, supermarkets, local shopping streets; city planning as economic policy; La Ville

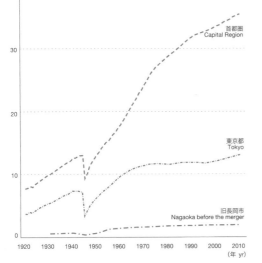

108 東京都と長岡市の人口の推移
Changes in the populations of Tokyo and Nagaoka City

1. モダンの都市

モダンの都市の特徴を、「工業化」、「公平な都市」、「健康な都市」、「交通都市」、「百貨店・スーパー・路線商店街」、「経済政策としての都市計画」、「輝く都市」というキーワードで素描する。

政府の主導

都市の世界史を概観すると、都市の全部または一部が全面的に新しく作られることがときどき起こることに気づく。その一つは、新しい政権が、権勢を示すために首都を建設する場合である。明治政府も例外ではなかった。明治維新は政治革命であると同時に大規模な文化革命である。結果的には、明治政府は東京を完全に作り直すことはしなかったが、すくなくとも意図においては、過去の都市を旧弊の象徴として否定し欧風の意匠で覆うとした。明治の改革者たちにとって、木造平屋建ての建築で埋め尽くされた日本の都市は、根本的に改革されるべきであった。明治政府による東京の都市計画は、機能の近代化と外観の欧化という二つの目的をもった都市改造であった（→110）。

こうした変革の多くは政府の手で進められたものも多いが、それ以上に政府が指し示した方向に民間が呼応して民間の力で立ち上げたものも多かった。封建制の時代に地方都市にも蓄積された資本や知識、外部世界への関心が一挙に吹き出したように見える。長岡でも、多くの近代的施設や制度が民間人の力で立ち上がった。

日清、日露の戦争と第一次世界大戦の勝利で東アジアに植民地を得て、海外の植民地に新都市を作る機会が日本の都市計画官僚や建築家に訪れた。植民地では、古代ローマも南米におけるスペインも、アジアにおけるイギリスも植民者の優越を示すために本国の様式で植民都市を作った。ところが日本は、極めて奇妙なことに、台湾にも朝鮮半島にも満州にも日本の欧化政策を持ち込み欧風街区と西欧様式の建築物を作った[20]。近代日本の帝国主義と欧化政策の間に奇妙なねじれがあったことは、日本の都市の近代化のもっとも大きな特徴である

110 錦絵に描かれた銀座煉瓦街
Ginza Rengagai (Ginza Brick Town) illustrated in a color print

明治政府は、明治5年（1872）に起きた銀座、築地の大火をきっかけに、銀座の表通り沿いの建物を煉瓦化して防火と欧化を同時に行った。
二世歌川国輝画(1873)

After the devastating fire in the Ginza and Tsukiji area in 1872, the Meiji government planned the construction of brick buildings along Ginza's main street as both a fire-proofing and modernizing effort.
Illustration by Utagawa Kuniteru (1873).

Radieuse.

Government Initiative

A survey of the history of the cities of the world shows that occasionally a city will be newly created, either in its entirety or in part. One situation in which that occurs is when a new government constructs a capital to demonstrate its authority. The Meiji government was no exception. The Meiji Restoration was not only a political revolution but a large-scale cultural revolution as well. As it turned out, the Meiji government did not completely remake Tokyo, but its intention at least was to reject the city of the past as a symbol of antiquated ways and cover it up with Western designs. Meiji reformers saw the Japanese city filled with one-story wooden buildings as something requiring a fundamental transformation. The city plan for Tokyo developed by the Meiji government was a restructuring project with two objectives: the modernization of functions and the Westernization of the city's external appearance (→110).

Many of these reforms were undertaken by the government itself, but the private sector, following the direction indicated by the authorities, played an even bigger role. Even in local cities, there was a sudden outpouring of capital and knowledge accumulated in the feudal period and curiosity about the outside world. Nagaoka too saw the establishment of many modern facilities and institutions by the private sector.

Victories in the Sino-Japanese War, the Russo-Japanese War and World War I resulted in colonies in East Asia and opportunities for Japanese city planning officials and architects to create new cities in them. Throughout history, colonial powers from ancient Rome to Spain in South America and England in Asia have created outposts in their own styles to demonstrate their superiority. Strangely, Japan brought with it to Taiwan, the Korean peninsula and Manchuria its policy of Westernization and constructed Western-style urban districts and Western-style buildings.[20] The curious dissonance that existed between the imperialism of modern Japan and its policy of Westernization was the key characteristic of the modernization of Japanese cities (→111).

Perhaps residential suburbs can also be regarded as newly created cities. Japanese cities began to develop

(→111)。

郊外住宅地も一種の新都市と見なすことができるかもしれない。日本の都市の郊外化は1920年代頃から首都圏と関西圏で始まる。「殖産興業」の名で産業革命が進み、給与生活をする中産階級世帯が増え、煤煙や病疫や犯罪などで都市が汚染されたからである。彼らは、できたばかりの鉄道を使って都心を離れて郊外に移り住んだ。郊外の住宅地開発を先導したのは民間鉄道事業者であった。国も自治体も量は少なかったが別の大きな役割を担った。郊外化初期でいえば、関東大震災(1923)の復興事業の一翼を担った同潤会は、日本に最初の本格的な都市の共同住宅を導入した。第二次世界大戦で、日本のほとんどの都市が連合軍の大規模な空爆を受けて壊滅的な破壊を被り、再びゼロからの出発が必要になったが、その復興も官僚主導であった。地方都市の戦災復興計画には、海外植民地での都市計画を担った専門家[21]や大学に籍を置く都市計画家も参加した。たとえば長岡は東京大学の高山英華が、広島市と前橋市を同じく丹下健三が、長崎市と呉市を早稲田大学の武基雄が担当した。計画の中心は、区画整理手法を駆使した道路網の整序と拡幅や緑地の充実であった[22]。

戦災復興が一段落して、日本経済も戦前のレベルを超えると、民主主義の自由な空気のなかで大建設時代が到来した。これまで軍事費に使われていた税金が建設投資に振り向けられたと言ってよいだろう。第二次世界大戦後のベビーブームは、日本では1947年から1949年の3年間に集中し、後に「団塊の世代」と呼ばれる世代グループを形成する。この人口構造が、1950年代後半から1960年代にかけての高度経済成長の基礎を築いた。戦後になって戦前の家族制度は法的には廃止されたが、実質的には古い家族観が残り、家督を相続できない農家の次男三男を中心に大量の人口が職を求めて大都市圏に流れ込んだ。そのために大都市における住宅不足は深刻な問題となった。大都市では、住宅不足を補うために、日本住宅公団、各自治体の公営住宅や住宅供給公社などの公的組織が都心と郊外で大小の住宅団地を開発し[23]、近代的な生活様式の普及の役割を担った(→112)。同じような人口移動が、長岡と周辺の農村間でも起こるのだが、人口が増える時期が首都圏より遅れ、1960年代から郊外住宅地が開発され始めた[24]。

111 大連広場(現中山広場)の風景
View of Zhongshan Square (Dalian)

大連は、新京(現在は長春)、奉天(同じく瀋陽)とともに植民地経営の重要な拠点都市であった。

Dalian was an important hub in the management of colonized areas, along with Hsinking (present-day Changchun) and Fengtian (present-day Shenyang).

suburbs in the Capital and Kansai regions in the 1920s. As the country industrialized under a policy encapsulated in the slogan "Increase Production and Promote Industry" (shokusan kōgyō), middle-class households of salaried workers increased, and cities became troubled by pollution, epidemics and crime. Thanks to newly-built railways, those workers were able to move from the center of town to the suburbs. Private railways took the initiative in developing suburban residential areas. The central government and local governments played a separate role that was small in scale but important nonetheless. Dōjunkai or the Mutual Profits Association, which played a part in reconstruction after the Great Kanto Earthquake (1923), introduced the first authentic urban apartments in Japan. In World War II, practically all cities in Japan were heavily damaged by aerial bombing conducted by Allied forces and had to be completely rebuilt. That reconstruction effort was led by government bureaucracy. Experts who had been responsible for city planning in overseas colonies and city planners teaching at universities took part in the reconstruction of local cities.[21] For example, Takayama Eika of the University of Tokyo was in charge of the city planning of Nagaoka; Tange Kenzō, also of the University of Tokyo, planned the cities of Hiroshima and Maebashi; and Take Motoo of Waseda University planned the cities of Nagasaki and Kure. These projects were focused on rearranging and widening the road system and creating green areas, using the Land Readjustment Law.[22]

Once the first phase of postwar reconstruction was completed and the Japanese economy had recovered and surpassed prewar levels, the era of major construction began in an atmosphere of democratic freedom. Tax revenues that had hitherto been funneled into military expenditures were now invested in construction. In Japan, the great majority of baby boomers after World War II were born in the three years between 1947 and 1949. This population structure served as the foundation for the intensive economic growth that took place from the late 1950s through the 1960s. The prewar family system was abolished as law, but conservative views regarding the family survived. In an agrarian household, the eldest son continued to enjoy the right of succession; as a result, there was a mass movement of people, many of them younger sons from such households, to metropolitan

同じ時期に大都市圏では、都市として完結したニュータウンが近郊の農地や山林を造成して開発された。関西圏では千里ニュータウン、名古屋圏では高蔵寺ニュータウン、首都圏では多摩、港北、千葉の三大ニュータウンが開発された。筑波研究学園都市もこの時期の開発事業であった[25]。1970年代になると日本の各地で自家用車所有が行きわたり、広汎な郊外化に弾みが付く。首都圏では、東京とその周辺の主要都市である横浜市、川崎市、千葉市のそれぞれの周辺に郊外住宅地が広がり、それらが繋がり、東京首都圏は世界で最大の人口を擁する巨大コナベーションにまで成長し、現在は1都3県で約3300万人の人口を抱える。一方、首都圏より少し遅れて長岡でも、1973年から長岡ニュータウン建設計画が始まった[26]（→113）。このように戦後の都市計画を振り返ると、首都圏が地方都市の発展モデルとして位置づけられていたとも言えるが、都市の規模の大小にかかわらず各地の都市に同じ開発モデルが適用されたとも言える。

工業化

明治政府の政策の中心に殖産興業と欧化があった。明治政府は八幡製鉄所をはじめとして国営でさまざまな近代工場を設立し、欧米から技術者を招聘して日本の工業化を促進した。日本は短期間のうちに西欧技術を輸入し、咀嚼し、自家薬籠中のものにし、極東の島国をアジアで随一の工業国にまで成長させ、同時に軍事大国として東アジア各地に植民地を獲得した。首都圏での工業化は、主に東京湾の埋め立てによって工場用地を拡張して進められた。一方、内陸側では、東京の東北側が工業用地と労働者階級の住宅地に向けられ、北半球の都市の常として中産階級の住宅地は湾岸を除く西南側に広がった。この基本的なゾーニングを保持しつつ、東京は、ほぼ層状に同心円的に拡大した。

明治維新以来の国家主導の産業革命の推進とその後の戦争経済体制は、大都市と小都市あるいは中央と地方の支配関係を強めた。地方都市は農業生産地として大都市に食料を供給し、農村は余剰人口を大都市圏に労働力として送り出す一方で、都市に基盤を置く企業の市場となり、観光地として都市住

112 赤羽台団地（東京都北区1962）
Akabanedai Danchi (Kita Ward, Tokyo, 1962)

113 長岡ニュータウン（2011年撮影）
Nagaoka New Town (photographed in 2011)

regions in search of employment. This caused a serious housing shortage in large cities. To offset this shortage, public agencies such as the Japan Housing Corporation and the housing corporations and housing supply corporations of local governments developed housing projects of all sizes in the middle of cities and the suburbs[23] and assumed the responsibility of disseminating a modern lifestyle (→112). A similar population movement occurred between Nagaoka and its surrounding farming communities, though it took place somewhat later than in the Capital Region; the development of the suburbs there only began in the 1960s.[24]

It was at this time that agricultural and wooded areas in the suburbs of metropolitan regions were developed into new towns which were complete cities in themselves: Senri New Town in the Kansai region, Kōzōji New Town in the Nagoya region and Tama, Kōhoku and Chiba New Towns in the Capital Region. Tsukuba Academic New Town was also developed in this period.[25] In the 1970s, automobile ownership became commonplace throughout the country, and suburbanization gained momentum. In the Capital Region, suburban residential areas spread on the fringes of Tokyo and the main cities around it—Yokohama, Kawasaki and Chiba—and merged. The Tokyo Capital Region grew into an enormous conurbation with the biggest population in the world, and at present the four prefectures, Tokyo, Kanagawa, Saitama and Chiba, have a combined population of about 33 million. Nagaoka lagged slightly behind the Capital Region, but a project to construct Nagaoka New Town began in 1973 (→113).[26] With respect to postwar city planning, one could argue that the Capital Region served as a development model for local cities; however, it may have simply been a case of the same development model being used in all cities, large and small (→114).

Industrialization

The central policies of the Meiji government were the promotion of industries and Westernization. In promoting the industrialization of Japan, the government established various modern factories that were state-administered, beginning with Yahata Steel Works, and invited technical experts from the West. In a short period of time, Japan imported, absorbed and mastered Western technology and

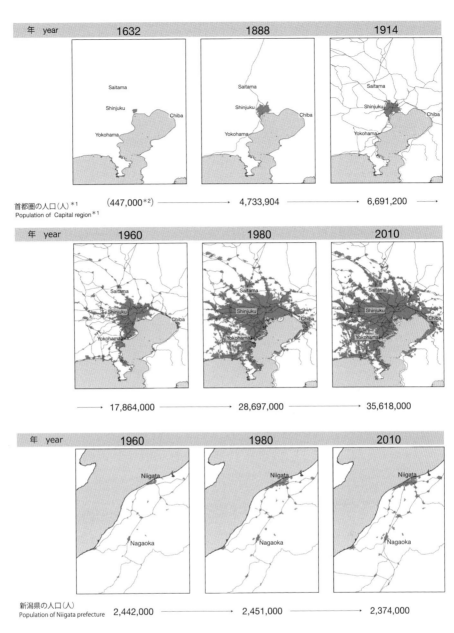

114 首都圏と新潟県西部における市街地の変遷
Changes in urbanized areas of the Capital Region and the western region of Niigata Prefecture

developed into the most industrialized nation in Asia and a military power with colonies in various parts of East Asia. The industrialization of the Capital Region was carried out mainly by expanding factory sites through reclamation projects on Tokyo Bay. Further inland, the northeastern side of Tokyo was turned into land for industrial uses and residential areas for the working class, and as was so often the case with cities in the Northern Hemisphere, residential areas for the middle class developed on the southwestern side except along the bay. In accordance with this system of zoning, Tokyo expanded concentrically in layers.

The promotion of an industrial revolution under the leadership of the state since the Meiji Restoration and the subsequent adoption of a war-time economic regime reinforced the dominance of large cities over small cities and of centers over localities. Local cities, as areas of agricultural production, supplied large cities with food, and farming villages sent their surplus populations to metropolitan regions to provide labor; at the same time, they became markets for corporations based in cities and tourist areas for city-dwellers seeking relaxation. To correct this situation, a Comprehensive National Land Development Plan (Zenkoku Sōgō Kaihatsu Keikaku (or Zensō)), to be carried out over a period of eight years, was established in 1962, and among its objectives was the industrialization of local cities.[27] In response to Zensō,

民を癒してきた。このような状況を是正するために、1962年に計画期間を8年間として「全国総合開発計画（全総）」が策定され、地方都市の工業化が図られた[27]。新潟県は全総を受けて1963年に「新潟県総合開発計画」を策定し、長岡市は三ヵ所の工業団地を作って工場を誘致した。長岡は周辺の大農業地帯を控えた商業都市であるが、少量の石油と天然ガスを産することから機械工業関連の技術基盤があった[28]。この時期に、長岡市をふくむ新潟県中部を選挙地盤とする政治家田中角栄が、全国の均衡ある発展を目標として国土開発を提唱する政権構想「日本列島改造論」を発表し、首相の座を射とめる。1983年には、政府は先端技術産業を中核として工科系大学などの学術研究、そして居住地域を含む都市構想を打ち出し、長岡は、その一つに選ばれた。長岡は信濃川の左岸に副都心的な地区を計画し、音楽ホール、美術館、文科系大学、研究施設、大型病院を集め、さらに巨大なショッピングモールを誘致して、自動車時代に対応して広い道路と広い駐車場を整備した（→114）。

建築レベルに着目して近代化を見ると、工業技術は防災技術を軸に発展し、これが建築形態そして都市形態に大きな影響を与えた。日本の近世都市は、頻発する火事に悩まされたため、明治政府は、重要な建物には煉瓦造を取り入れ（→110）、市街地の一般の建物には瓦葺き土蔵造りの壁を求めた。本格的な防火措置は、1920年に制定された市街地建築物法のなかに防火構造として盛り込まれた。防火構造とは、隣家から延焼を防ぐ目的で木造の建物を不燃性の外装材で包む方式で、今日でも木造建築の主要な防火対策であるが、近世までの日本建築の基本的な外観であった真壁方式を駆逐し都市景観を一新した[29]。1923年の関東大震災では耐火の切り札であった煉瓦造の建物が壊滅的な被害を受けたので、耐火耐震を兼備した構造方式として鉄筋コンクリート造が普及した（→115）。戦後になって、耐震研究はさらに進み、超高層建築を可能にし、1968年には国内初の超高層オフィスビルである霞ヶ関ビルが竣工する[30]。再び日本の都市景観を一新することになる。かように、工学技術の発展が日本の都市景観に大きな影響を与え続け、日本の建築教育が工学中心に組織化されることに繋がった。

115 関東大震災で被災した東大法学部講堂（八角講堂）
University of Tokyo Faculty of Law Auditorium (Hakkaku Hall) damaged by the Great Kanto Earthquake

東京帝国大学の初期の校舎は、お雇い外国人の建築学科教授のジョサイア・コンドルによって設計された。
The first school buildings at Tokyo Imperial University were designed by Josiah Conder, a government-employed foreign professor in the architecture department.

116 日本の鉄道網の拡大
Expansion of the railway network in Japan

明治維新によって封建制から解放された地方の資本家によって、全国に多数の民間鉄道会社が誕生した。全国を覆う鉄道網の必要性を認識した政府は、その多くを買収して国営の鉄道網を樹立した。第二次世界大戦前は、全国を覆う路線網は国が経営し、地域ごとの鉄道を私鉄が経営した。戦後もこの体制が継承されたが、1987年には6つの地域鉄道会社と1つの貨物鉄道会社に分割されて民営化された。

A great number of private railway companies emerged throughout the nation due to local investors released from the constraints of the feudal system by the Meiji Restoration. Acknowledging the need for a railway network stretching across the nation, the government bought out many of these companies to found a national railway network. Prior to WWII, railway networks stretching across the country were operated nationally, while those in each local region were managed by private railway

Niigata Prefecture established a Comprehensive Development Plan for Niigata Prefecture in 1963. Nagaoka City created three industrial estates and invited factories to locate there. Nagaoka is a commercial city near a major agricultural belt but possessed a machine industry-related infrastructure[28] because it produces a small amount of petroleum and natural gas. It was in this period that Tanaka Kakuei, a politician and soon-to-be prime minister whose stronghold was in central Niigata Prefecture including Nagaoka, announced his plan for the remodeling of the Japanese archipelago advocating a balanced development of the entire country. In 1983, the government set forth a concept for cities where advanced technology industries, institutes of technology for academic research and residential areas would be developed. Nagaoka was selected as one of those cities. Nagaoka created a sub-center on the left bank of Shinano River, with a concert hall, an art museum, university dedicated to the humanities, research facility and large hospital. It also invited an enormous shopping mall to locate there and built wide roads and extensive parking areas to adapt to the age of automobiles.

If we focus on the effect of modernization on architecture, the development of industrial technology, particularly fire-preventive technology, greatly influenced the forms of both buildings and cities. The Meiji government introduced brick construction for important buildings and promoted the use of tiled roofs and wattle-and-daub construction for ordinary buildings in urban areas because feudal cities in Japan had suffered from frequent fires. "Fireproof construction" (bōka kōzō) included in the Urban Area Buildings Law (Shigaichi Kenchikubutu Hō) enacted in 1920 was the first authentic fire-preventive measure taken. Fireproof construction is a method of construction in which a wooden building is wrapped in an exterior finish that is non-flammable in order to prevent fire spreading to neighboring houses and is still the main fire-preventive measure for wooden buildings today. It replaced walls made in shinkabe zukuri (a method of construction in which walls were built between posts, leaving the posts themselves exposed), which had determined the exterior appearance of Japanese architecture through the feudal period, and completely altered the townscape.[29] Brick construction, introduced for its fire resistance, did not

交通の革命

公共交通が近代都市の形態に与えた影響は計りしれない。1868年に樹立した明治政府の文明開化策として最初に行なったことの一つが鉄道敷設であった。日本で最初の鉄道路線は1872年に東京の市街地の南端の新橋と最初に外国船に対して開かれた港のある横浜の間に敷かれた。2年後には関西で神戸大阪間に鉄道が敷かれ、それから25年後に東京と大阪が鉄道で結ばれた。同時進行的に全国各地に官民による大小の鉄道路線が敷かれ、それが相互に繋げられて国土を覆う鉄道網が広がり、19世紀の間に日本列島をほぼ縦断した[31]（→ 116）。

鉄道はそれまでの地域間の人流、物流の主要な手段であった舟運の地位を奪い、多くの人の往来を可能にし、郵便の速達性を高め、新聞の全国展開を可能にし、文化を運び広める流路となった。それは、国民国家の一体感を醸成するうえで大きな役割を担った。国土を縦貫する鉄道の存在は軍隊の作戦行動を容易にしたので、日清戦争（1894-1895）と日露戦争（1904-1905）でも大きな役割を果たした。第二次世界大戦後になっても、国鉄は意欲的で、1964年には新幹線の営業運転を開始し、世界に先駆けて弾丸鉄道の時代を切り拓いた。

一方、大都市で中産階級のための郊外住宅地が広範に成立しえたのは、誰でも簡単に郊外と都心を行き来できるようになったからであり、それもまた鉄道網の力であった。大阪を本拠地とする阪急電鉄の小林一三は、1900年代から20年代にかけて郊外線の開設とそれに並行して郊外住宅地開発を進め、都心側駅に百貨店を置き、郊外側の端に遊園地を置くという郊外居住者の消費生活をすべてパッケージ化するビジネスモデルを確立し成功を収めた（→ 117）。このモデルはやがて首都圏を含めて日本の大都市に広まった。私鉄は大都市の都市空間の形成に大きな影響力を現代に至るまで持ち続け、1970年代に自家用車時代を迎えた後も大都市の郊外は鉄道郊外であり続けた。

東京の郊外は、社会階層からみるとかなり均質である[32]。しかし、鉄道によって枠組みが与えられた東京の郊外では、沿線ごとに消費社会的階級性が存在し、都心までの所要時間は変わらないのに地価は大きく違う。しかし、その違いは、部外者に

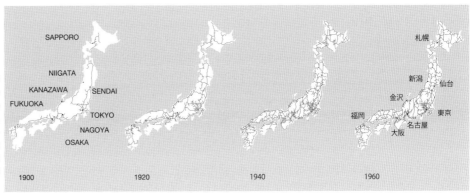

companies; and this system continued to be maintained after the war. However, in 1987, Japanese National Railways was divided and privatized by six local railway companies and one freight company.

prevent many buildings constructed by that method from being destroyed in the Great Kantō Earthquake of 1923 (→ 115). This led to the dissemination of reinforced concrete construction, a method that is resistant to both fires and earthquakes. Further progress was made in research into seismic resistance after World War II, making possible the construction of skyscrapers; the Kasumigaseki Building, the first supertall office building in Japan, was completed in 1968.[30] The Japanese townscape was completely remade once more. In this way, advances in engineering technology continued to have a major influence on the Japanese townscape and led to the engineering orientation of architectural education in Japan.

Revolution in Transportation

Public transportation has had an immense effect on the form of the modern city. One of the first measures taken by the Meiji government after its establishment in 1868, under the slogan "Civilization and Enlightenment" (*bunmei kaika*), was the laying of railways. The first railway line in Japan was laid in 1872 between Shinbashi, at the southern end of Tokyo's urban area, and Yokohama, one of the first ports to be opened to foreign ships. In the Kansai region, a railway was laid between Kobe and Osaka two years later, and 25 years later Tokyo and Osaka were linked by rail. Public and private railway lines, both large and small, were laid throughout the country; these became linked, and a railway network crisscrossing the entire archipelago developed in the nineteenth century (→ 116).[31]

Railways replaced boats and ships as the main channels for transporting people and goods and made possible the movement of large numbers of people, the speedy delivery of mail, and the growth of newspapers on a national scale. They became the route by which culture was disseminated. The presence of railways crisscrossing the country facilitated the deployment of troops and played a major role in both the Sino-Japanese War (1894-1895) and the Russo-Japanese War (1904-1905). Japanese National Railways set an ambitious agenda after World War II. In 1964, it began to operate the Shinkansen and launched the era of bullet trains. Meanwhile, suburban residential districts for the middle class expanded in large cities because a network of railways made it possible to commute between the suburbs and the centers of cities. Kobayashi Ichizō of Hankyū

117 阪急電鉄の沿線開発戦略
Development strategy along the Hankyū Railway

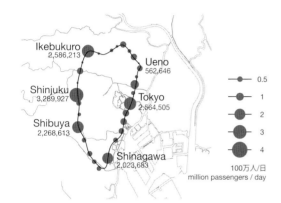

118 山手線の乗降客数(200X年)
Number of passengers on the Yamanote line (2009)

東京の活動を支え、東京の空間認識の参照枠となっているのが1周34.5km29駅からなる環状鉄道線の山手線である。最新のデータでは、各駅での乗降客と乗り換え客の合計は1日当たり約4200万人に上るという。

The Yamanote line, a 34.5km-long railway loop line consisting of 29 stations, supports the activities of Tokyo and acts as a frame of reference for spatially understanding the city. According to the most recent data, the total number of incoming and outgoing passengers per station, including transfers, amounts to approximately 42 million per day.

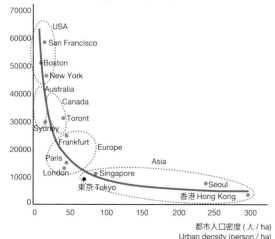

119 人口密度と一人当たりの交通エネルギー消費
Population density and per capita energy consumption in transportation sector

はほとんど察知できないほど微妙な差異である。

　東京の鉄道網は都心を囲む環状のJR山手線とそれに接続する私鉄郊外線と都心の細やかな地下鉄網を特徴とする。多くの郊外線は地下鉄と相互乗り入れをして継ぎ目のないネットワークを形成し、朝の時間帯に東京都心部に流入する交通の約3/4は鉄道が担っている(→118)。その結果、東京首都圏は世界一の人口を抱え、巨大な経済活動の場でありながら、驚異的な低炭素都市なのである(→119)。

　日本の大都市の郊外における鉄道駅の重要性は、それが住宅地の生活の中心になっていることである。郊外の各駅ごとに鉄道駅を起点として線路に直交する線状の商店街が発達し、東京の地図をみると、鉄道と商店街が十字型のパターンが散在していることが認められる(→120)。

公平な都市

　近代都市計画は、産業革命がもたらした社会の歪みの是正という使命を背負っていた。その中心には、衛生観念と平等主義があった。劣悪な環境と化した都心と悲惨な生活を強いられ

Railway, which was headquartered in Osaka, developed suburban residential areas while establishing suburban lines from the 1900s to the '20s. His success was founded on an innovative business model that packaged all goods and services needed by suburbanites including a department store at the urban end of a line and an amusement park at the suburban end (→117). This model was eventually adopted in metropolises throughout the country including the Capital Region. Private railways have continued to exercise an enormous influence on the formation of urban spaces in metropolises to the present day, and railways have remained the backbone of metropolitan suburbs even after the ownership of automobiles became widespread in the 1970s.

　Tokyo's suburbs are fairly uniform from the perspective of social class.[32] However, in those suburbs that are structured around railways, differences in class of consumer do exist from one line to the next, as do differences in the price of land even though travel times to the center of the city may be similar. Those differences, however, are subtle and may not be recognizable to outsiders.

る労働者階級を救済するには、十分な日当たりと新鮮な空気が必要であり、その意味で郊外こそ近代社会の正しいライフスタイルを実現する場と考えられた(→112、121)。なかでも、公共集合住宅は近代都市計画の主要な都市戦略であり、日本の戦後の公共住宅政策の主役として取り入れられた。特に1950年代の戦災復興期には、公共住宅が全住宅供給量の2割を超える時期もあったが、60年代になると1割程度に低下し、その後は漸減している。しかし、生活スタイルへの影響ははなはだ大きく、戦後の日本の住まい方の選択肢に中層集合住宅を加え、扉で区切られる個室、椅子式の生活、そしてダイニングキッチン方式など戸建て住宅の間取りにまで及んだ。また、公共組織が居住様式のプロトタイプ開発を担い、それを民間デベロッパーや住宅メーカーが後追いして商品化するという日本の公共政策の典型的な過程が成功した例でもある。

一方、経済的側面では、累進課税などの再分配政策も有効に働き[33]、首都圏ですら、特権的な階級の住区というものがほとんど見当たらない極めて公平な風景が、日本の都市に出来上がった。

120 商店街と鉄道駅
Shopping street and railway station
小さな線分が路線型商店街を表している。
Small line segments represent shopping streets.

121 L. ヒルベルザイマーの「高層建築都市」(1924)
L. Hilberseimer's "High-Rise City" (1924)
ルードヴィッヒ・ヒルベルザイマー、の提案の特徴は、空中歩廊と地上面の自動車道路の分離、歩廊の上に平行に均等に配置された板状の集合住宅と地上に配された商業業務建物である。
Hilberseimer's "High-rise City" proposes a multi-level complex that segregates pedestrian traffic in the air from automobile traffic on the ground. South-facing slabs house equally furnished residential units on the upper levels and commercial/business facilities on the lower levels.

Tokyo's rail network consists of a fine mesh of subway lines servicing the center of the city, the JR Yamanote Line looping around that center, and private suburban railway lines that connect to that loop line. In many cases, where a suburban line meets a subway line, through service is provided, eliminating the need for transfers and forming an uninterrupted network. Private railways are responsible for approximately three-fourths of all transportation flowing into the central districts of Tokyo in the morning (→118). As a result, despite having the biggest population in the world and being a place of enormous economic activity, the Tokyo Capital Region is a remarkably low-carbon city (→119).

Railway stations in the suburbs of Japanese metropolises are important in that they are the centers of life in residential areas. At each station in the suburbs, a linear shopping district has developed at right angles to the railway. A map of Tokyo shows a proliferation of such cruciform patterns formed by railway lines and shopping streets (→120).

経済策としての都市計画

日本では1950年代半ばから1970年代前半までを高度経済成長時代と呼ぶ。この間 (1956~1970) 年度平均9.6%の実質経済成長率を記録している (→122)。一度は灰燼に帰した都市が短期日のうちに甦り、世界第二の経済規模に到達した時期である。その中心的政策に、国家主導による産業育成政策、そしてケインズ政策すなわち不況が到来すると公共投資で有効需要を創出する政策があった。ケインズ政策による公共施設整備は基盤整備が不十分である段階では非常に有効である。社会基盤の建設は経済活動を直接的に刺激し、国民の生活水準を向上させる。また、不況のときは建設費が安いので割安に建設ができ、経済が好転すると、この基盤が次の経済成長の跳躍台となるという好循環を生み出す。しかし、一方で、この政策の成功が、この国では、都市計画を都市施設の建設計画と取り違える原因となったのである。つまり、都市施設の新築偏重体質を助長し、都市空間の管理技術を未熟なままに放置するという問題を残した。

もう一つの日本の都市政策の大きな特徴は、持ち家を奨励

City of Equality

Modern city planning had a mission—to correct the social distortions caused by the Industrial Revolution—that was based on ideas of health and sanitation and on egalitarianism. Ample sunlight and fresh air were needed to relieve members of a working class forced to lead miserable lives in city centers whose environment had been allowed to deteriorate, and the suburbs were seen as the very places to achieve a lifestyle appropriate to a modern society (→121). Apartment buildings were the main urban stratagem of modern city planning and played a central role in the public housing policy of postwar Japan. Especially in the period of postwar reconstruction in the 1950s, public housing accounted for more than 20 percent of the total housing supply, but by the 1960s that share fell to approximately 10 percent, and has fallen even further since then. However, public housing had an enormous influence on lifestyle. Thanks to it, the medium-rise apartment building became a housing option in postwar Japan, and with their separate bedrooms, each equipped with a door, combined dining room and kitchen, and style of life premised on the use of chairs, they even influenced the

し公的住宅ローンで支援する政策を選び、公共住宅供給は戦後の住宅不足が解消するや最低限の線に後退したことである。持ち家政策は人々の勤労意欲をかき立てるが、都市環境の質を維持するのが容易ではない。ただし、結果をみると、日本が特別に持ち家率が高いわけではない。それは1/3程度で、アメリカやイギリスに近く、フランスやドイツより低い。また東京の持ち家比率は、アジアの大都市、上海やムンバイやシンガポールよりも低い。

消費の拡大

近代を象徴する小売り形態は、前期においては百貨店と路線型商店街であり、後期においてはこれにスーパーマーケットが加わる。百貨店は、小売業の用語で言えば「買い回り品」とよばれる耐久消費財や趣味品などを販売し、消費都市の象徴として長らく都心を飾った。日本の百貨店には二種類の系譜があり、一つは江戸時代から続く呉服商が発展したもの、そしてもう一つは大都市の郊外化とともに鉄道会社によって都心の郊外線のターミナル駅に作られたものである。

一方、庶民の日常的買い物、小売り業の類型で言えば「最寄り品」と呼ばれる食料品や日用雑貨などの販売は、地場の路線型の商店街が担った。この機能の一部は後にスーパーマーケットが引き継ぎ、郊外住宅地の生活の実質を支えた。商店街の発生は、第一次世界大戦後に遡るという。産業革命が成功し工業化が順調に進展したことと経済不況が大量の余剰人口を農村から都市に押し出し続け、その多くが零細な商店を構えることになる。彼らは商業者としては未熟であり資本も貧弱であったが、それを超える需要があった。都心の強力な百貨店に対抗するために、商店主たちは商店組合を形成した。それは垂直の百貨店に対して「横のデパート」を目指したという[34]。ほとんどの大都市の郊外住宅地の商店街は、戦後に鉄道駅を核に発展した。小売り店だけではなく喫茶店や家族の晴れの日の食事をするレストランなどもでき、郊外コミュニティの中心的空間として機能した。

モダン都市の雛形

都市の形態は、合理的で分析的な思考から生まれるというよ

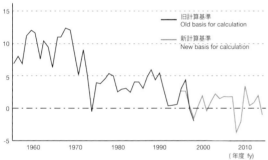

122 第二次世界大戦後の日本の経済成長率の推移
Changes in post-WWII economic growth rate in Japan

123 ル・コルビュジエの「現代都市」(1922)
Le Corbusier's "Ville Contemporaine" (1922)

ル・コルビュジエとピエール・ジャンヌレの提案になる300万人の人口を擁する「現代都市」の都心部の透視図。公園的な空間に超高層建築が離れて建っている。現代都市の全てがここに描かれた。

Perspective of the "Ville Contemporaine," proposed by Le Corbusier and Pierre Jeanneret for a population of 3 million. Skyscrapers stand a good distance apart from one another in a park-like space.

layouts of detached houses. This development by public agencies (particularly the Japan Housing Corporation) of a housing prototype that was subsequently commodified by private developers and housing manufacturers was a successful instance of a process typical of public policy in Japan.

On the economic front, measures for the redistribution of wealth such as progressive taxation worked effectively,[33] and a landscape characterized by extreme equality in which there are practically no enclaves for the privileged class has developed in Japanese cities, even in the Capital Region.

City Planning as Economic Policy

In Japan, an 18-year period from the mid-1950s to the mid-1970s is referred to as the period of intensive economic growth. During that time, the average real economic growth rate was 9.6 percent (→ 122). Though most of its cities had been reduced to ashes by the war, Japan developed in a short period of time into the second biggest economy in the world. The key to this achievement was a policy of government-guided development of industries and a Keynesian policy, that is, the creation of effective demand through public investments in times of recession. The improvement of public facilities as a stratagem of Keynesian policy is extremely effective when infrastructure is still not fully developed. The construction of social infrastructure directly stimulates economic activity and improves living standards. Moreover, this gives rise to a favorable cycle—construction costs are low in times of recession, making construction relatively economical, and once the economy picks up, the new infrastructure provides a springboard for the next stage of economic growth. On the other hand, the success of this policy led to the mistaken belief that city planning meant planning the construction of urban facilities. That is, it encouraged those in charge of policy to place too great an importance on the new construction of urban facilities and to neglect the development of techniques to manage urban spaces.

Japan's urban policy is also distinguished by the choice made to encourage homeownership and support public housing loans; the supply of public housing was reduced to a minimum after the postwar housing shortage was eliminated. The homeownership policy increases people's

リモデル的発想から演繹的に導かれることが多い。つまり形が啓示を与えるのである。たとえば古代ローマとその植民都市、7世紀の唐の冊封体制のもとでの冊封国の都市、オースマンのパリと同時代の欧州の首都などを見れば、古代ローマや長安やパリが雛形になっていたことがわかる。モダンの都市では何が規範の地位に着いたのだろうか。モダンの都市にとっての雛形となった現実の都市はなく、代わりに理念と図に描かれた都市が規範となった。それはル・コルビュジエが「300万人の現代都市」(1922)と名付けて世に問うたモデルである(→ 123)。これは近代の価値である工業化、公平、経済、交通、衛生（健康）を体現している。後年、彼は、このプロジェクトを含む初期の都市プロジェクトで示した原理をまとめ、それを「輝く都市(La Ville Radieuse)」と呼んだ。

一方、郊外住宅地のモデルを探すことは難しい。リバーサイド[35]、レッチワース、ブロードエーカーシティ[36]など郊外のイメージを形成した計画や開発は多数あるが、「輝く都市」ほど決定的な影響を郊外住宅地に与えた空間モデルはなかったのではないだろうか。日本では、塀が連なる近世の武家住宅地が規範の位置にあったように思える(→ 124)。

2. ポストモダン都市

1973年と1979年の二度のオイルショックは、世界を大きく変えた経済事件であった。これは、後進的な地域から原油などの資源を安く調達して経済的繁栄を追求するモデルがもはや成立しなくなったことを意味していた。実際、1973年度までの日本の経済は驚異の成長を示し、1968年にはGDPで世界第二位になるのだが、その後は、1974年度からの経済成長率は平均3.8％に下がり、1990年度から2007年度になると、さらに1.3％まで下がる。2011年には遂に世界二位の経済国家という地位を中国に譲ることになる。このような経済成長の鈍化は、生活文化からデザインの方向性まで広く影響を与えた。かつてのような社会民主主義的メンタリティをもった官僚が主導する都市政策は採られなくなり、都市経営は市場経済の一部門としてみなされ、都市空間は資本の流れを加速させるターボチャージャーとしての役割が期待されるようになる。

will to work but makes it difficult to maintain the quality of the urban environment. However, for all that, the rate of homeownership in Japan is not particularly high. It is approximately one in three, which is close to that of the United States or England and lower than France or Germany. The rate of homeownership in Tokyo is lower than rates in other Asian metropolises such as Shanghai, Mumbai and Singapore.

Expansion of Consumption

Forms of retailing symbolic of the first half of the modern era were department stores and local shopping streets (consisting of shops fronting directly on streets), and to these were added supermarkets in the second half. Department stores sell what is referred to in retailing as "shopping goods," that is, durable consumer goods or goods that appeal to tastes; they were long a symbol of urban consumption and a feature of the central districts of cities in Japan. There are two types of Japanese department stores, depending on their origin. One type developed from dealers in kimono fabrics whose businesses can be traced back to the Edo period, and the other was created by private railways in the city terminals of the lines they laid to serve the suburbs of metropolises.

Local shopping streets, on the other hand, sell everyday goods to ordinary people, that is, food and everyday goods that are referred to in retailing as "convenience goods." This function was later taken over in part by supermarkets, which became the mainstay of life in the suburbs. Shopping streets are said to have originated after World War I. The success of the Industrial Revolution, advances made in industrialization and economic recession drew excess population from farming villages into cities, and many of these immigrants set up small stores. These people were inexperienced in trade and had little capital, but demand was great enough to offset those disadvantages. Storekeepers formed associations to compete against powerful department stores in central districts. Their aim can be said to have been the creation of "horizontally deployed department stores" as opposed to department stores proper which were vertical and multi-floored[34]. Nearly all shopping streets in the suburbs of metropolises developed around railway stations after the war. Including not only retail stores but coffee shops and restaurants

このような新しい状況は広くポストモダンと呼ばれている。この時代の都市空間の再編の特徴を「稼げる都市」、「歴史性」、「都市空間全体の商業化」、「観光」、「都市の地政学」、「輝く都市とマンハッタンの結婚」というキーワードで捉えてみよう。

稼げる都市

1970年代、80年代を通して民間資本の蓄積が徐々に進む一方で公共財政は逼迫する。そこで、政府は都市空間の再組織化に民間資金を利用することを思いつく。理念は高邁だが金のかかるモダン期の都市政策の代わりに、金を稼げる都市が求められた。その先鞭はイギリスがつけた。近代化の先頭を走ってきたイギリスは第二次世界大戦後には経済的に停滞した。マーガレット・サッチャーは、官僚主義の低い生産性、手厚い福祉政策、強すぎる労働組合が原因だと主張して首相に就任する[37]（1979〜1990）。サッチャーは、近代市民社会の基本概念である自由と平等の秤を自由の側に大きく傾け、モダニズムの都市政策を推進してきたGLC（大ロンドン議会）を解体した。このような政策はアメリカ合衆国のロナルド・レーガン大統領そして日本の中曽根康弘首相の政策とともに新自由主義政策と呼ばれている。その後、新自由主義はベルリンの壁の崩壊の余勢を駆って、グローバル資本主義へ拡大し、都市の経営にも市場原理を行き亘らせるように要請する。この頃には、モダニズムの都市を支えてきた西欧先進諸国の製造業部門の競争力が衰え始めていた。その代わりに都市を支える産業部門として情報と金融が浮上してくる[38]。

イギリス、アメリカそして日本と順次新自由主義的政策に傾斜し、その波は中国にまで至る。都市空間への投資の対象は郊外から都心に戻って来た。平等の理念に基づいた市民社会を実現する手段としての都市計画という発想が退けられただけではなく、都市を「計画する」こと自体が市場への無用な公的介入として嫌われ、都市建設を公共政策としてではなく、市場経済の一部門として見なすべきだと主張されるようになる。都市づくりの主役を降りた政府の役割は、モダニズムの都市政策を特徴付けたさまざまな規制を骨抜きにすることに変わったのである。言い換えれば、都市空間は流動化する資本の流れを加速させるターボチャージャーとして期待されたのである。

124 東京の鉄道郊外の風景（杉並区）
View of a Tokyo railway suburb (Suginami Ward)

where families dined on special occasions, shopping streets functioned as the central spaces of suburban communities.

Template of the Modern City

Instead of being a product of rational analysis, an urban form is often derived from a model. Under such circumstances, the form is intended to be revelatory. For example, if one looks at ancient Rome and cities built in parts of the Roman Empire, Chang'an and cities of tributaries of the Tang dynasty in the seventh century, and Haussmann's Paris and other European capitals of the same period, it is evident that ancient Rome, Chang'an and Paris served as templates in their respective spheres of influence. Then what served as the model for the modern city? The model was not an actual city but instead an ideal illustrated by Le Corbusier in a series of projects such as "Contemporary City for Three Million Inhabitants" (1922) that embodied modern values, namely industrialization, equality, economy, transportation and sanitation (or health) (→ 123). Later in life, Corbusier integrated the principles suggested in these projects and called that vision "La Ville Radieuse."

Meanwhile, it is difficult to discover the model for suburban residential areas. There are many projects and developments that contributed to the formation of the image of the suburbs such as Riverside,[35] Letchworth and Broadacre City,[36] but no spatial model for it seems to have been as influential as La Ville Radieuse was for the modern city. In Japan, the residential quarters of feudal-period samurai, with each estate surrounded by a wall, arguably served as a model for the suburbs (→ 124).

2. Postmodern Cities

The energy crises of 1973 and 1979 were economic events that greatly altered the world. They signaled that the old model for the pursuit of economic prosperity, based on resources such as petroleum obtained at low cost from undeveloped regions, would no longer work. Japan's economy grew at a sensational rate until 1974 (and in the process became the second biggest in the world as measured by GDP in 1968). However, from 1974 the rate of growth declined to an average of 3.8 percent, and from

商業化する都市空間

70年代になると東京の23区内には新たな住宅開発の余地が無くなり、郊外住宅地はさらに外部に向かって拡大し、それまでの鉄道郊外とは異なり、周縁部は大衆化し始めた自動車に対応した空間構造を持つようになった。他の先進諸国と同様に、脱工業化が進展し、東京湾沿岸に集中していた重工業は次々と地方都市や海外に出てゆき、後には大規模な空地が残された。こうした場所は都心に近く、しかも敷地規模が大きく自由な開発が可能である。工業地帯のための鉄道線は旅客線として位置づけ直され都心と結ばれ[39]、高速道路や高規格の道路が敷かれ、その周りに大規模な公園や遊園地、都心近くにはオフィスと高層集合住宅とショッピングモール、都心から離れた地区には集合住宅団地と戸建て住宅地などの複合的な大規模再開発が行われた (→125)。こうしたかつての工業地帯を開発して労働者の街が高等教育を受けた優雅なヤッピーの街に変えることは、脱工業化を果たした先進諸国の共通の都市政策である。

ポストモダン都市を特徴付けるのは過剰消費と都市空間の商業化である。小売り形式では、ショッピングモールとコンビニエンスストア(コンビニ)が代表する。スーパーマーケットからショッピングモールへの転換は、自動車社会が後押しをした。ショッピングモールはスーパーマーケットをテナントとの一つとして飲み込み、最寄り品と買い回り品の両方を揃え、さらにレストランやシネマコンプレックス、子供の遊び場なども組み込み、市民の週末の過ごし方を提示した。戦前の小林による郊外モデルが鉄道交通に対応していたのに対して、ショッピングモールは自家用車交通に対応し、空間的には孤立した「島」状をなすことが特徴である。「島」戦略は、駐車場がなく品揃えも古くさい地方都市の商店街に最後のとどめの一撃を加えた。

ショッピングモールの「島」のなかはテーマパーク手法で演出される。この戦略は都市の一地区全体に広げられ、やがて、都市空間のテーマパーク化ということになる、都市デザイナーは都市という巨大な売り場のインテリアデザイナーというわけである。そして、このような戦略を取れないような都市は、都市間競争から振るい落とされてしまうことになる。

他方、近隣商店街が担っていた住宅地の近所の小売店の役

125 東京湾岸の埋め立てと土地利用の転換
Land reclamation in Tokyo Bay and conversions in land use

1990 to 2007, it declined further to 1.3 percent. In 2011, China finally overtook Japan and became the second biggest economy in the world. The slowing of economic growth in Japan had a widespread effect on everything from lifestyle to the direction of design. Urban policies of the kind once guided by bureaucrats with a social democratic mentality were no longer adopted. Urban management came to be regarded as a branch of market economics, and urban spaces were now expected to accelerate the flow of capital.

These new conditions are widely referred to as "postmodern." The following phrases help to convey the characteristics of the reorganization of urban spaces in this period: "cities that bring in money," "historical character," "commercialization of urban space as a whole," "tourism," "urban geopolitics" and "the marriage of La Ville Radieuse and Manhattan."

Cities That Bring in Money

Through the 1970s and '80s, as private capital accumulation continued unabated, public finances became increasingly tight. Governments hit on the idea of using private capital for the reorganization of urban spaces. Instead of the urban policies of the modern era whose ideals had been noble but which had cost money, the policy now is to demand that cities bring in money. In this, the United Kingdom was the pioneer. The country that had led the way in modernization experienced economic stagnation after World War II. Asserting that low productivity tolerated by a bureaucratic mindset, generous welfare policies and powerful labor unions were to blame, Margaret Thatcher became prime minister (1979-1990).[37] She tipped the balance hitherto maintained between freedom and equality in modern civil society in favor of freedom and dissolved the Greater London Council which had promoted modernist urban policies. Thatcher's policies as well as the policies of Ronald Reagan in the United States and Nakasone Yasuhiro in Japan were referred to as neoliberal. Invigorated by the fall of the Berlin Wall, neoliberalism developed into global capitalism, and the application of market principles was demanded even in the administration of cities. It was around this time that the manufacturing sectors of Western developed countries that had helped support the cities of modernism began to lose their competitiveness. In their place, communication and finance emerged as the industrial sectors underpinning cities.[38]

One by one, the United Kingdom, the United States and Japan gravitated toward neoliberal policies, and that wave eventually reached even China. Investments in urban spaces, directed for a time to the suburbs, came to be channeled to central districts once more. Not only was the idea of city planning as a means of realizing a civil society based on the ideal of equality rejected, the planning of cities was itself condemned as a useless public intervention in the market; according to this view, urban construction should be regarded, not as public policy, but as a sector of the market economy. No longer a major

割はコンビニに取って代わられる(→126)。コンビニは、POSシステムを採用して消費者の最新の選好を反映させた品揃えをし、同時に製造業で開発されたジャストインタイム式の配送システムによって店舗の極小化と出店の機動性を身につけた小売り形態であり、通信と物流のネットワーク時代の産物である。コンビニの商圏は、大都市では徒歩5〜10分程度、2000人に一軒といわれ、首都圏はコンビニの商圏にほぼ覆われている(→127)。コンビニは営業時間を24時間に延ばしただけではなく、単なる小売業を超えて公共料金の徴収代行、宅配便の受け渡しと発送まで業態を広げ、さらに防災拠点に位置づける動きもある。コンビニは今や日本社会のなかで地域コミュニティの総合的拠点として確たる位置を築きつつある。これほどの重要性を帯びると、コンビニがあるかどうかが、住宅地の評価を左右すらするようになってきている。

　高齢化やフードデザートが広がる現代の日本では、良き近隣社会を代表する都市施設として路線型商店街が郷愁をもって眺められ、高齢社会での復活を期待する議論もあるが、これまでの小売り商業の歴史を冷静に見れば、従来の路線商店街のような形態での復活は大都市の一部を除いてありえないと考えるのが合理的である(→128)。

歴史的連続性

　ポストモダン期の新自由主義的性格を強調するだけでは、この時代を正確に説明したことにならない。近代主義の神話の崩壊は、同時に価値観の多様化を推し進めた。市場の支配が強まるのに対抗するかのように、市民たちが都市空間の再組織化に直接関わることを要求するようになった。また近代主義の進歩主義の前で居場所を奪われていた歴史性や場所の固有性にも関心が向けられ、市民が自分の居住地や慣れ親しんだ場所の環境の改善や景観や歴史の保全や復原が新たな権利として主張されはじめた。それは、60年代末ごろからのマンション紛争において主張された日照権運動や各地の歴史的な街並保存運動であった。その結果、1976年には建築基準法の改正で日影規制が盛り込まれ、1975年には文化財保護法改正により伝統的建造物群保存地区(伝建地区)が設けられた[40]。特に後者は画期的で、それまでは圧倒的に寺社が多く「民家」と言

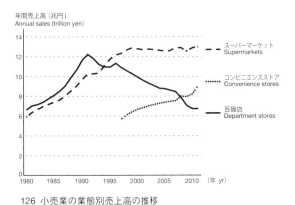

126 小売業の業態別売上高の推移
Changes in sales per retail outlet

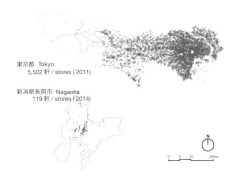

127 東京都と長岡市におけるコンビニの立地
Locations of convenience stores in Tokyo and Nagaoka City

player in urban development, government was now expected to pull the teeth of various regulations that had distinguished the urban policies of modernism.

Commercialization of Urban Space

In the 1970s, there was no room for new residential development in the 23-ward areas of Tokyo. Suburban residential areas expanded further; whereas suburbs in the past had been structured around railways, peripheral areas now had spatial structures adapted to automobiles, ownership of which had become widespread. As in other developed countries, society became more post-industrialized; one after another, the heavy industries that had concentrated along the coast of Tokyo Bay moved to local cities or overseas, leaving large, unused areas of land (→ 125). These places were located close to the center of the city, and their large sizes made unrestricted development possible. Large-scale, comprehensive redevelopment projects were implemented. Railway lines that had served the industrial belt were turned into passenger lines linked to the center of the city;[39] expressways and high-standard roads were laid; large

parks and amusement parks were located on the periphery, office buildings, high-rise housing and shopping malls were built near the center of the city, and housing estates and detached houses were built in districts at a greater remove from the center. Developing a former industrial belt and transforming a city of workers into a city of elegant yuppies who have received higher education is an urban policy common to advanced countries that have achieved post-industrialization.

The postmodern city is characterized by excessive consumption and the commercialization of urban space itself. Its representative forms of retailing are the shopping mall and the convenience store. The shift from the supermarket to the shopping mall was facilitated by the automobile. The shopping mall has swallowed up and made a tenant of the supermarket. Offering both convenience goods and shopping goods and incorporating restaurants, a multiplex, and a children's play area, it suggests ways people might spend much of their weekend. Whereas the prewar suburban model established by Kobayashi Ichizō was adapted to rail transportation, the shopping mall is adapted to automobile transportation; spatially, it is

えば農家どまりであった文化財行政に都市住宅である町家が加わり、しかも、保存対象が単体から街並に広がったこと、そしてこの動きが地方都市から起こったことも革命的であった（→ 129）。

「地域性」と観光

現代では、「自然」と同じくらいに、「地域性」は肯定的に捉えられている。なぜだろうか。

モダニズムは普遍主義であり進歩史観に裏打ちされているので歴史性や場所の固有性に関心を抱かなかっただけではなく、それらを排撃した。一方で、高速移動の大衆化と情報革命によって、海外旅行は大衆化し、さらにお茶の間にいてもテレビやインターネットを通じて古今東西を体験できるようになった。だれでも地球を俯瞰するような視点を持つようになり、自分の街もよその街も一望でき、比較できるようになる。「地域性」の誕生である。しかし、高速移動の大衆化と情報革命が進めば、「地域性」が消滅してゆくのは自然である。なぜなら人々は、この数十年のあいだに同じような物を見、同じようなものを着、同じような物を食べるようになったからである。人は失くした物を懐かしむが、地域性の消滅が不可避な帰結である以上それは仮構としてしか存在しえない。俯瞰的視点は旅行者の視点でもある。旅行者たちは訪れる土地に独特さを要求するものである。そもそも、近代人は、土地から切り離され、動き回るなど本質的に旅行者的な価値観をもっている。

現代資本主義は、大衆の旅行者的視線に新たな商売の可能性を見いだした。それは、都市を観光資源に変え、都市観光を一大産業として成長させただけでなく、さらに都市空間を消費の欲望を刺激する神話生成装置にまで高めた。その一つの都市戦略が、都市空間のテーマパーク化である。テーマパークは直接的には消費者を刺激し、同時に、都市どうしを競わせ都市自身にユニークであることを欲望させる。

「地域性」の希求は、大きい流れが支配する現代においては人々の感情の自然な発露なのだが、実は資本によって巧妙に仕組まれた販売促進活動のスローガンでもあることを忘れてはならない。

128 衰退する商店街（北海道夕張市）
The declining shopping street (Yūbari City, Hokkaidō)

129 町家の街並（川越市幸町）
Landscape of townhouses (Saiwaichō, Kawagoe City)

distinguished by its isolated, island-like form. Shopping malls were the final blow that finished off the shopping streets in local cities which lacked parking areas and offered only unfashionable sales items. Inside the island that is a shopping mall, techniques originally developed for theme parks are employed. Indeed, the same strategy is now used over entire districts in cities, transforming urban spaces into theme parks. Urban designers have become interior designers of what are essentially enormous sales areas. Cities that are unable to adopt such a strategy drop out of the competition among cities.

Meanwhile, convenience stores have taken over the role, once played by local shopping streets, of neighborhood retail stores in residential areas (→ 126). Convenience stores represent a form of retailing—and a product of the age of networks in communication and distribution—that uses the point-of-sale (POS) system to offer a selection of merchandise reflecting the latest preferences of consumers; they are able to minimize the size of each store and to open a new store quickly through the use of a just-in-time style of distribution. The target area of a convenience store lies within a five to ten-minute walking distance in a metropolis and has a population of 2,000; the Capital Region is almost entirely covered by the target areas of convenience stores (→ 127). Convenience stores not only operate 24 hours a day but offer, besides retail services, payment service for public utilities and courier service; there are also places where they have been designated bases for disaster-prevention activities. Convenience stores have become so important that their presence or absence nearby can affect the valuation of residential land.

In contemporary Japan with its aging society and extensive food deserts, the local shopping streets of the past are remembered nostalgically as symbols of more cohesive communities, and there is talk of reviving them. However, one can only conclude from a dispassionate examination of the history of retailing that, excepting certain parts of metropolises, the revival of the local shopping streets of the past is impossible (→ 128).

Historical Continuity

An accurate portrait of the postmodern period would require more than an emphasis on its neoliberal character.

都市の地政学

　1970年代には、産業構造の脱工業化を追いかけるように、交通環境の変化が日本の地方都市を大きく変えた。都市間の時間距離の短縮と移動の個人化は、都市の勢圏を大きく変えた。都市間の移動がたかだか時速100キロ程度の鉄道に限られていた時代には一つの県に商業・文化の中心的都市が複数あったが、時速200キロの時代になると、それらの多くが没落し、代わりに札幌、仙台、東京、横浜、新潟、金沢、名古屋、京都、大阪、広島、福岡などの限られた都市だけが繁栄し、数県に跨がる広域を商圏として支配するようになる（→130、131）。

　1990年以降の、経済と文化のグローバル化は、この傾向に拍車をかけた。特に東京を含む首都圏は、時速900キロで活動する企業や、国際機関、教育・学術活動、創造活動などの拠点として選ばれることを求めて国際的な競争の中に曝される。都市の魅力は優秀な人材と資金を集める鍵であり、これらがグローバル時代の大都市の繁栄を左右するということが認識されるようになる。

ポストモダン都市のモデル

　都心では、モダン期の都心の雛形であった「輝く都市」は、ニューヨークのマンハッタンと結ばれ、その間にできた子供が新たな雛形の座につく。母のマンハッタンは資本主義の城であり、近接性と自己主張の強い超高層建築の遺伝子をもち、父の「300万人の都市」は近代の衛生思想と工業技術の結晶であり、建築と建築、建築と街路の分離を遺伝子としてもつ。理想主義と現実主義を両親とする都市モデルは最強であった。パリのデファンス、アメリカ大陸の都市、中国の大都市、そして東南アジアの大都市、産油国の都市など世界中の都市開発のモデルとなった。日本でも、1965年に始まる東京の新宿副都心を皮切りに全国の大都市の埋め立て地区の開発のモデルとなり、近年は郊外の開発にまで及んでいる（→170）。

　大衆的なレベルで展開される日本の住宅地の雛形は、ポストモダン期になっても捉え難いが、現代都市風景のなかで一番似ているのは何かと問えば、それは住宅展示場の風景ではないだろうか。日本には、実物の住宅を展示し、美麗なカタログを用意して販売員が顧客に応対する住宅展示場が全国にある。これ

The collapse of the myth of modernism also promoted a diversification of values. The increasing dominance of the market was met by increased demand by people for direct involvement in the reorganization of urban space. Concern grew for historical character and the distinctive character of place, factors hitherto dismissed by modernism with its progressivist slant, and people began to call for the improvement of the environment and the preservation or restoration of the landscape and history of the places in which they lived or with which they were familiar. The construction of condominiums toward the end of the 1960s touched off "right-to-sunlight" movements and movements to preserve historic townscapes in various parts of the country. As a result, regulations governing shadows cast by new construction on neighboring lots were introduced into the 1976 revision of the Building Standards Law, and the 1975 revision of the Cultural Properties Law established Preservation Districts for Groups of Traditional Buildings.[40] The latter in particular was a landmark event; most buildings designated as cultural properties had hitherto been temples and shrines, with farmhouses (*nōka*) being the only traditional houses (*minka*) accorded such protection. Now, however, townhouses (*machiya*) were added to the types of buildings eligible for consideration. What is more, instead of individual buildings entire townscapes were now to be preserved. The fact that this movement began at the grassroots level in local cities was also revolutionary (→129).

Local Character and Tourism

　Today, "local character" is considered something almost as positive and worthy of support as "nature." Why? Modernism, with its universalist perspective and belief in historical progress, did not simply ignore historical character and the distinctive character of place, it actively sought to eliminate those qualities. Meanwhile, the availability of high-speed modes of transport made travel,

130　ブランドCが店舗を置く都市
Cities with Brand C shops

domestic and overseas, a mass phenomenon, and the information revolution enabled people to experience in their living rooms through television or the internet any period of history and any place on earth. People were able to step back and get a broader view—a bird's-eye view—of things, that is, to see their towns within a larger context and compare one place with another. That was how "local character" was born. However, mass utilization of high-speed means of transport and the information revolution will ultimately lead to the disappearance of local character. That is because over the last several decades, people have all come to see similar sights, wear similar clothes and eat similar foods. People are nostalgic about what they have lost, but the disappearance of local character is unavoidable; local character can only continue to exist in the imagination. A bird's-eye view is the viewpoint of the traveler. Travelers demand uniqueness of the places they visit. Uprooted and mobile, people of the modern era have had from the start the mindset of travelers.

　Contemporary capitalism has discovered in the travelers' perspective of the general public the potential for new business. It has not only changed cities into tourist

は自動車のショールームと同じ販売方法であり、大企業による工業化住宅が市場で成功しているからこそできることである。自動車のショールームとの違いは、複数の住宅メーカーが展示場を共有することである。住宅展示場には、複数の住宅メーカーが差異を競って北欧風から和風、アメリカ西海岸風、フランクロイドライト風などさまざまなデザインの住宅の実物を並べて展示している。そこには、まとまりのある外部空間や意図をもって形成された風景が認められない。ただ、商品が相互に無関係に並んでいるだけである (→132)。それが現代の日本の住宅地にそのまま再現されている。

日本の自家用車に依存する郊外の景観を特徴付けるのは、住宅展示場に似た住宅地と、その間に挟まる細切れの農地、それに工場のような外観をもつ巨大なショッピングモール、そして散在する小さなコンビニである。

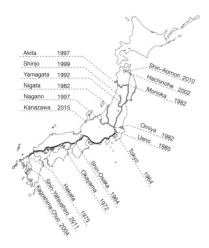

131　新幹線網の延伸拡大
Expansion and spread of Shinkansen network

resources and developed urban tourism into a major industry but transmuted urban spaces into myth-generating devices that stimulate the desire for consumption. One urban stratagem is to transform urban spaces into theme parks. Theme parks stimulate consumers directly; at the same time they inspire cities to compete against each other and instill in them a desire to be unique.

A longing for local character is the natural expression of people's feelings at a time like today that is dominated by Big Flow, but one should not forget that it is also something deliberately and cleverly exploited by capital to promote sales.

Geopolitics of Cities

In the 1970s, the post-industrialization of the industrial structure was followed by changes in the transportation environment that greatly altered local cities in Japan. The shortening of travel time between cities and the increasing tendency to move, not in groups, but individually, greatly changed the spheres of influence of cities. When movement between cities was limited to railways traveling about 100 kilometers per hour, multiple cities serving as commercial and cultural centers could coexist within a single prefecture, but when travel speeded up to 200 kilometers per hour, many of those cities fell by the wayside. Only a limited number of cities such as Sapporo, Sendai, Tokyo, Yokohama, Niigata, Kanazawa, Nagoya, Kyoto, Osaka and Fukuoka have prospered; each dominates a large commercial sphere of influence that extends beyond prefectural borders (→ 130, 131).

The globalization of the economy and culture since 1990 has accelerated this trend. In particular, the Capital Region including Tokyo seeks to attract corporations whose activities are premised on travel at 900 kilometers per hour, international agencies, and educational, scientific, and creative activities, and in that pursuit competes against other international cities. It has come to be recognized that the appeal of a city is key to assembling a pool of talented people and raising funds and that these factors can determine the prosperity of metropolises in the age of globalization.

132 住宅展示場の風景（埼玉県志木市）
View of a housing gallery (Shiki City, Saitama)

Template of the Postmodern City

The offspring born of the union of La Ville Radieuse, the model of the central district in the modern era, and Manhattan, the stronghold of capitalism, has become the new template for the center of a city. The City of 3 Million Inhabitants, a crystallization of the modern idea of hygiene and industrial technology, is the "father" whose genes produce a clear separation of building from building and building from the street. Manhattan is the "mother" whose self-assertive skyscrapers are packed closely together. An urban model born of these two visions, one embodying idealism and the other realism, has proved all-powerful and served as the template for urban development throughout the world including La Défense in Paris, the cities of the American continent, the metropolises of China, the mega-cities of Southeast Asia, and the cities of petroleum-producing countries. In Japan too, it has been the model for the development of reclaimed areas in cities throughout the country, starting with the Shinjuku Subcenter of Tokyo in 1965. In recent years, its influence has been felt even in the development of the suburbs.

The template for ordinary residential areas of Japan is difficult to pinpoint even in the postmodern era, but the landscape of a "housing center" may come the closest. In Japan, places where actual houses are arranged and sales personnel are on hand to show them to potential customers are to be found throughout the country. This is the same sales technique used in automobile showrooms and is possible only because industrialized houses by large corporations do well in the marketplace. A housing center is different from an automobile showroom in that different house manufacturers display their products side by side. In a housing center, houses in diverse designs, suggestive of, for example, Scandinavia, traditional Japan, the West Coast of the United States and Frank Lloyd Wright, compete for the visitor's attention. There is no coherent outdoor space or deliberately formed landscape. Products are simply arranged next to one another without any discernible relationship (→ 132). The result perfectly recreates contemporary residential areas in Japan.

The townscape of the automobile-based suburbs of Japan are characterized by residential areas that resemble housing centers interspersed with narrow strips of farmland, enormous shopping malls that have factory-like exteriors, and small, scattered convenience stores.

われわれは、最初に縮小という現実を確認した。続いて20世紀の都市の二段階の発展史を振り返ったばかりで、いささか性急ではあるが、21世紀の縮小する都市はどうあるべきかに議論を進めよう。

[1] 自然と民族と未来からの収奪に依存することなく、当面の縮小、そしてその先の定常状態にあるべき社会と都市を構築すること

　部分的や一時的な縮小はときには起こることであり心配ないが、全面的かつ長期に縮小が続けば、やがてその社会は消滅してしまう。社会構造の変化の調整としての縮小が過ぎたあとは定常状態に落ち着かなくてはならない。縮小の原因は、鉱物資源と地球環境からの収奪と他地域（民族）からの収奪そして未来からの収奪の三つの収奪の限界、先進諸国における低い出生率と高齢化、そして欲望の飽和であった。それゆえ21世紀の世界では、少なくとも先進地域においては「縮小」は不可避

第三章：21世紀の都市のための10箇条

Chapter 3. Ten Articles for the Development of Twenty-First Century Cities

We began by confirming the reality of shrinkage. Now, having reviewed the history of the two-stage development of cities in the twentieth century, we shall briefly discuss what shrinking cities of the twenty-first century ought to be like.

[1] Construct society and cities that are appropriate for the process of shrinkage we face now and the "steady state" we hope to achieve eventually, without exploiting nature, other peoples or the future.

　A partial or temporary shrinkage occurs from time to time and is not a cause for concern, but if a total, long-term shrinkage continues indefinitely, society will eventually wither away. Once sufficient adjustments have been made through shrinkage, society needs to settle into a "steady-state" condition. The causes of shrinkage are the limits to the three forms of exploitation—namely the exploitation of mineral resources and the global environment, the exploitation of other regions (or peoples), and the exploitation of the future—the low rate of birth and the aging of the population in advanced countries, and the

である。「不可避」の意味は、客観的状況として他に選択肢がないということ、この地球世界を存続させるためには、先に発展し「収奪」の恩恵を味わった地域は率先して「縮小」を自ら選択すべきという倫理的な要請の両方である。ただし、後者は、苦難を耐え忍ぶべきだということではなく、経済成長がなくても人々が満足できるような価値転換を図り、少ない資源投入で多くを得る方法を編み出し、拡大と成長を前提条件に組み立てられた近代都市計画理論に代わる新しい都市理論を構築することに向けて叡智を集めることである。

　産業革命が都市に大きい衝撃を与え、近代都市を定式化したように、「縮小」も都市と建築に大きい衝撃を与えることは間違いないことである。

[2] 自然と人工物を等価に位置づけ、人間の居住地を大きい環境システムの一部とすること

　前項の目標を補完するためには、自然の征服をもって人知の偉大さを誇示するという19世紀的な思考をやめなければならない。自然と人工物を協働させることによって人間の環境を作る方法を追求しなければ「自然からの収奪」を止めることはできず、将来の世代に不当な付けをまわすだけであることをしっかり認識しなければならない。

[3] 都市と農林漁業地域との健全な相互依存関係を確立すること

　より具体的には、都市と農村漁村を人間の居住域のシステムとして調和的に再組織化しなければならない。日本の制度に顕著に見られる、都市行政と農村行政を区分けするという官僚的習慣と専門分野間の無関心や敵愾心を排することが第一歩である。そして、第一次産業は地域の環境保全と食料自給など重要な役割を担っているにもかかわらず、後継者不足に悩まされグローバルな競争にも曝されている過酷な状況にあることを知らなければならない。その解決のためにも、これらの産業における世襲的な事業継承から脱し、意欲のある人なら誰でも就労できる産業に変わるべきである。

satiation of desire. Accordingly, shrinkage is unavoidable in the twenty-first century, at least in advanced regions of the world. "Unavoidable" means not only that no other alternatives from an objective viewpoint exist, but that ethically, regions which developed first and enjoyed the benefits of "exploitation" ought to take the initiative and choose shrinkage for themselves. However, choosing shrinkage does not mean having to endure hardships; instead, it is to embrace different values that bring people satisfaction without economic growth, to devise ways of obtaining more from the investment of fewer resources, and to marshal intellectual forces toward the building of a new theory for cities to replace a modern city planning theory premised on expansion and growth.

　The Industrial Revolution had an enormous impact on cities and established their modern form. Without a doubt, shrinkage will also have an enormous effect on cities.

[2] Place equal value on nature and artifact and make the human habitat a part of the larger environmental system.

　To help achieve the objective intended by Article 1, we must discard the nineteenth-century idea that the conquest of nature demonstrates the greatness of human knowledge; unless we develop a method of creating the human environment from both nature and artifact, we cannot stop exploiting nature and making future generations pay unfairly for our actions.

[3] Establish a sound relationship of interdependence between cities and localities dedicated to agriculture, forestry and fishing.

　More specifically, cities and agricultural and fishing villages must be reorganized in a harmonious way as a system of human habitation. The elimination of the bureaucratic practice of dividing urban administration from the administration of agricultural villages and the mutual indifference or hostility of different fields of specialization evident in Japanese institutions is a first step. Awareness is necessary of the severe conditions in which primary industries must operate—the shortage of successors and the exposure of these industries to global competition—despite the important role they play in the preservation of

[4] 自然と社会の多様性を尊重すること

21世紀の社会を考えるときに最も重要な概念の一つが多様性であろう。高速交通と大容量高速通信の発達は、都市に荒々しい多様性を持ち込むが、同時に、弱い地域文化が強い地域文化に曝され、人々が落ち着いて判断する間もなく容易にしかも短時間に支配される危険性も増す。生態系は多様性を失うと脆弱化するが、これは人間社会にもあてはまる。都市のなかにある力の弱い文化を許容することこそが都市の創造性を刺激し、生産性を高める。

[5] 過去の物的環境をできる限り継承し、改修して使い続けること

多様性のなかには、時間的多様性も含めなければならない。つまり、進歩主義を排し、異なる時代の思考と物が共存することが重要である。より具体的に言えば、歴史との連続性を保ち、再利用を促進することが必要である。建築を平均30〜40年しか使ってこなかった日本では歴史的構造物を改修して使い続ける習慣が官民ともに乏しい。真の地域性は、現在と過去の対話のなかにしかない。ただし、そこに残っているだけの歴史的遺物は過去の抜け殻でしかない。改修・再利用は現在と過去との対話であり、新しい地域文化の創造に繋がる行為である（→133）。

[6] 人間の居住地のあり方すべてを市場原理だけに任せてはならない

多様性を尊重しようとすれば、経済的強者にのみ優先権を与える市場万能主義に足かせを掛けねばならない。居住地の多様性、自然の適切な管理、健康な食の確保、地域文化の継承、弱者の人権の保障などは都市にとって欠かせない重要なことである。これは市場機能だけでは達成することができない。

133 高島屋日本橋店（1933、1952-1965）
Takashimaya Nihonbashi Store (1933, 1952-1965)
髙橋貞太郎設計による折衷様式の正面部分に、村野藤吾設計になる近代様式の増築部で構成された百貨店建築。2009年に国の重要文化財指定を受けた。

This department store incorporates a façade in an eclectic style designed by Takahashi Teitarō together with a modern addition by Murano Togō. It became a nationally designated Important Cultural Property in Japan in 2009.

the local environment and self-sufficiency in food. To solve these problems, these industries must themselves change, doing away with the custom of hereditary succession and accepting any worker with ambition willing to work in them.

[4] Respect the diversity of nature and society.

One of the most important concepts in considering the society of the twenty-first century is diversity. Progress in high-speed transportation and high-capacity, high-speed communication introduces diversity into cities, but at the same time there is increased danger that weak local cultures are exposed to powerful local cultures and that people are swayed before they can make up their own minds dispassionately. Ecosystems are weakened by loss of diversity, and the same is true of human society. It is precisely the toleration of weaker cultures in cities that stimulates urban creativity and leads to higher productivity.

[5] Accept, repair and continue to use the physical environment of the past as much as possible.

Diversity must include temporal diversity. That is, it is important to reject progressivism and permit the coexistence of ideas and things from different periods. More specifically, it is necessary to preserve historical continuity and promote recycling. In Japan, where buildings on average have been used only thirty to forty years, neither the public sector nor the private sector is accustomed to repairing and continuing to use historical structures. True local character only exists in the dialog between present and past. Historical remains that have simply survived are no more than the lifeless shells of the past. Repair and recycling represent the dialog of the present with the past and leads to the creation of a new local culture (→133).

[6] Refuse to let all aspects of human habitation be governed by the market principle.

Reliance on the market to resolve all issues, an approach that awards priority only to the economically powerful, must be kept in check if diversity is to be respected. Diversity of habitation, appropriate management of nature,

[7] 大都市はグローバルな文脈での魅力的な場を提供し、中小都市は地域で循環する経済を打ち立てること

　日本の都市における多様性の危機は、中小都市の衰退に現れている。大都市と中小都市の共存は国や地域の活力と持続可能性にとって重要である。大都市は他の国際都市と張り合える魅力を持たなければならない。一方、交通網や情報網が整備されるとグローバル資本や全国基盤の資本の支配力が強くなり、地方都市は草狩り場にしかならず、地域での消費が地域に還元されず地域から流出してしまう。里山資本主義を提唱している藻谷浩介は、「大都市につながれ、吸い取られる対象としての「地域」と決別し、地域内で完結できるものは完結させよう……」[41]と主張する。人口減少は好機だと考える広井良典は「経済の中にあった互酬性や相互扶助の要素……をもう一度回復していくこと、つまり経済をもう一度コミュニティや自然とつないでいく」[42]と主張している。いずれも、地域で循環する経済の重要性を言っているのである。

[8] だれもが自分の住む場所の向かうべき方向が感じられ、その決定に参加ができること

　民主主義は最善の方法と言えないかもしれないが、現在われわれが手にしている制度のなかでは、構成員の尊厳をもっとも尊重できる政治制度であり、その維持と発展に向けて努力を怠ってはいけない。都市に関ることで言えば、市民が自分の住む場所のあり方に如何に関われるかが重要である。成長経済下では人任せであっても待てば良くなったが、縮小期の都市経営は成長期に比べると各段に難しい。住民が納得ずくで都市の縮小に対応しなければ、やがて住民に付けが回ってくる。

　第二次世界大戦後半世紀間は、多くの先進諸国では民主主義がよく機能し、国民の基本的人権を守ろうという熱意にあふれていた。しかし、社会の性格が変わり新しい状況や集団が生まれると、社会は平等や自由の概念をそこにどう適用すればよいのかわからなくなる。倍増する高齢者、労働参加が求められる女性や高齢者、限界集落などの衰退地区そして外国人労働者や移民などが「新しい状況や集団」である。新しい状況のも

availability of healthy foods and regard for the rights of the weak are all vital urban objectives that cannot be achieved solely through the workings of the market.

[7] Ensure that metropolises offer attractive places within a global context and that small and medium-size cities build economies in which money is circulated locally.

　The crisis of diversity in Japanese cities is evident in the decline of small and medium-size cities. The coexistence of metropolises on the one hand and small and medium-size cities on the other is important to the vitality and sustainability of the country and localities. Metropolises must possess the appeal to compete with other international cities. Meanwhile, once transportation and communication networks develop, global capital and capital that has a nationwide groundwork gain dominance; local cities become nothing more than places from which many outsiders seek to profit, and profit from local consumption is not reinvested in that locality but flows out instead. Motani Kōsuke, an economist and proponent of *satoyama* or "rural" capitalism, urges localities to cut ties to metropolises insofar as those ties allow metropolises to feed off them; "a locality ought to be complete in itself where possible."[41] Hiroi Yoshinori, who believes the decline in population offers an opportunity, advocates "the restoration once more of elements of reciprocity and mutual aid that once existed in the economy, that is, to connect the economy once more to community and nature"[42]. Both assert the importance of circulating money within a locality.

[8] Enable everyone to participate in deciding the direction the place they live in will take.

　Democracy may not necessarily be the best political system, but it is the one that most respects the dignity of its members; we must always strive to maintain and develop it. With respect to the city, the way in which citizens are involved in determining the nature of the place they live in is important. In a growing economy, conditions improved even if work was left to others, but in a period of shrinkage, urban administration is far more difficult. People must understand, and be committed to, measures taken to deal with urban shrinkage or they will be the ones to pay a

とで、新しい住民構成なのかで、自分たちの町の方向を決めなければならない。

[9] すべての人が好きなときに好きな所に移動できる機会が得られること

都市計画の大きい使命は、市民の基本的人権を保障することであるが、その項目として忘れられがちなのが公共交通である。移動は基本的人権の問題である（→ 134）。

[10] 縮小の時代の都市空間の再組織化は政府、市場、住民の三者の共同作業によること

これらの目的を目指しながらも、同時に居住者人口の減少や人口の高齢化に直面せざるをえないのが日本の都市である。居住地としての質を維持し活力を保ち激化する地域間競争に勝ち抜かなければならない（→ 135）。しかも、放置すれば事態は深刻になる。それを誰が舵取りするのだろうか。成長期の主役であった地方自治体は、人口減少で税収が減り、高度経済成長期以来作られ続けた大量の公共施設が2020年頃から徐々に更新時期に差し掛かり出費が増える[43]。また、今後大都市は人口の高齢化に直面し介護と医療の費用が嵩む。縮小期こそ自治体の役割が期待されるが、多くを期待するのは難しいだろう。80年代以降、都市開発の主役を張ってきた民間企業にとって、今後地価が長期的に下落、低迷すれば不動産業はこれまでと違って魅力的な産業部門ではなくなり、この産業から撤退する企業も増えるだろう。

民間事業者も自治体政府もいずれも単独で、都市空間の再組織化の主役を張れないとなると、住民自身が自分の居住地経営に乗り出さざるをえないだろう。ただし、集落の集団移転や区分所有マンションの解散など当事者だけでは解決が難しい課題も増える。というのは、住民は当事者なので、住民間で利益が相反することもあるからである。また、地域を超えた広域に関わる問題の判断を住民に期待すべきでもない。つまり、住民の参加は不可欠であるが、住民が単独で地域を切り盛りして合理的な方向に導けるというものでもない。

134 自転車タクシー（八王子市）
Bike taxi (Hachioji City)
大規模な団地である館ヶ丘団地内での移動を助ける活動をNPOが行っている.
An NPO works to assist mobility in the large-scale housing complex of Tategaoka.

135 老朽化する町家（徳島県美波町）
Deteriorating townhouses (Minami-chō, Tokushima)
近世の伝統を受け継ぐ庶民の住まいの多くは老朽化し消滅しようとしている.
Many commoners' houses that inherit the traditions of the early modern period are starting to deteriorate and are facing extinction.

penalty of some kind.

For about a half-century after World War II, democracy functioned well in many advanced countries, and the basic human rights of citizens were zealously guarded. However, when the character of society changes and new conditions or groups come into being, society has difficulty understanding how to apply concepts of equality and freedom to those new conditions or groups. Here, by "new conditions or groups," I mean a twofold increase in the number of seniors, women and seniors in the workforce, areas in decline such as so-called "critical villages" (*genkai shūraku*) that have become so depopulated that they can no longer function as communities, and foreign workers and immigrants. People must determine the direction their towns will take under new conditions and with new community composition.

[9] Give everyone the opportunity to move when and where they like.

A major mission of city planning is to guarantee the basic human rights of citizens, and one such right that is apt to be forgotten is the right to public transportation (→ 134). Movement is a question of basic human rights.

[10] Make the reorganization of urban spaces in the period of shrinkage a collaborative endeavor on the part of the government, the market and the public.

Even as they work toward these objectives, Japanese cities will have to confront a decline in and the aging of their populations. They must not only maintain their quality as habitats but retain their vitality and survive the intensifying competition among different localities. If neglected, the situation will become serious (→ 135). Who will guide them? Local governments that played major roles in the period of growth have seen their tax revenues shrink with the decline in their populations but will need to increase their expenditures as the need to rebuild or renovate public facilities that continued to be constructed in great quantities from the period of intensive economic growth gradually makes itself felt from around 2020.[43] Moreover, metropolises will confront the aging of their populations, and the costs of nursing and health care will

可能性のある途は、自治体政府、民間企業、そして住民の三者がそれぞれ責任をもって地域の経営にあずかる協同的な主体を構想しなければならないだろう。そして、新しい担い手に適した都市空間の再組織化手法と制度の開発が期待される。

このような展望を述べると、必ず、日本の市民の成熟度の低さが問題にされる。確かに、官僚による父性主義的政策が日本の市民としての成長を妨げたところはあるが、明治維新の尋常ではない開化の速度の裏には、むしろ、力強い市民力があったことを忘れてはならない。長岡を例にとれば、明治時代の最初の10年の間に、長岡市民は、銀行を設立し、病院と学校を開き、汽船の運航会社を作り、育英会を組織し、信濃川に橋を掛けた[44]。これらが皆、地域の発意で民間資金で進められたのである。封建制の身分制のなかで資本主義の企業家精神が育っていたことを伺わせる。ところがその後の近代化の過程で地方都市の創発力は中央政府に掠め取られてしまった。高度成長期以降、都市プロジェクトを立案するのは国であり、地方自治体はそれに対して関心を表明するだけの立場になってしまった。政府が自治体を選び、補助金が支給される、というのが現在の姿である。しかし、こうした政策は、国は財政難で今後は取りづらくなっていくだろう。再び市民の力を発揮するときである。

mount up. Much will be expected of local governments in the period of shrinkage, but it will be difficult for them to meet such expectations. Private enterprises have been key to urban development since the 1980s, but in the future, as land prices fall and stagnate, many of them are likely to find the real estate business less appealing and to withdraw from it altogether.

If neither a private developer nor a local government is by itself capable of playing the main role in the reorganization of urban spaces, residents may have no alternative but to embark on the administration of their own habitat. However, residents by themselves will find it difficult to solve certain problems such as the collective relocation of a village or the dissolution of the ownership arrangement of a condominium building. That is because the residents in such cases are interested parties, and what may benefit one may be disadvantageous to another. Moreover residents cannot be expected to judge on a problem that has an impact beyond the immediate area in question. Thus, though the participation of residents is indispensable, the management of a community and the guidance of that community in a rational direction cannot be left solely to the residents. The only possible course is to devise a collaborative body that enables all three parties—the local government, private developer(s) and residents—to participate in the management of the area. Methods and systems for the reorganization of urban spaces suitable for such a body then need to be developed.

When such prospects are explained, the immaturity of Japanese citizens is often cited as a problem. The paternalistic policies of bureaucrats have certainly impeded the growth of the Japanese as citizens, but the fact that powerful civic forces were behind the extraordinary speed with which enlightenment was achieved in the Meiji Restoration ought not to be forgotten. To take Nagaoka as an example, in the first ten years of the Meiji period, the citizens of the city established a bank, founded a hospital and schools, created a steamship company, organized a scholarship society and built a bridge over the Shinano River.[44] These were all carried out at the suggestion of people in the area and with private funds. This indicates that despite the social class system imposed in the feudal period, a capitalist, entrepreneurial spirit had already developed. However, in the subsequent process of modernization, the central government robbed local cities of their creative initiative. Since the period of intensive economic growth, it has been the central government that has devised urban projects, and local governments have only been allowed to express interest in such projects. At present, the central government awards subsidies to selected local governments. However, the central government, faced with financial difficulties, has trouble continuing to adopt such policies. It is time that citizens demonstrate their power once more.

ポストモダン期もいよいよ大詰めを迎えて、日本の都市も時代の変り目にある。日本の都市の潜在力を診断してみよう。これは、第二部で、日本の都市の将来像を考える上での設計条件でもある。

1. 隙間に息づく自然

かつて、日本の都市も建築も自然との距離が近かった。日本の都市には、西欧の都市と違って、至る所に隙間があり、そこから自然が顔を出している。西欧人の専門家が日本の街並に出会って真っ先に不思議がるのが都心の建物と建物のあいだにある特に用のない隙間である。隙間は人工物の裂け目である(→136)。半世紀前の多くの家庭の台所は、湿っていて、暗く、すすけていて、幾分黴び臭く、ナメクジが常連客であった。それは土が露出する地面も同じで、半世紀前は未舗装の道はそこら中にありふれていた。道際には雑草が生え、雨が降ればぬかるむ。こうした隙間や裸地は近代的感覚の嫌悪の対象となり、台所はシステムキッチンとなり、道路は完全に舗装し尽くされた。

136 縁の下
Empty space under the raised floor

第四章：日本の都市の診断書

Chapter 4. A Diagnosis of Japanese Cities

The postmodern period is at last coming to a close, and Japanese cities are at a turning point. Let us examine the latent power of Japanese cities. The conditions that are identified are also the design conditions for conceiving the future image of Japanese cities in Part 2.

1. Nature in Gaps

Japanese cities and buildings were once close to nature. Unlike Western cities, Japanese cities have gaps everywhere, and nature appears in those gaps (→ 136). On first encountering Japanese townscapes, the things that most puzzle Western experts are the gaps that exist between buildings in the middle of a city and that have no particular use. Gaps are interruptions in man-made objects. Half a century ago, many household kitchens (*daidokoro*) were damp, dark, stained with soot, slightly musty and apt to harbor slugs and insects. The same was true of areas of exposed earth; fifty years ago, unpaved roads were commonplace. Weeds grew by the side of a road, and rainfall would turn earth into mud. Such gaps and exposed earth came to be loathed by anyone with a

身の回りの小さな自然があらかた人工物で覆い尽くされてしまうと、われわれの記憶の深部にある自然の近さを好む心性がうずき始める。神社の境内や、都心に奇跡的に残る渓流、護岸が施されていない岸辺、放置されて雑草が繁茂する空き地、倉庫の裏のコンクリート舗装のひび割れから伸びる雑草、巨木、崩れかけの石垣、それらに、多くの人は自然の息吹を感じ、共同体の神話や個人の神話の世界に引き戻される（→137）。人の感受性は移り変わるのだが、深部には集団的な記憶が宿っている。東京の古層の探訪を行なった宗教学者中沢新一の本が多くの読者を惹き付けた理由はその辺りにあるだろう。『アースダイバー』という表題の通り、中沢は、現代の人工物の下に潜む古層を求める都市民俗学的な旅の案内役である。人は、身の回りに複数の時間軸を見いだすとき充実感を感じる。

古代の都市計画理念は幾何学的秩序を好む中国の強い影響下にあったが、日本人の地理的な空間意識にまでは影響しなかったように思える。むしろ、伝統的な日本の都市には、人間の住む領域は主に自然物で定義されるべきであり、人工物はそれを補う程度でよしとする傾向があった。文献と実地調査から伝統社会の日本人の領域意識を研究した樋口忠彦の研究では、7種類の領域の型を構成する要素はすべて自然の山や川である。平時の自然がそれほど過酷ではない日本では山並にいだかれた里という概念が生まれたとしても不思議ではない。

江戸時代の浮世絵からわかるように、江戸の町筋の焦点には山が描かれている（→138）。都市史の研究者である桐敷真次郎は都市図を分析して、江戸の主要街路の方向は一見無秩序に見える、江戸を包囲する山々の一つである富士山や筑波山などが道の先に見えるように決められたという仮説を提出した[45]。日本の伝統的な社会では、山は神道において神性に満ちた場所として祀られていた。視覚構造の焦点として選ばれたのが、一つでなく複数であるところはバロック都市と似ているが、バロック都市の主要道路の焦点に置かれるのは建築モニュメントであるのに（→139）、江戸では都市の境界をなす山が焦点になっているところが対照的である。

137 東京の名所と線状緑地
The major sights of Tokyo and its linear green belt

138 焦点に自然がある軸線で構造化された都市
A city organized on an axis with nature at its focal point

前後の町筋を合わせても長さ500 m余の駿河町の町筋が100 km先の富士山を見据えている。短い直線街路と焦点にある山が江戸の都市計画の特徴である。歌川広重画「名所江戸百景する賀てふ」(1856-8)

This small piece of road named Surugachō less than 500 meters, even counting the streets leading to and from it. The road looks towards Mount Fuji, located 100km

modern sensibility; in time, *daidokoro* were replaced by system kitchens and roads were completely paved.

A deep-seated desire to be close to nature begins to assert itself as the immediate environment becomes increasingly artificial. We sense the power of nature and are drawn back to the world of communal myths in the precinct of a shrine, a woodland stream that by some miracle survives in the middle of the city, a riverbank that has not yet been covered by a concrete embankment, an untended vacant lot overgrown with weeds, flowering plants growing from cracks in the concrete paving behind a warehouse, an enormous tree or a half-collapsed stone wall. Human sensibility is subject to change, but deep down inside group memory lives on (→137). That may be why so many readers were attracted to a book by Nakazawa Shin'ichi, a scholar of religious studies who investigated the old layers of Tokyo. As its title, *Earth Diver*, suggests, Nakazawa is our guide on a journey into urban folklore in search of the layers from the past that lie concealed under the contemporary man-made environment. One has a greater sense of the richness of life when one discovers multiple time frames in one's immediate environment.

China, which was partial to geometrical order, heavily influenced Japanese city planning ideas in the ancient period, but that influence does not appear to have extended to the Japanese consciousness of geographical space. Instead, in traditional Japanese cities, there was a tendency to have natural objects define domains of human habitation; in this, man-made objects had merely an auxiliary role. According to Higuchi Tadahiko, who studied the consciousness of territory in traditional Japanese society through an examination of literature and first-hand inspection, seven types of territory were composed entirely of natural elements, that is, mountains and rivers.[45] It is not strange that in Japan, where nature is usually mild, the idealized image of home has traditionally been a village cradled by mountains.

In Edo-period woodblock prints, mountains are the focal points of streets in Edo (→138). Kirishiki Shinjirō, a scholar of urban history, analyzed maps and put forward the hypothesis that the main streets of Edo, which at first glance seem to have been laid without any underlying principle, were in fact oriented so that one of the mountains surrounding the city such as Mt. Fuji or Mt. Tsukuba could

2. 線形性好み

日本には、都市、文学、視覚芸術すべてにおいて線形の形式が好まれる傾向があった。巻紙の手紙は中国から伝わった形式であり、日本でも由緒正しい形式であるとされた。切り紙と違って文旨とは関係のないところで文章が切れる心配がなく、手紙を連続体として書くことができる形式である。絵巻形式も同じく無限に続く絵画形式である。絵巻物の祖形とみられるものに奈良時代に制作された『絵因果経』がある[46]。平安時代になると、王朝文学の物語、説話などを題材として『源氏物語絵巻』、『信貴山縁起』、『鳥獣人物戯画』などが制作されるようになった（→ 140）。中国にも唐の『五牛図巻』、五代の『韓熙載夜宴図』、北宋の『清明上河図』や『千里江山図巻』などがある。

詩歌では、連歌であろう。多人数の連作で作る文芸的遊びであり、鎌倉時代に興り、南北朝時代から室町時代にかけて発展した。参加者は、直前に詠まれた句の内容や調子を踏まえて自分の句を作る。各参加者は全体の流れを意識しつつひと続きの集団的即興詩を作る。

日本の伝統的都市空間の特徴の一つは西欧的な意味での広場空間を持たないことである。しかし、そのことは、日本の都市に市民が集まる公共空間がなかったということではない。その役割を道が果たした。西欧の教会前の広場がそうであるように、近世の日本では、仏教寺院や神社の参道に沿って、お茶を飲みながら話をする茶店や酒場や店や旅館などが並び建ち、市民が集まる場所となった。

大都市の構成原理が小集落に由来することはままある。日本の農業集落は山と水田の境に一列に並ぶ。また商業集落も街道の両側に狭い間口で奥行きの深い敷地に建つ町家が軒を接して並ぶ。いずれも線形である。中国経由で渡来した仏教より古い歴史をもち、農業祭祀との結びつきが強い神道の神社は、神体山を背にして置かれることが多く集落のはずれに立地する。つまり、日本の集落では、共同体の精神的支柱である神体山が空間的には周縁にあるという奇妙な転倒が起こり、線形性が生まれる源になっている（→ 141）。線形の形態は外界との接触面が長いので防衛には適さないが、周辺の自然との接触性は高い。外国からの侵略にさらされることが少なかった

away. Such short, linear streets converging on a view of a mountain were characteristic of urban planning in Edo. Utagawa Hiroshige, *A Hundred Famous Views of Edo: Suruga-chō* (1856-8).

139 焦点にある人工物で軸線で構造化された都市
A city organized on an axis with art at its focal point

ウイーンの郊外にあるシェーブルン宮殿では、都市の街路も庭園の園路も一つの軸の上にあり、すべてが宮殿に集まっている。

At Schönbrunn Palace, located in a suburb of Vienna, the streets in the city as well as the garden are located along a single axis centered on the palace.

be seen at the far end of each[45] (→ 138). In Japan's traditional society, mountains were worshiped by the Shinto religion as sacred places. Edo resembled a Baroque city in that it had multiple focal points in its visual structure, but, whereas monumental structures served as the focal points of the main streets in the latter, mountains defining the boundary of the city functioned as such in the former (→ 139).

2. A Predilection for Linearity

In Japan, a predilection for linear forms was to be found in cities, literature and all visual arts. The custom of writing letters on rolled paper, introduced from China, was regarded as a time-honored form. It enabled one to write a letter as one continuous text; one did not have to be concerned that the writing might be interrupted unintentionally, as might be the case with cut sheets of paper. Handscrolls are a form of painting that also continues endlessly. These *emakimono* can be traced back to works such as the *Kako genzai inga kyō emaki* (Illustrated Sutra of Past and Present Cause and Effect) from the Nara period (794-1185).[46] In the Heian period, they included works that took as their themes narratives and stories from court literature such as *Genji monogatari emaki* (Tale of Genji Scrolls), *Shigisan engi emaki* (The Legends of Shigisan) and *Chōjū jinbutsu giga* (Scrolls of Frolicking Animals and Humans) (→ 140). In China, illustrated handscrolls include works such as *Five Oxen* from the Tang dynasty, *Night Revels of Han Xizai* from the Five Dynasties and Ten Kingdoms era, and *Myriad Miles of Rivers and Mountains* and *Along the River During the Qingming Festival* from the Northern Song Dynasty.

In literature, there was linked verse (*renga*), a form of poetry that flourished in the Kamakura period (1185-1333) and developed further in the Muromachi period (1333-1568). Two or more persons would take part, with each poet writing his or her stanza on the basis of the content and style of the preceding stanza. It was an improvised, collaborative undertaking in which participants had to be always aware of the overall flow of the sequence of verse.

One of the special characteristics of the traditional urban spaces of Japan is the absence of open squares or plazas in the Western sense. However, that did not mean that the

日本の集落は防衛性より接触性を選んだと考えられる。

近世の大名庭園の多くは回遊式と呼ばれる形式に則っている（→142）。その名前の由来は、池の周りを巡る園路が空間構成の基本になっているからである。池の周りを散策すると、庭木によって池畔の風景が巧みに見え隠れし、次々と変化して見える。移動の経路を前提として継起的に現れるさまざまな風景を繋いでいくことを期待する庭園デザインである。これもまた線形性の一形態である。

3. 人工物と自然の混成系

江戸は、西部の山岳地帯に繋がる多摩丘陵が終端する縁に君主の座す城を築き、その下に広がる沖積平野の低地に庶民の住区、丘陵地を武士階級の住区にするというゾーニング制を前提に、いくつかの土木工事を行って首都の基盤整備をした。大掛かりな土木工事の一つは、お茶の水の切り通しである（→143）。それ以前は現在の神田あたりを通って、日比谷入り江に流れ込んでいた平川（神田川の旧名）の河道を城の東側に誘導し隅田川に直接放流することで、洪水の被害を防ぎつつ江戸の街を掘り割りで囲繞しようという計画が作られた。そのために、本郷台地の端部を切り開いた[47]。現在は、切り通しでできた崖の表面を樹々が覆い、武蔵野の古自然が露出したかのように見えるが、実は人工的介入の結果である。渓谷の縁を中央線の通勤電車が走り、切り通しを貫くように地下鉄丸の内線のトンネルが顔を出し神田川を跨ぐ様は、自然と人工が掛け合いをするダイナミックな景観となっている。

東京に限らず大都市の風景は、自然に対する大小の人工的介入を通して形成され、自然と人工が渾然一体となって、どこまでが自然で、どこからが人工とは言えない混成系をなす。お茶の水の景観はその好例である。東京の景観で、もっとも大きい混成系は海浜の埋め立て地である。江戸時代以来、干拓や埋め立てが行われ、東京近郊には既に自然海岸は1割程度しかない[48]。一方、内陸では住宅地を拡大するために、崖地に擁壁を作ってひな壇化する造成が広く行われている。

近代主義に対する景観的批判のなかに、人工物による自然の蹂躙という論点があり、高架構造物はそのやり玉にあげられ

140 平治物語絵巻　三条殿夜討巻（部分、鎌倉時代・13世紀後半）
Night Attack on Sanjō Palace, Heiji Monogatari Emaki (partial image from scroll, Kamakura period, second half of the 13th century)

ふつうには絵画は、写真のように時間の流れのなかの一瞬を切り取ったと考えられるが、絵巻物では、連続する時間に起こった出来事を無限に続く巻き紙に順番に描く。

While paintings, like photographs, are generally thought to capture a single moment in time, picture scrolls sequentially depict events that have occurred along a continuous stretch of time on an endless scroll.

residents of Japanese cities had no public spaces in which to gather. Roads served such a purpose. In feudal Japan, tea shops where customers could engage in talk while drinking tea, sake houses, shops and inns lined approaches to Buddhist temples and Shinto shrines. Like plazas in front of Western churches, they functioned as places where people could gather. The organizational principles of large cities sometimes originated in small villages. The typical farming village of Japan was arranged as a row of houses at the boundary between the mountains and the paddy fields. In a commercial village, townhouses were arranged side by side on either side of a highway, on deep lots with narrow frontages. In each case, the form was linear. Shinto has an older history in Japan than Buddhism, which was introduced through China, and close ties to agricultural rituals. Its shrines are often located on the outskirts of villages, in front of mountains that are their objects of worship. Thus, linearity was the result of the location of the worshiped mountain on the edge of the village that provided the community with spiritual support (→ 141). A linear form is disadvantageous from a defensive point of view because the village is exposed to the outside world over its entire length but creates close contact between the village and nature. Japan rarely faced the threat of invasion by a foreign country, and that fact conceivably led villages in this country to give priority to contact with nature over defense.

Many of the gardens of residences for lords of the feudal period were designed in the so-called "tour" style (→ 142), a term derived from the way a garden was structured by a path circling a central pond. As an observer strolled around the pond, the landscape was by turns hidden and revealed by strategically arranged trees and shrubs. The design was premised on movement on the path; the sequence of different views created in the observer a sense of anticipation. This too was a linear form of expression.

3. A Composite System of Man-made Objects and Nature

In Edo, the castle for the ruler was constructed on the eastern edge of the Tama hills which extended westward toward mountains. Premised on a zoning system in which the alluvial plain that spread below the castle was assigned to commoners and the hills to the samurai class, a number

ることが多い。日本の都市の多くに鉄道と高速道路の高架構造物がある。当然、鉄道高架の方が歴史が古く上野から新橋までは煉瓦で作られた明治時代の高架構造物が今も残り、構造物の下が商業施設として利用されている。長岡にも新幹線の高架構造や関越自動車道路の高架構造があり、都市域の外に直線的に風景を切り裂いて地域の景観構造に大きな影響を与えている。高架構造物に対する批判的観点からの政策転換として、ボストンの高速道路の問題、横浜の高速道路の地下化、ソウルの清渓川（チョンゲチョン）の高速道路撤去と河川再生などがある。これらは巨大構造物を悪とみて、隠したり撤去したりしている。最近、これらの事例とは別の態度が表れている。構造物は残して新たな使い道を探るプロジェクトが各地で行われるようになったのである。自然には存在しない高架構造物も人工と自然の混成系と看做そうという態度といえる。パリにはヴィアデュック・デザール（→144）と呼ばれる廃線の高架鉄道構造物の緑道化プロジェクト(1994年再生)がある。散歩やジョギングに使われている。横浜の「山下臨港線プロムナード」は、港湾貨物線の高架構造物を遊歩道に転用して一般開放された

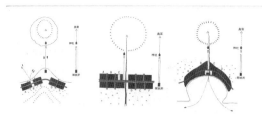

141 神代による日本の集落の祖型（1975）
Prototypes of the Japanese village, according to Kōjiro

神代雄一郎は、景山春樹の研究に示唆を得てこの空間図式を作成した。
Kōjiro Yuichi formed this spatial diagram with suggestions from Kageyama Haruki

142 六義園（東京文京区、1702）
Rikugi-en (Tokyo, Bunkyō Ward, 1702)

六義園は、徳川幕府の重臣柳沢吉保が下屋敷に作った回遊式庭園である。
Rikugi-en is a circuit-style garden established in a villa by Yanagisawa Yoshiyasu, a chief vassal of the Tokugawa shogunate.

of civil engineering projects were undertaken and the infrastructure for the capital was developed. One major civil engineering project was the excavation of Ochanomizu (→ 143). A plan was devised to divert Hirakawa River (now called Kandagawa), which until then flowed through an area in the vicinity of the present-day Kanda district and into Hibiya Cove, to the east side of Edo Castle and into the Sumidagawa River, so as to prevent flooding and surround Edo with waterways. The edge of Hongō Plateau was excavated for this purpose.[47] Today, the surface of the bluff created by the excavation is covered with trees and seems natural, but it is in fact the result of human intervention. Commuter trains on the Chūō Line travel along the edge of the ravine, and the Marunouchi Subway Line is revealed where it crosses the ravine. The result is a dynamic landscape in which nature and man-made objects interact.

The landscape of Tokyo or indeed any metropolis is formed through human interventions, large and small, in nature; nature and man-made objects are in harmony and form a composite system in which it is impossible to say where nature ends and man-made objects begin. The landscape of Ochanomizu is a good example. The largest

143 神田川とお茶の水駅（東京）
Kanda River and Ochanomizu Station (Tokyo)

（2002年再生）。マンハッタンのハイラインは、廃線で放置されていた高架鉄道構造物を再利用した緑道で、今や当地の名所になっている（2009年再生）。これらはいずれも市民や旅行者から歓迎されている。

4. 芝生・スポーツ・ショッピング・アメリカ

芝生

　建築様式と造園様式の両方を持つことによって初めて、ある時代は楽園を夢想することができる。大衆的な性格をもつ近代には庭園様式がないように見えるが、実は、芝生に覆われた公園こそが近代の庭園ではないのだろうか。実際、近代の住宅の名作として名高いサボワ邸（→ 145）もファンズワース邸もいずれも芝生のうえに建っているし、産油国の富者は石油より高い水を撒いて芝生の庭園を持つ。

　東京の西郊にある小金井公園は面積約70 haで都立公園随一の広さである（→ 147）。大半は起伏のある地形に森と芝生が広がっているだけの場所である。早朝のラジオ体操や太極拳

144　ヴィアデュック・デ・ザール（1994）
　　　Viaduc des Arts (Paris, 1994)

145　ル・コルビュジエ サボワ邸（1931）
　　　Le Corbusier's Villa Savoye (1931)

パリの郊外に1931年に竣工したサボワ邸は近代建築を代表する名建築としての誉れが高い。

The Villa Savoye, completed in 1931 in a Parisian suburb, is a renowned piece of modern architecture.

146　亀倉雄策による1964年東京オリンピックのポスター（1962）
　　　1964 Tokyo Olympics Poster designed by Kamekura Yūsaku（1962）

composite systems in the landscape of Tokyo are reclaimed land in the bay. Land reclamation has been undertaken since the Edo period, and only about ten percent of the seashore in the area around Tokyo is now natural.[48] Further inland, residential areas have expanded, and retaining walls have been constructed on hillsides to form terraced developments.

One of the points made in criticizing the scenic effect modernism has had is that man-made objects intrude on nature, and elevated structures are often singled out for condemnation. Elevated railways and expressways exist in many Japanese cities. Elevated railways obviously have the longer history. The elevated structure constructed of brick in the Meiji period from Ueno Station to Shinbashi Station still survives, and the space under the tracks is used by commercial facilities. Nagaoka also has elevated structures, in its case for the Shinkansen and the Kan-Etsu Expressway, which cut directly through the landscape outside the city region and have a huge impact on the scenic composition of the region. Examples of change made in policy in response to views critical of elevated structures include the halt to construction of expressways in Boston, the decision made to place an expressway in Yokohama belowground, and the dismantling of an expressway and the restoration of the stream in Cheonggyecheon in Seoul. In these instances, enormous structures, regarded as detrimental, were concealed or removed. More recently, a different approach has been taken. Projects to leave structures and to give them new uses have been implemented in various places. The approach is one that sees elevated structures which do not exist in nature as composite systems combining artificial and natural elements. In Paris, a project called Viaduc des arts (1994) has turned the elevated structure for a discontinued line into a green road used for strolls and jogging (→ 144). The Yamashita Rinkō Line Promenade (2002) in Yokohama is the result of a conversion of the elevated structure for a line used by freight trains serving the harbor. The High Line (2009) in Manhattan is a green road that makes use of an elevated rail structure for a discontinued line and has become a well-known attraction. In these cases, the elevated structures are seen as parts of nature. These projects have all been welcomed by residents and tourists.

からはじまって、昼間は幼児を連れた若い家族や、ピクニックをするグループ、ラクロスのサークルの練習、夕方には、サッカーをする小学生、老人夫婦のウォーキング、楽器の練習をする若者、学校帰りの高校生のカップル、朝と夕方に集まる大型犬を連れた人たちなどなど早朝から深夜まで、春夏秋冬多様な利用がなされている。この多様さの秘密は包容力豊かな芝生にある。

米作地帯では雑草は稲の生育を妨げる厄介者である。農民は黄金色の収穫は喜んでも緑色の芝生を特別なものとして見てはいなかった。近代以前の都市では王侯貴族の宮殿や宗教施設にしか庭園はなく、街路樹という概念はなかった。ところが近代の都市の衛生思想は緑地に「都市の肺臓」という栄えある称号を贈り[49]、近年の地球環境問題は「緑色」を特別の象徴に押し上げた。環境保護を訴える政党は緑の党と称し、建物は草で覆われ、緑化は揺るぎない正義である。植物は、健康、生命、成長、清浄、庇護を体現する。芝生でくつろぐ家族は近代家庭の幸せの図である。緑を増やすことは善であり、樹々を伐採する開発はもっとも恥ずべき行為である。人間が蹂躙した自然は回復されなければならないので、現代都市デザインの主な任務は都市に緑を持ち込むこととなる。

植物はもはや単なる機能を超えた近代社会の象徴であり、芝生は近代の楽園の敷物である。

スポーツ

かつては、スポーツは有閑階級だけに許された暇つぶしであり、有能な軍人を育てる身体鍛錬であった[50]。現代社会では大衆的な娯楽と健康維持のための主要な活動となり、市民的な善のシンボルの座に就いた。

スポーツは美の基準を提供する。古代ギリシャではスポーツで贅肉を落とした肉体は美しいとされ、今や世界的に普遍的な価値観となっている。ルノワールの描いた豊満はもはや豊かさの象徴ではなく、現代では不摂生の象徴である。健康な生活を目指すのなら、薬を飲む前にスポーツで体を鍛えるべきである。スポーツ選手が公正な選挙や正直な納税申告キャンペーンのイメージキャラクターに起用されるのは、彼らが単に体が強いだけではなく、忍耐心と克己心があり、思慮深い戦略家であり、思いやりがあり、勇気があり、正義を重んじる人格者であるとみな

147 小金井公園（東京都小金井市）
Koganei Park (Koganei City, Tokyo)

4. Lawns, Sports, Shopping, the United States

Lawns

Paradise can be envisioned only in an age that possesses both an architectural style and a style of garden landscape. The modern era, with its mass cultural character, may not appear to have a style of garden, but a park with a lawn is precisely that. Indeed, the Villa Savoye and the Farnsworth House, famous as modern masterpieces of residential architecture, both rise above grass, and the rich in oil-producing countries in the Middle East flaunt lawns maintained with water, a resource that is more expensive in that region than oil (→ 145).

Koganei Park, which has an area of approximately 70 hectares and is located in a western suburb of Tokyo, is the biggest metropolitan park in the capital. Most of it is simply trees and grass on undulating land (→ 147). The park is used in diverse ways depending on the time of day and the season: people exercise to instructions on the radio or practice *t'ai chi* shortly after dawn; young couples with infants dawdle, groups picnic and a lacrosse club practices during the day; elementary-school students play soccer, elderly couples stroll, youths play musical instruments and young couples return from high school in the afternoon; and people walk large dogs morning and evening. The secret to this diversity is the all-accommodating character of grass.

In rice-growing regions, weeds are a nuisance that may impede the proper growth of rice plants. Golden harvests are welcome, but green grass in itself is nothing special. In premodern cities, gardens only existed in the palaces of royalty and the aristocracy or in religious facilities, and the custom of lining streets with trees was not widely practiced. However, as modern ideas of health and sanitation took hold, green areas came to be regarded as the "lungs" of the city,[49] and as such, having a salutary effect. In recent years, global environmental issues have elevated the color green to the status of a symbol; political parties calling for environmental protection are referred to as "green parties;" buildings have been covered with grass; and greening has become an indisputably righteous act. Plants embody health, life, growth, purity and protection. A family relaxing on a lawn is the picture of a happy modern household. Increasing greenery is good, and development

されているからである。彼らは、現代のヒーローである。芸能関係のスーパースターが性と金のスキャンダルにまみれ、強欲と自己顕示と独占つまり世俗を象徴するのに対してスポーツ選手は聖性を帯びている。日本の武芸は技術を超えた精神的鍛錬を求める「道」であると考えられ、人格涵養を目指す。そのような高潔なアスリートが集合する国際的なスポーツ大会は、世界が民族や国民国家間の競争であることを、美的に様式化する祭典である（→ 146）。だから、スポーツは多くの人々を愛国者に変える。クーベルタン男爵は、賢明にもスポーツを平和の祭典として讃え、スポーツの価値をさらに高めた。

　植物もスポーツもともに生命の発露であり健康を約束する。植物を育てることは正しく、植物を愛する人は好人物であるとされる。また、植物の形は調和の源である。スポーツも植物も、国技、国花、県木、潜在植生、など地域性や共同体と結びつけられる。実際は有史以来さまざまな植物が国境を越えて移動しているにもかかわらず、近年は地域の植物の固有種を守るためには格別の努力が払われている。

　植物もスポーツも都市のなかに特別な意味を与える上で大きな役割を果たす。東京のもっともファッショナブルな地区とされる青山から原宿地区は、東京オリンピックが行われた代々木や神宮外苑に近いことと無縁ではない。そのうえ、それらのスポーツの聖地が、代々木公園、明治神宮の森、表参道の欅の並木道、神宮外苑の森と続く東京都心の最大の緑地帯と密接な関係にある。

ショッピング

　近代社会以降、社会が豊かになるにつれ、消費は市民にとって重要な余暇活動となる。単に必要を満たす行為ではなくなり、時代が降るに従って、消費の重要性は益々大きくなる。ところが、コルビュジエから丹下健三まで、新時代の都市を構想した近代建築のパイオニアたちは、消費に大した関心を示さなかった。彼らの関心は伸び盛りの工業に集中し、その後の商業の圧倒的な力を予見するまでには至らなかった。

　消費は生活に必要なモノやサービスを買うことが第一義にありそうであるが、実際には人は本当に必要なものは値切っても、必要性の低いものに多額の出費をする。不条理なのにそれが

that cuts down trees is shameful. Nature into which humans have intruded must be restored; therefore, the main task of contemporary urban design is to introduce greenery into the city.

Plants symbolize a modern society that transcends functions, and the lawn has become the modern vision of paradise.

Sports

In the past, sports were pastimes permitted only to the leisured classes or forms of physical discipline for training capable soldiers.[50] In contemporary society, sports have become major activities for popular recreation and the maintenance of health—they have come to epitomize civic virtue.

Sports provide standards for beauty. In ancient Greece, a body that had lost excess weight through sports was considered beautiful, and today that has become a universal point of view. It is an aesthetic that originated in the West. The voluptuousness depicted by Renoir is today a symbol, not of abundance, but of disregard for one's health. If one's objective is to lead a healthy life, one ought to get into shape through sports instead of taking pills. Athletes are often called upon to be the faces of campaigns exhorting citizens to abide by election laws or to file income tax returns honestly and expeditiously because they are regarded as not only physically strong individuals but persevering, abstemious, thoughtful strategists and caring, courageous persons of character with a sense of justice. They are contemporary heroes. Whereas superstars of the entertainment world are caught up in sex and money scandals and symbolize greed, vanity and privilege, that is, all that is worldly, athletes have a spiritual aura. The martial arts in Japan are considered "ways" that go beyond matters of technique and require spiritual discipline; their aim is to build character. International meets where noble athletes gather are aesthetic stylizations of the worldwide competition among peoples and states (→ 146). That is why sports turn so many people into patriots. Baron de Coubertin astutely praised sports as a festival of peace and further raised the value of sports.

Both plants and sports promise health and symbolize life. Growing plants is considered a worthy endeavor, and someone who loves plants is considered a good, honest person. Moreover, the forms of plants are marvels of harmony. Both sports and plants have been linked to locality or community, as in such concepts as national sports, national flowers, prefectural trees, and potential natural vegetation (PNV). Despite the fact that plants have in fact moved and crossed borders throughout recorded history, considerable energy has been expended in recent years to protect plants considered distinctive to certain localities.

Both plants and sports play an important role in endowing places in the city with special meaning. The Aoyama-Harajuku district is considered the most fashionable district in Tokyo, and the proximity of Yoyogi National Gymnasium and the National Stadium in Jingū Gaien, two venues for the Tokyo Olympics of 1964, and the woods surrounding Meiji Shrine has a great deal to do with that image. Moreover, those sports facilities are a part of the biggest green belt in Tokyo consisting of Yoyogi Park, the woods of Meiji Shrine, the zelkova trees lining Omotesandō, and the woods of Jingū Gaien.

Shopping

In modern times, as a society becomes affluent, consumption becomes an important leisure activity. It is not simply a way to satisfy a need; as time passes, the importance of consumption becomes ever greater. However, the pioneers of modernism who conceived cities for the new age, from Le Corbusier to Tange Kenzō, had

格別の快楽だからである。消費は蕩尽性に本質がある。百貨店はかつて消費の中心にあった。人々は百貨店に出かけるときには身なりを整えて「お出かけ」をしたものである。日本の百貨店の全盛期は、店数でいえば1997年に最多となり、日本全国に432店の大小の百貨店があったが、小売りにおける売り上げの占有率でみれば1970年代が最盛期であった（→ 148）。百貨店の重要な商業戦略の一つに、子供に照準を当てたことがある[51]。遊園地を屋上に備え、子供服売り場を備え、御子様ランチを用意した。子供が小さな大人ではなく、慈しんで育てる対象と見なされるのは近代家族の大きい特徴である[52]。百貨店は、他に先駆けてシステムキッチンを売り出し[53]、洋食が食べられるレストランや美術館や劇場を併設することで、新時代の都市生活者生活様式と価値観を提供する啓蒙の中心となった。こうして百貨店は、それまでは存在しなかった新しい欲望を作り上げたのである。長岡に本格的な百貨店ができたのは1954年で、当時の感覚であれば、長岡もこれで真の都市に格上げされたのである。その時代の日本の百貨店を代表する三越百貨店は、1960年に日本橋本店の中央にある大吹き抜けに高さ11mの極彩色の天女像を据えた。これが商業戦略的に見て、いかほどの効果があったかは怪しいが、ともかく消費のカテドラルの威厳を示したのである（→ 149）。

百貨店がカテドラルであったことは、長岡の目抜き通り面して1958年に開店した百貨店の大和（ダイワ）が2009年に店を畳むというので長岡の人たちにとって大事件として意識されたことに表れている。百貨店は、それがなければ都市と名乗れないほどに栄光に包まれた存在であった。大和の撤退は百貨店という業態の問題なのだが、街の人にとっては誇りの問題であった。

百貨店の全盛期を過ぎた今、それを引き継ぐのは何だろうか。ブランドショップは確かに商品の価格と神話的な雰囲気において所持者に威信を与えるから正当な継承者と言えるが、全盛期の百貨店が提供した世界の縮図性、そして家族の祝祭の舞台という点では遥かに及ばない。百貨店には大衆性と愛の理念があった。そしてその凋落は日本の家族の一つの時代の終わりを意味する。

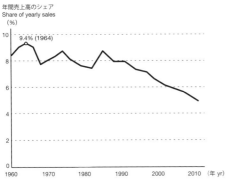

148 衰退する百貨店
The declining department store

149 三越百貨店本店の天女の像
Statue of a celestial being in the main store of Mitsukoshi Department Store

little interest in consumption. Concerned instead with the emergence of industry, they could not foresee the overwhelming power that commerce would subsequently wield.

Purchasing goods and services necessary to everyday life might seem the most important objective of consumption, but in fact people are more likely to spend large sums on things they do not especially need even when bargains on things they truly do require are available. That is because such a course of action, however absurd, is pleasurable. Profligacy is the essence of consumption. Department stores were once centers of consumption. Going to a department store was an occasion, one for which people dressed up. The total number of department stores, large and small, in Japan peaked at 432 in 1997, but from the point of view of retail's share of sales, the golden age of department stores in this country was the 1970s (→ 148). One important commercial strategy of department stores was to target children.[51] They set up amusement parks on roofs, set aside sales areas for children's clothes and offered specially-prepared meals for children in cafeterias. It is in the nature of the modern family to regard children, not as small adults, but as individuals to be loved and nurtured.[52] In introducing system kitchens for sale[53] and installing on their premises restaurants offering Western-style meals, art museums and theaters, department stores became sources of enlightenment presenting urban lifestyles and values for a new age. In this way department stores created new, hitherto unknown desires. The first department store in Nagaoka was built in 1954, and by the yardstick of the day it upgraded Nagaoka to the status of a full-fledged city. The most famous Japanese department store of that era was the main store of Mitsukoshi Department Store, located in NIhonbashi, Tokyo. In 1960, a richly colored, eleven-meter-tall sculpture of a celestial being was installed in its atrium and lent it a majesty befitting a temple to consumption, though the effectiveness of the measure from the point of view of commercial strategy was questionable (→ 149).

The status accorded department stores was demonstrated when a branch of the Daiwa Department Store, which opened in 1958 on the main street of Nagaoka, closed in 2009. This was seen by the people of the city as a highly regrettable event. A department store

アメリカ

芝生、スポーツ、百貨店とみてくると、すべての背後に多かれ少なかれアメリカが控えていることに気づく。アメリカは世界にとって20世紀の最強のシンボルであった。日本にとっては殊更である。明治維新はペリー提督率いるアメリカ海軍の4隻の軍艦の来航で始まるし、第二次世界大戦の後の連合軍の占領は実質アメリカによる支配であり、戦後の日本の社会体制の基本路線を敷いた。アメリカは日本の近代化の後見人の立場にあった。だから、サンフランシスコ講和条約が成立するまで占領部隊が駐留したワシントンハイツに近い表参道は特別だったのである[54]。同様に、銀座が古い目抜き通りから、現代的なショッピングストリートとして生き残っているのは、そこは外国人居留地があった明石町に接する場所だからである。

5. 土地神話の崩壊

日本では、明治以来人口の増加と経済発展によって、一時的な例外状態を除いて地価は上がり続けたので、資産としての土地に対する信頼は絶大になり、土地本位制とまで言われる経済システムを作り上げた。しかし、長期の人口減少社会では地価の下落は避けられないし実際に地価の下落は始まっている。土地神話が完全に無くなるには土地が上がり続けた時代に青年期を送った団塊の世代が消えるのを待たなければならない。そのとき、資産保全手段としての土地の役割は終わり、今まで経験したことがない地平が開かれるだろう。カール・ポランニーが言うよう[55]に、地価は資本主義経済下では土地という商品の価格であるが、土地は移動が不可能であり自然の一部であり、そもそも本来の商品とは異なるものである。土地の資産価値が減れば、そこにある建物や植栽や利用の実績などを評価することになろう。建築市場からみれば中古市場の成熟や改修文化の展開などこれまで日本にはなかった新しい動きの下地を作ることになろう。

6. 新築依存症

日本は国家としての歴史が長く、古くから洗練された文明を

was a marvelous presence, and a city was not worthy of the name without one. The withdrawal of Daiwa from the city was, for the company itself, merely an acknowledgment of problems having to do with the department store as a business, but for the people of the city, it was a blow to their pride.

Now that the golden age of department stores is over, what will take their place? Will it be brand stores? Certainly they can be said to be the legitimate successors to department stores in that their prices and ambience confer prestige on their patrons, but brand stores are not nearly the presences that department stores were at the height of their influence, when they offered the world in miniature and a festive stage for the entire family. Department stores were intended for the general public and embodied an ideal of familial love; their decline signifies the end of an era for the Japanese family.

The United States

It was their association with the United States that made lawns, sports and department stores so influential. The United States was the most powerful symbol of the twentieth century for the entire world but especially for Japan. The Meiji Restoration began with a visit by a squadron of US warships under the command of Commodore Matthew C. Perry, and the occupation of Japan after World War II by the Allied Forces was in fact implemented by the US, which laid the groundwork for the country's postwar social system. The United States was in effect the guardian under whose watchful eye Japan modernized. Omotesandō, located near Washington Heights where Occupation forces were stationed until the San Francisco Peace Treaty was signed, was special for that reason.[54] Similarly, Ginza was able to survive and make the transition from an old main street to a contemporary shopping district because it was adjacent to and associated with Akashichō, where the foreign settlement was once located.

5. The Collapse of the Myth of Land

For the most part, land prices continued to rise in Japan from the Meiji period because of the rise in population and economic development. As a consequence, the Japanese

築いてきたが、残念ながら往時の建造物はあまり残っていない。東京にも長岡にも歴史的地区らしい地区というものがない。その理由は、木造建築が中心で、腐朽しやすく火災に非常に弱いところに地震や台風などの激烈な自然災害が頻発し、さらに戦火に遭い何度も灰燼に帰したからである。こうした外的条件に加えて、日本では、近代に目覚ましい経済的、技術的発展が100年以上も続き、それが成功したため、古いものは遅れたものであり、新しいものは優れているという確信が一般化した。古いものは新しいもので置き換えるべきだという、まさに近代主義の信条を、市井の人までが信じてきた。つまり、日本の都市の物的脆弱性が近代日本人の進歩主義を補強し、進歩主義がますます都市の更新を加速したのである（→ 150）。こうしてできた東京の風景も長岡の風景も、常に変動し続ける活動のエネルギーを示す代理指標となり、東京の風景はあまりに早く変わり続けると半ば嘆息と半ば感嘆をもって言われ、長岡は風景が変わらないと嘆かれるばかりである。

確かに、戦後の経済復興からの奇跡的な成長は、社会基盤に対する公共投資政策なくしてはなかっただろう。いち早く弾丸列車を成功させて国土の時間距離を縮めて人材の効果的な再配置を実現し、少しの公共住宅によって現代的生活像を効率的に示して消費を喚起し、建設労働で農業の余剰人口を吸収し、高速道路網の完成により高速で緻密な物流網を築いたのである。しかし、副作用も大きかった。建設産業は肥大化し、その旺盛な食欲を満たすために常に新たな建築工事を必要とするようになってしまった。そのため、世界でも有数の建築寿命が短い国になり、建築の再利用や再生が根付かない（→ 151）。

1980年代に入ると、専門家はフローの経済からストックの経済への転換が必要だと言い始め、政府の政策にも書き込まれるようになった[56]。しかし、現実はなかなか変わらなかった。成功が華々しければ華々しいほど成功譚を人々は忘れることができないのも世の常である。

7. 日本橋と二条城

歴史否定は近代主義の中核的な思想である。20世紀の都市

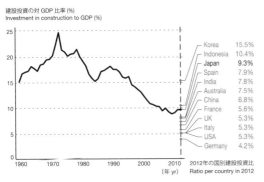

150 建設投資の推移と国際比較
Changes in construction investments and an international comparison

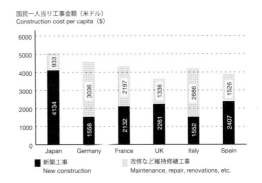

151 新築と改修の国際比較
International comparison of new construction and renovations

came to have enormous faith in land as an asset, so much so that it was suggested that the country adhered to a "land standard," that is, that the economic unit of account in Japan was based on land. However, a fall in land prices is inevitable in a society whose population is in decline over the long term, and in fact land prices have begun to fall. The myth of land will not be completely dispelled until the generation of baby boomers in whose youth land prices continued to rise itself disappears. When that occurs, the role of land as a means of safeguarding assets will end and an entirely new horizon will open up. As Karl Polanyi states, under capitalism, land prices are the prices of a commodity called land, but land, being immovable and a part of nature, is different from a true commodity.[55] If the asset value of land declines, then the buildings and plants that exist there and the use to which land is being put will likely become the yardstick by which its value is measured. From the perspective of the building market, this will probably initiate new trends such as the maturing of the market in used buildings and the emergence of a culture of remodeling.

6. A Dependence on New Construction

Japan is a country with a long history and a civilization that has been refined since ancient times. Unfortunately, not many buildings from the past survive. There is nothing like a historic district in Tokyo or Nagaoka. That is because most structures being of wooden construction have been subject to decay and vulnerable to fires; in addition, violent natural disasters such as earthquakes and typhoons occur frequently, and wars have repeatedly reduced many buildings to ashes. In addition to such external factors, there is also in Japan a widespread conviction that old things are out-of-date and new things are superior, arising from the remarkable economic and technological development of the country for over 100 years in the modern era, that is, a belief even among many ordinary people in what is essentially the tenet of modernism that the old ought to be replaced by the new. Thus the progressivism of the modern Japanese, reinforced by the physical vulnerability of Japanese cities, has accelerated the continual renewal of cities. The townscape has become an index of activity—a sign of the level of energy—in a city.

計画思想は、基本的に今あるものを全部白紙に戻して新しい色で描き直そうとする。なぜなら、過去のものは劣り、新しいものは優れているという考えるからである。彼らは歴史的遺物には懐古的な趣味以上の価値があるとは思いもよらない。その考えは、1922年に発表されたコルビュジエの「現代都市」の現実の都市への適用であるヴォアザン計画に明瞭に現れている。パリのルーブル美術館の北側の中世の香りが残る街区を壊して、合理的で新しい超高層群で置き換えようという提案であった（→152）。

　20世紀の物質的な発展の速度は人類史上で最大であり、「日進月歩」の表現が誇張ではないほどに急激にかつ広範囲に進展したことはよく知られている。実際、明治維新以来、今日まで、多くの日本人は、幼少のころの生活と年老いてからの生活を比べると別世界と言っていいほどの大きい変化を経験してきた。だから、進歩に対して人々は絶大な信頼を寄せてきたのである。しかし、近代主義に対する懐疑は1960年代から徐々に広がってきているが、人々に染み付いた近代主義的思考法はそんなに簡単には変わらない。二つの事例を考えてみよう。

　まず、日本橋と上を跨ぐ首都高速道路（シュトコウ）である。これは景観的に不人気である[57]。首都高を撤去すべきだと主張する人は、日本の都市景観が惨憺たるのは日本人が歴史を大事にしないからだと難じる。しかし、歴史を振り返ればわかるとおり、日本橋ができた明治44年ころは、周辺は川越に残るような重厚な関東町家並ぶなかにところどころ西欧様式建築が混じる街並だったのだから（→153）、この日本橋自体が景観破壊の張本人であった。それに、祖父の時代の仕事を残せと主張するのであるなら、現在からみた父の時代の仕事を残しておくべきであろう。今の父の時代の仕事は次の世代からみれば祖父の仕事だからである。首都高は、完成した当時は優れた土木技術と未来的な景観で絶賛を博した父の時代の立派な仕事であった[58]。

　ここには近代以来の日本人の矛盾する心性が表れている。一つは、父の時代の首都高をけなす裏には、父親を乗り越えようという近代日本の原動力となった進取の気性が見て取れる。明治の人は江戸を封建的となじり、大正の人は明治を江戸に片足を突っ込んだと言い、戦後の人は戦前を軍国主義と否定し、

Therefore, it is often remarked upon, with a mixture of admiration and exasperation, that the landscape of Tokyo continues to change at too rapid a pace, just as it is often lamented that the landscape of Nagaoka is unchanging.

Certainly, the miraculous growth of the country after its postwar economic reconstruction would not have been possible without a policy of public investments in social infrastructure (→150). The success of the bullet trains shortened distances in the country and produced an effective redeployment of human resources; a limited supply of public housing was able to effectively suggest an image of contemporary life; construction work managed to absorb the surplus population in rural areas; and the completion of a system of expressways produced a high-speed, closely-woven network of distribution. However, there were harmful side effects. The construction industry grew enormously, and a constant supply of new construction projects was now needed to satisfy its prodigious appetite. For that reason, buildings in Japan are among the shortest-lived in the world, and the recycling of buildings has not taken root in this country (→151).

In the 1980s, experts began to state that a shift from an economy of flow to an economy of "stock" was necessary, and such statements also found their way into expressions of government policy.[56] However, reality was slow to change. The more brilliant success has been, the more difficult the task of getting past it is.

7. Nihonbashi and Nijōjō

Rejection of history is a core principle of modernism. The approach to city planning in the twentieth century was basically to wipe the slate clean and start from scratch in the belief that the new was superior to the old. It did not occur to modernists that historical remains had any value beyond antiquarian appeal. That is evident in the Plan Voisin, an application of the Contemporary City idea to an actual city that was presented in 1922 (→152). The proposal called for demolishing the area north of the Louvre in Paris where traces of the medieval past still remained and constructing a cluster of rational skyscrapers.

Material development occurred at a faster pace in the twentieth century than at any other time in human history. In fact, ever since the Meiji Restoration, many Japanese have experienced such radical changes in their lifetime that in looking back in their old age the world of their childhood has seemed incredibly remote. That is why they have had such great faith in progress. Inured to a modernist way of thinking, people have found it difficult to change, despite doubts about modernism gradually beginning to spread in the 1960s. Let us consider two examples of this.

First, there is the case involving a stretch of the Metropolitan Expressway that passes over a bridge in Tokyo called Nihonbashi. This is unpopular for scenic reasons.[57] Those who argue that the Metropolitan Expressway ought to be removed assert that the reason Japan's townscapes are so wretched is that the Japanese do not place enough value on history. However, as is evident from looking back on history, when the bridge was constructed around 1911, the surrounding area had a townscape that was a mixture of buildings in Western styles and the imposing traditional townhouses (machiya) characteristic of the Kantō region (examples of which still survive in Kawagoe, Saitama Prefecture) (→153). Thus, Nihonbashi itself originally disrupted an existing scenic environment. If those in favor of removing the expressway argue that a work dating from our grandparents' days ought to be left undisturbed, then what do they propose we do about a work dating from the days of our parents, who will inevitably become the grandparents for the next generation? The Metropolitan Expressway is a superb work

現代人は高度成長時代を馬車馬の如くと非難するのが常である[59]。

もう一つは、特段優れたデザインでもない西欧風の橋を現代工学の成果である土木構造物より文化的に価値があるとすることである。その裏には、西欧に対する劣等意識があるように思える。ストックの時代だとか街のコンテクストが大事だといわれて久しいが、ストックの時代を本気に考え、歴史が堆積する街を厚みのある都市としたいなら、まずは目の前に存在する構築物を事実として受け入れる覚悟が要る。パリの都市景観を誉める人は多いが、それなら19世紀の「首都高」であるエッフェル塔を残した寛容さを見習うべきではなかろうか。

二番目の例は、日本政府が2006年に京都御所内に新築した京都迎賓館である。近代和風[60]の建物で、サミットのような国際会議をするための施設である。日本では、このような施設を新築することに誰も疑問を挟まないが、もし欧州に倣って発想するならば、京都でのサミットは二条城で行なうべきである（→ 154）。現に第一回サミットは、パリの近郊にある1375年に建ったランブイエ城で行なわれている。二条城は1603年に建っ

152 ル・コルビュジエによる「ヴォアザン計画」(1925)
Le Corbusier's "Plan Voisin" (1925)

ル・コルビュジエは、「現代都市」(1922)に続いて、このプロジェクトを発表した。パリの中心部を作り替える提案である。ヴォアザンは支援者の自動車メーカーの名前である。

Le Corbusier published this project following his "Ville Contemporaine" (1922). A proposal for rebuilding the center of Paris, the name Voison is adopted from an automobile manufacturer that was one of the project's supporters.

153 竣工当時の日本橋とその界隈
Nihonbashi and its surroundings at the time of its completion

現在見る日本橋は、1604年に掛けられた第19代になり、1911年（明治44年）に竣工した。

The present-day Nihonbashi was completed in 1911 (Meiji 44) as the 19th reconstruction of the bridge first built in 1604.

from the 1960s that was praised for its marvelous civil engineering technology and futuristic appearance at the time it was completed.[58]

This reveals the contradictory mentality of the Japanese evident since the advent of the modern era. For one thing, the enterprising spirit—the determination to surpass one's parents—that was the driving force of modern Japan can be glimpsed in the disparagement of the Metropolitan Expressway built in the days of our parents. People of the Meiji period scorned Edo, calling it feudalistic; people of the Taishō period (1912-1926) saw the Meiji period as having one foot still in the Edo period; people of the postwar period rejected prewar militarism; and contemporaries always criticize the period of intensive economic growth, regarding it as a time of people who slaved away.[59]

A second interesting thing about the case of Nihonbashi is the implication that a Western-style bridge of undistinguished design has greater cultural value than a civil engineering structure that is the fruit of contemporary technology. This seems indicative of a feeling of inferiority toward the West. It has long been said that this is an age in

154 二条城二の丸御殿（京都市、1603）
Nijōjō Castle Ninomaru Palace (Kyoto City, 1603)

二条城は徳川家康が上洛時の滞在場所として建てた。現在国宝に指定されている。

Nijōjō is a castle built by Tokugawa Ieyasu as a place to stay during his visits to Kyoto, where the emperor resided.

た。二条城は堀で囲まれセキュリティも万全であり申し分ない。少しばかりの対応は必要であるがランブイエ城だって、なにがしかの対応をしたはずである。二条城にも現代的な家具と機器を持ち込めばよい。海外の要人も数年前にできたばかりの高級温泉旅館風情の会議場と17世紀の古城の大広間とどちらで開催されるサミットを喜び、どちらの会議が日本文化に対する尊敬を勝ち取れるか、一目瞭然なのだが、そのような検討をしたと聞いたことがない。見識のある人であれば地球環境を守るためには既存の施設の再利用を進めるべきだと主張するだろうが、誰もサミットを二条城で行なおうと言わない。一方、欧州では14世紀の城でサミットを行なって当然だとされ、誰も現代のリゾートホテルのような国立の会議場を作ろうとしない。これらの違いは合理性の問題ではなく思考習慣、つまり、文化の問題である。そしてこの思考習慣にも、近代以来の日本人の矛盾する心性が表れている。一つは、日本のだれにとっても新築こそ最先端であり古いものは価値がないという素朴な進歩主義がある一方、日本の伝統の真価は日本人にしかわかるはずがないという日本文化特殊論がある。

縮小の時代の都市を実現することの難しさの核心はまさにここにある。思考習慣という檻を一つ一つ打ち破らない限り、新しい時代を作ることはできない。サミットを二条城で行なおうと皆が自然に発想するようになったとき、初めて建物の再利用が日本の建築文化の一部になるのである。

8. ハコモノ

かつて、どこの地域でも建築は、それほど多数の型を持っていなかった。基本的には宗教建築と住宅と倉庫だけで成り立っていたといっても過言ではない。日本の近世で言えば、寺社の他は、町家と武家屋敷形式で構成されていた。庶民の施設は町家型で作られ、劇場も巨大な町家形式であった。一方、武家屋敷型建築は行政庁や学校として使われた。

19世紀になると、欧州都市ではさまざまな新しい活動が生まれ、それぞれに専用の建物が誕生した。オフィスや駅、温室、展示所、百貨店などである。ただ外観は駅であっても宮殿のままであった。20世紀の建築家たちは、これをさらに推し進め、用

which the existing stock of structures must be valued and that urban context is important, but if we are truly serious about setting store by the existing stock of structures and respecting what successive periods of history have produced in cities, then we must be prepared to accept the structures that exist now as a reality. Many people praise the townscape of Paris, but in that case they should learn from the example of the French who were sufficiently open-minded to allow the Eiffel Tower, the nineteenth-century equivalent of the Metropolitan Expressway, to remain standing.

A second example of the way a modernist way of thought continues to prevail is the Kyoto State Guest House, built in 2006 on the grounds of Kyoto Palace by the Japanese Government. Designed in a modern Japanese style,[60] this facility is intended to be used for international conferences such as summits. In Japan, no one questions the construction of such a facility, but if the Japanese were to follow the example of Europe, a summit would be held in Nijōjō (→ 154). As a matter of fact, the first summit was held in the Chateau de Rambouillet, which was built on the outskirts of Paris in 1375. Nijōjō was built in 1603. Surrounded by a moat and completely secure, it would be perfect. Some measures would be necessary to adapt it to such a use, but adaptations of some kind were no doubt required for Rambouillet as well. Contemporary furniture and equipment could be introduced into Nijōjō too. Which would foreign dignitaries prefer as a venue for a conference and win greater respect for Japanese culture, a conference hall recently built in a style suggestive of a high-class inn at a hot spring resort or the large reception hall of a seventeenth-century castle? The answer is obvious, but as far as I know such a possibility was never studied. Persons of good judgment routinely argue that existing facilities ought to be reused to protect the global environment, but no one advocates holding a summit in Nijōjō. In Europe on the other hand, a fourteenth-century castle is considered an appropriate venue for a summit and no one is in favor of building a national conference center designed in the style of a contemporary resort hotel. This difference is not a problem of rationality but a problem of habit of thought, that is, of culture. This habit of thought is also indicative of the contradictory mentality of the Japanese since the advent of the modern era. The first case is an expression of a naive progressivism, a belief that something newly constructed is advanced and something old is without value. The second is an expression of the idea that Japanese culture is somehow special and the value of Japanese tradition can only be understood by the Japanese.

This is the essence of the difficulty we face in realizing cities for the age of shrinkage. We cannot usher in a new age unless we break our habits of thought, one by one. The recycling of buildings will become an accepted part of Japan's architectural culture only when everyone finds the idea of holding a summit in Nijōjō perfectly natural.

8. White Elephants

In the past, not many different building types existed anywhere. People made do basically with only religious buildings, houses and storehouses. Townhouses (*machiya*) and samurai estates (*buke yashiki*) were the two types in cities in Edo-period Japan aside from temples and shrines. Facilities for commoners, even theaters, were of the townhouse-type. Meanwhile, the samurai estate-type was used for government agencies and schools.

In the nineteenth century, various new activities developed in cities in the West, and buildings dedicated to each came into being including office buildings, railway stations, hothouses, exhibition halls and department stores. However, the development of new building types

途ごとに一番相応しい間取りと、他の用途の建物と区別できる独特の外観があるべきだという機能主義的美学を主張した。20世紀になると、新たに飛行場、コンピューターセンター、オフィス、ショッピングモールなどが加わった。

こうした型の分化は、少なくとも日本では、公共建築の分野で顕著に起こり、公共サービスの種類ごとに独特の型を発展させた。そこには、建築学の一分野である建築計画学の取り組みと官僚主義があずかり、ある特定の公共サービスを提供するためには一定の基準を満たした専用の建築施設を用意することが制度化された。また、各都市が提供する公共サービスのレベルが公共建築の量で評価されるようになり、政治家は公共施設を作ることが実績になる政治風土が生まれた。公共サービスと強く結合した公共建築は、利用人口が減ったときに施設規模を縮小しようにも融通が利かなくなる弊害を生む。しかし、本当のところは、建築は、それほど厳密に機能と形態が対応しているわけではなく、過不足なく面積を利用し尽くそうと考えさえしなければ、学校を事務所として使うことも、美術館を郵便局として使うこともほとんど問題なくできる。

9. 孤立と互恵性

いま、日本では、高齢者だけでなく全世代で単身者が増えている(→ 166)。単身世帯は、やがて家族形態の最多の類型になると予測される。一方、グローバル経済と新自由主義的政策によって収入格差が拡大し平等な社会が崩壊し始めていると議論されている。単身者は経済的弱者であることが多い。日本では国際的に比較しても自殺率が高いことに注目しなければならない。これは、人々が孤立しやすく無力だと感じる社会構造が一因だと言える。日本社会は、他人との接触が少なく社会的紐帯が弱く、他人からの精神的ならびに物質的支援を受けられない孤立した個人が増える「寂しい社会」に向かっているのかもしれない。

「寂しい社会」化に対抗して、近隣の互助関係の復活を願う人たちが増えている。たとえば、昨今のコンパクトシティにも暖かい近隣関係の復活の期待が込められているし、建築界におけるシェアハウスの流行にも建築形態が親密な関係を醸成する期待がある。しかし、現代都市住民が、期待通りに近隣社会で

was still incomplete, and a station, for example, might outwardly resemble a palace. The architects of the twentieth century took this further and promoted a functionalist aesthetic—the idea that there ought to be an optimal layout and a distinguishable exterior for each function. In the twentieth century, airports, computer centers and shopping malls were newly added to the list of building types.

This differentiation of types was conspicuous, at least in Japan, in the field of public buildings, and a distinctive type developed for each variety of public service. Building planning, a field of architectural studies, and the bureaucracy institutionalized the idea that each specified public service required its own dedicated building facility. Moreover, the level of public service provided by a city came to be judged by the public buildings that city possessed, and a political environment came into being in which getting a public facility built was a feather in one's cap for a politician. The strong link between public services and public buildings becomes an impediment when facilities need to be reduced in scale in response to a decline in population. In truth, however, the correspondence between function and form in architecture is not that strict. As long as one is not intent on using every last bit of space, there is virtually no problem in using a school as an office building or an art museum as a post office.

9. Isolation and Reciprocity

Not just seniors but people of all generations in Japan are increasingly living alone. Single-person households are expected to constitute the majority of households eventually. It is argued, meanwhile, that as the result of the global economy and neoliberal policies, the gap between the rich and the poor is expanding and social equality is beginning to disappear. Many persons living alone are economically weak. It should be noted that, compared with other countries, the rate of death by suicide is high in Japan. Arguably, one reason for this is that the social organization is such that people can easily become isolated and feel powerless. Japanese society, where contact with others is limited and social bonds are weak, may be turning into a "lonely society" of isolated individuals unable to receive psychological or material support from others.

Increasingly people desire a revival of reciprocal neighborhood relationships to combat such a trend. For example, the proposal now being made for Compact City has been developed in the hope that warm neighborhood relationships might revive, and the hope that architectural form might promote the growth of close relationships is behind the popularity of shared houses in the architectural world. However, whether neighborhood relationships will flourish as contemporary citydwellers hope is open to question. The bonds of neighborhood relationships are maintained through small, everyday interactions by those who have been trained in appropriate social behavior. In the past, people were trained in such behavior within extended families and through experience in society, but those who have grown up in a "lonely society" are not likely to have experienced even neighborhood relationships much less relationships in families or society. For example, it may be necessary to arrange the inclusion of an attendant who can provide guidance to the uninitiated in the ways of getting along with a neighbor.

Another practice that is apt to be forgotten in contemporary society is reciprocity or mutual assistance not involving money that takes place among neighbors. Over the long term, help given and help received among neighbors in a settled community are likely to balance out, but in contemporary society where people are frequently

うまくやっていけるかどうか疑問が残る。近隣関係の紐帯は、日常的な些細なやりとりの集積のなかで蓄積され、訓練された参加者によって維持される。かつては拡大家族のなかで訓練され、社会のなかで体験できたが、「寂しい社会」のなかで育った人は良き近隣関係そのものを体験していないからである。たとえば、隣人との付き合い方を指導してくれる介助者を組み込む工夫などが必要ではないだろうか。

もう一つ忘れられがちなことは、近隣社会による金銭を介さない扶助は互酬的ということである。定住を前提にすれば、人生は長いので世話をしたり世話をされたりと帳尻は合うが、居住地の移動が当たり前の現代社会では悪意がなくても、タダノリをされる危険性が高い。近代社会は、この問題を金銭と公共サービスという方法で解決したが、金銭だけを介すれば経済的弱者がサービスを受けられないし、公共サービス頼みは公共財政が逼迫する縮小社会では十分なことができない。

他人のお世話という本来的に金銭換算が難しい行為をどう評価し、互酬的関係を地域に作り出すかは極めて困難な問題である。未だ実験的ではあるが、地域通貨のように、移動社会を前提とした互酬性の仕組みを構築する必要がある。

10. お一人様支援技術

通販は、2013年度の日本の小売業界の売り上げの1割を占め、個人農家が消費者に新鮮な農産物を直接届ける新しい流通形態を可能にし、外出がおっくうな高齢者世帯に毎日のお惣菜を届けてくれ、重い荷物を運べない人にとってはポーターの役割もしてくれる。通販の隆盛の裏には、宅配便と呼ばれる家庭向け小口配送サービスがある。1976年にヤマト運輸株式会社によって始まり、いまでは複数の企業が参入し、いまや日本全国ほとんどの地域に1〜2日で届く。しかも2〜3時間刻みで配達時間の指定ができ、しかも受取人が不在のときには追加料金無しで再配達をしてくれる。また、冷蔵庫を備えたトラックで冷凍食品も配達してくれる。通販での購入者にはさまざまな支払い形態の決済の代行もしてくれる。

遠隔医療とよばれるITを利用した診療はまだ普及していないが、やがて「無医村」を無くすだろう。単身世帯や高齢者世帯の

moving from one place to another, the possibility of people freeloading, intentionally or unintentionally, is quite real. Modern society solved this problem by means of either services rendered for a fee or public services. However, when services can only be received when fees are paid, the economically weak are at a disadvantage, and in a society that is shrinking, financial constraints limit the scale of public services.

Helping someone is an act whose monetary value is difficult to calculate. Evaluating such acts and creating reciprocal relationships in a community are quite difficult things to do. However, there is a need to build a mechanism for reciprocity premised on a society with a mobile population, one that is perhaps similar to the "community currencies" that have been introduced in various parts of the country.

10. Party-of-One Support Technologies

Mail-order sales accounted for ten percent of all retail sales in Japan in 2013. They have made possible a new form of distribution that delivers fresh farm produce directly to consumers from individual farmers and everyday meals to seniors who find going out too much trouble and incidentally serves as porters for people unable to carry heavy loads. The home delivery service system has made it possible for mail-order sales to flourish. This service was begun in Japan in 1976 by Yamato Transport Company, Ltd., and a number of other companies have since entered the field as well; today, a package can be delivered to practically any region in Japan in one to two days. A customer can designate not only the date but a two- or three-hour window when delivery is to take place. When the recipient is not at home, the package will be redelivered at no additional expense. Refrigerated trucks are available for delivering frozen foods. Several different ways of settling accounts for the purchase of a mail order are available.

Telemedicine, that is, clinical health care provided at a distance using IT, has not yet become commonplace but will eventually eliminate so-called doctor-less villages. The increase in single-person households and senior households has led to a shortage of doctors, and demand for telemedicine is growing. Such service technologies,

増加は医師の不足を招き、遠隔医療市場の需要を拡大する。

これらのサービス技術を「お一人様支援技術」と呼べば、これは生活基盤の新しい類型になりつつあると言ってよい。今後、単身者世帯が増え、都市郊外部においても限界集落的状況の発生が懸念される日本では「お一人様支援技術」に対する期待は高い。農村に残る近世的地縁社会の桎梏を逃れて都市に向かうことが日本人にとっての近代化であったが、実際には、都市にも共同体的束縛はある。「お一人様支援技術」はそれすらも解いてくれ、誰からも干渉されず、どこにいても同じサービスを受けることを可能にするまさに近代の技術と言える。

しかし、近代黎明期の村落においては封建的地縁関係しか選択できなかったことが苦痛であったように、一人でいることしか選択できないとしたら、それもまた別の苦痛である。「お一人様支援技術」が席巻し他の選択肢を排除することもまた問題である。

11. 遠く・速く・大量に

近代の交通は「遠く・早く・大量に」を目標に進歩してきた。これを「大きい交通」と呼ぶならば、新幹線はその代表である。日本政府は、1961年の東京オリンピックに当たって、戦災からの復興ぶりを世界各国の人々に印象づけようと、さまざまなプロジェクトに取り組んだ。その筆頭が東京都内では首都高速道路の建設、都市間では東京と大阪を結ぶ新幹線であった。特に、時速210キロの営業運転は当時世界最速であり、「遠く・速く・大量に」を実現する高速鉄道時代を切り拓いた。今日では、弾丸列車[61]は世界の都市戦略の必須項目になっている。もう一つの「大きい交通」は都市間自動車専用高速道路である。最初は1964年に開業したのは名神高速道路で、1968年には東京まで延びた。太平洋ベルト地帯の都市群は二本の「大きい交通」で繋がれ、首都圏、中京圏、関西圏を連続的な経済回廊と見なす東海道メガロポリスが実体化された。同時に、航空路の国内線の拡充も進み、「大きい交通」が重層される。

「大きい交通」は、都市間の距離を縮め、中心の洗練された

which might be called "party-of-one support technologies," are becoming a new type of infrastructure. There are high expectations for single customer-support technologies in Japan, where single-person households are increasing and there are fears that conditions akin to those of so-called "critical villages" (i.e. villages that have difficulty maintaining their communities because of depopulation and aging) will begin to appear in the suburbs of cities. For the Japanese, modernization has meant escaping from the fetters of feudalistic customs in villages and heading for cities, but in fact communal constraints exist in cities as well. Party-of-one support technologies will lift even those constraints; these technologies can be said to be truly "modern" in that they enable everyone to receive the same services anywhere without interference from anyone else.

However, if it was agonizing in villages at the dawn of the modern era to have only feudal relationships based on territorial ties to choose from, then it can also be agonizing to have only the choice of being by oneself. If party-of-one support technologies win out and eliminate other choices, that would be a problem too.

11. Further, Faster and on a Larger Scale

The objective of transportation in the modern era was to carry people and things "further, faster and on a larger scale." The Shinkansen is representative of transportation of this kind, what one might call "Big Transportation." For the Tokyo Olympics of 1964, the Japanese Government undertook various projects intended to impress on the rest of the world that the country had rebuilt itself since the war. At the top of the list were the Metropolitan Expressway in Tokyo and the Shinkansen linking Tokyo and Osaka. The latter in particular with trains operating at 210 kilometers per hour, which made them the fastest in the world at the time, ushered in an age of high-speed railways and transportation that was further, faster and on a larger scale. Today, bullet trains[61] have become indispensable to urban stratagies throughout the world. The other class of Big Transportation is inter-city expressways. The first of these was the Meishin Expressway (between Kobe and Nagoya), which opened in 1964 and was extended to Tokyo in 1968. Three major economic spheres—the Capital Region, the Chūkyō region and the Kansai region—were now joined

文明を周縁地域に届け、都市の地政学を塗り替える。中心都市からみれば「大きい交通」は商圏的野望や政治的野望を拡大させ、領土的野心の実現手段となる。実際、東海道メガロポリスの形成は沿線の都市の繁栄以上に東京経済圏の支配権を強め関西の経済的地位を低くした。周縁的な地域では、中心の文明の影響力が強まることを「便利になった」と表現するが、便利と地域性はトレードオフの関係にある。「流れ」がある限り、強く魅力的な文明が他を席巻するのが文明史である。

「大きい交通」は強力で魅力的であるが、「小さい交通」と補い合わなければ人々の移動の欲求に十全に応えることができない。ところが現実には、「大きい交通」が「小さい交通」を駆逐しながら発展してきた。1930年代から1950年代には、全国の67都市に路面電車が走り[62]（→155）、バス網も細かく張り巡らされていた。この時代は、日本の地方都市が経済的にも文化的にも自立し繁栄していた時代である（→156）。それ以降、中小都市では路面電車は自家用車の普及に押されて縮小、撤去され、大都市では輸送力の大きい地下鉄に置き換えられていった。長岡にはいわゆる市電はなかったが、周辺の町村と長岡を結ぶ栃尾鉄道（1913年開業）と長岡鉄道（1915年開業）の二本の私鉄線が敷かれ地域の足として、また貨物輸送に活躍したが、栃尾鉄道が1975年に、長岡鉄道は1995年にそれぞれ廃された。それは東海道から約10年強の遅れで関越自動車道（1980年）と上越新幹線（1982年）が長岡まで延びた時期である。

12. 自家用車過依存

人が自分の意志で移動することは生活上必須なだけでなく生活の質を保つうえでも欠かせない。高齢になると、たとえ健康であっても移動の身体的負担は大きい。その意味で、自家用車ではさまざまな技術開発も進み、高齢者の移動を楽にし、生活を豊かにしてくれる。ただ、高齢になると交通事故の可能性が高くなるので運転を控えたいと考えるし、年金生活者を含めて自家用車を保有することが経済的に難しい世帯も増える。完全に自家用車での移動を前提にした都市では、スーパーマーケットも、ファミリーレストランもファーストフード店も皆ショッピン

155 路面電車の衰退と自動車の増加
The decline of the streetcar and rise of the automobile

together by two systems of Big Transportation, creating a continuous economic corridor, the Tōkaidō Megalopolis, on the Pacific coast of Japan. At the same time, airlines expanded their domestic services, adding a further layer to Big Transportation.

Big Transportation shrank distances between cities, brought the refined culture of the center to peripheral areas, and completely altered the geopolitics of cities. From the standpoint of central cities, Big Transportation heightened commercial and political ambitions and became a means of realizing territorial designs. Actually, the formation of the Tōkaidō Megalopolis reinforced the dominance of the Tokyo economic region instead of bringing prosperity to other cities situated along the transportation spine and led to a decline in the economic status of the Kansai region. In peripheral areas, the increased influence of the culture of the center is said to have made things more "convenient," but there is a trade-off between convenience and local character. Throughout history, the flow of people and things has enabled powerful and attractive cultures to overwhelm others.

Big Transportation is powerful and appealing, but it and Small Transportation must be complement one another to meet people's needs effectively. In reality, however, Big Transportation has developed by wiping out Small Transportation. Streetcars operated in 67 cities in Japan from the 1930s through the 1950s, and closely-knit bus networks also existed (→ 155).[62] This was a period in which local cities in Japan were still economically and culturally independent and prosperous (→ 156). Since then, in small and medium-sized cities, streetcar systems have been reduced in scale or removed entirely as automobiles proliferated, and replaced in metropolises by subways which have a greater carrying capacity. Nagaoka did not have a streetcar system, but there were two private railway lines linking it to surrounding towns and villages and transporting both passengers and cargo, Tochio Railway (which opened in 1913) and Nagaoka Railway (which opened in 1915). The former was discontinued in 1975 and the latter in 1995, around the time the Kan'etsu Expressway and the Jōetsu Shinkansen were extended from Tōkaidō to Nagaoka after a delay of approximately ten years.

グモールの中かバイパス沿いに集中して並んでいるから、自動車が使えないと食事にも食材にもありつけない地域が増える。英国では、これをFood Desert（食の砂漠）と呼び、日本では人に焦点を当てて「買い物難民」と呼んでいる（→ 157、158）。

自動車は、排気ガスによる環境汚染の原因となり、交通事故を皆無にすることができず[63]、さらに自動車を支える基盤整備にも相当な費用がかかる。こうした社会的費用のすべてが自動車の利用者によって負担されているわけではない。経済学者宇沢弘文や上岡直見は悪影響を金銭価格で評価して、両氏ともその額を1台あたり約1200万円と試算している[64]。日本の自動車ユーザーの負担は1台当たり400万円程度なので、残りは国民が広く負担していることになる。両氏とも自動車の利用者が適切なコストを負担していないことが過度な自家用車依存を存続拡大させていると主張している。

今後に増える高齢者のために、都市が何かをするとしたら、その一つは便利で充実した公共交通の整備である。介護施設も必要であるが、大半の老人は自分で生活をこなせ、ちょっとした助けさえあれば自立できる。介護も支援も要らない老人は

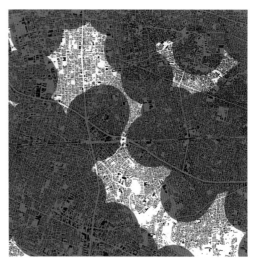

157 買物難民地図（東京杉並区上荻1丁目、2015）
Map of shopping refugees in a "food desert" (Kamiogi 1chōme, Suginami Ward, Tokyo, 2015)

濃い円は電話帳掲載の「スーパーマーケット」から歩ける範囲（500m）を示す。
Dark circles represent the walkable ranges (500m) from supermarkets, as listed in the telephone directory.

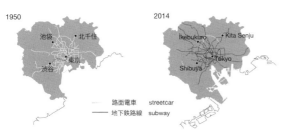

156 都電の衰退と地下鉄網の延伸（東京）
The decline of the metropolitan streetcar and the extension of the subway network (Tokyo)

12. An Overdependence on Automobiles

Being able to move about is not only essential in everyday life; it is crucial to maintaining quality of life. In old age, moving about becomes physically demanding, even when one is in good health. In that sense, automobiles, in which diverse technological advances have been made, facilitate moving for seniors and make their lives richer. However, the increased possibility of traffic accidents may make seniors think twice about driving, and the cost of maintaining an automobile may become prohibitive for seniors, particularly those living on pensions. In cities that are premised on the ownership and use of automobiles, all enterprises from supermarkets to family restaurants and fast food stores are concentrated in shopping malls or along bypasses. As a result, areas where people without the use of automobiles cannot get meals or foodstuffs are increasing. In England, such areas are referred to as "food deserts" (→ 157, 158).

The exhaust gases of automobiles cause environmental pollution, traffic accidents cannot be entirely eliminated,[63] and maintaining the basic infrastructure for automobiles is

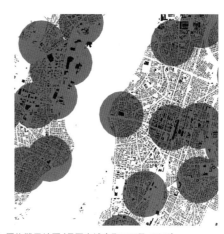

158 買物難民地図（長岡市城内町二丁目、2015）
Map of shopping refugees in a "food desert" (Jōnaicho 2 chōme, Nagaoka City, 2015)

濃い円は電話帳掲載の「スーパーマーケット」から歩ける範囲（500m）を示す。

Dark circles represent the walkable ranges (500m) from supermarkets, as listed in the telephone directory.

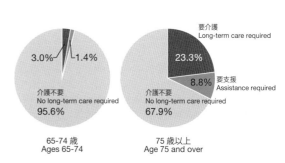

159 介護が必要な高齢者の割合
Ratio of elderly in need of nursing care

65歳以上75歳未満で約8割以上、75歳以上でも7割弱である (→ 159)。都市をコンパクトにすることができれば確かに公共交通の充実に資するかもしれないが、都市をコンパクトに改造する前に、超高齢社会はやってきてしまう。いますぐにも、自動車を使えない人も自由に動ける公共交通システムの整備に向けて舵を切らなければならない。それをしなければ、日本の地方都市は高齢者にとっては住みにくくなる。今や物と情報はどこにいても均等に届くので、大都市と地方中小都市を分け隔てるものは公共交通の充実度だけと言っても過言ではない (→ 161)。若者が地方都市に魅力を見いだせず離れてゆく傾向が続いているなか、高齢者までが地方都市を見放せば地方都市に未来はない。地方都市が生き残るためには、自動車依存症から脱することである。

13. 狭い道路

自動車が一般に普及する前は、道は狭くても何の問題もなかったが、たくさんの自動車がそこら中を走り回るようになると歩行者にとって危険になる。東京には歩道がない道が76.6%もある[65]。歩道があっても道路の端に電力会社などが電柱を立てるので、歩行者が自由に使える空間は益々狭くなる (→ 160)。

日本の都市が、自動車を受け入れざるをえなくなったとき、狭い道しか持たない都市を自動車交通に適応させる解決の方法にはいくつかあった。第一の解決法は、都市計画によって道路を広くすることである。日本の自治体が作る都市計画図には「都市計画道路」を示す赤い線が引かれている。しかし、実際に拡幅を行うためには道路に沿った宅地を買収しなければならないので、地主との交渉が必要であり時間も費用も膨大に要する。

第二の解決法は、市街地の外の田園に迂回路（バイパス道路）を作って、自動車交通を都心部に入れない方法である。この方法は多くの地方都市で標準的な都市計画として採用されている。第一の方法に比べると費用は遥かに掛からないし、市街地を拡大させたいという期待と自動車時代に適した便利な都市にしたいという期待に沿うが、結果的にはバイパスの沿道に内外の勢いのある商店が集まり、人々の生活をますます自動

160 狭い道路（東京都文京区）
Narrow street (Bunkyō Ward, Tokyo)

日本の法律では、道路は最低4mを要するが、現実にはそれより狭い道は多数あり、延焼の恐れや緊急車輛が入れないなど防災・安全上の問題がある反面、静かな住環境を提供していることは評価されている。

While Japanese laws regulate street widths to at least 4m, many streets in Japan do not meet this standard. Although this gives rise to disaster and safety concerns such as the spread of fires or obstructing the passage of emergency vehicles, these narrow streets do contribute to the creation of quiet living environments.

costly. These expenses to society are not borne entirely by automobile users. Economists Uzawa Hirofumi and Kamioka Naomi have evaluated the monetary cost of the harmful effects of automobiles; both calculate that this is approximately 12 million Yen per automobile.[64] The cost borne by automobile users in Japan is approximately 4 million Yen per automobile, which means the remainder is borne by a wide segment of the general public. In the opinion of both economists, the fact that automobile users are not paying an appropriate amount is helping to maintain and expand the excessive dependence on automobiles.

One of the things cities can do for seniors whose numbers will increase in the future is to develop a convenient and improved system of public transportation. Care facilities are also necessary, but the majority of seniors are able to manage their own lives if they have some help. Over 80 percent of seniors 65 or older but under 75 in age—and slightly less than 70 percent of those 75 or older—require neither care nor support (→ 159). Making cities compact will certainly contribute to the improvement of the public transportation system, but a society in which seniors make up a much greater percentage of the population than they do currently will arrive before cities are reorganized and made compact. The development of a public transportation system that enables people who do not have use of automobiles to move about freely is urgently needed. Without it, local cities in Japan will become difficult to live in for seniors. Today, things and information are delivered everywhere. It would not be an exaggeration to say that the only thing separating big cities from small and medium-sized local cities is the level of development of public transportation. Finding nothing appealing in local cities, youths are leaving for metropolises. If seniors too abandon local cities, then those cities truly have no future. If local cities are to survive, the dependence on automobiles must be overcome.

13. Narrow Streets

Narrow streets presented no problems before the proliferation of automobiles but become dangerous for pedestrians once traffic becomes heavy. In Tokyo, 76.6 percent of all streets have no sidewalks.[65] Even when there

車に依存させることになる。そして同時に、古い都心の活動に大きな打撃を与えて地元の古い商業者を苦境に立たせる一因にもなる。

　第三の解決法は、自動車を市街地から締め出して歩行者専用に変える方法である。これは1972年に北海道旭川市の商店街で始められた。その後、これに倣って、歩行者専用道路の整備や、週末などに時間を限定した運用での「歩行者天国」が日本中に広がり、今では都市のなかの道のあり方として市民権を得ている。天国という名前は、道がコミュニティの中心的公共空間になり、時に祝祭空間に変わる日本の伝統を、幾分の皮肉を交えて思い出させてくれる。自動車を都心部から締め出すような交通政策は、東京都などの首都圏の自治体、大阪府や兵庫県、愛知県などでディーゼルエンジン車やトラックを対象に窒素酸化物と粒子状物質の排出基準の制限で行なう間接的な管理だけである[66]。日本では、自動車使用の自由を主張する発言が強いところは、アメリカが銃の所持の主張に寛容なことと似ている。

　第四の解決法は、公共交通を充実させて自家用車への依存

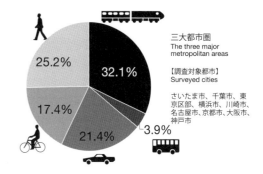

三大都市圏
The three major
metropolitan areas

【調査対象都市】
Surveyed cities

さいたま市、千葉市、東京区部、横浜市、川崎市、名古屋市、京都市、大阪市、神戸市

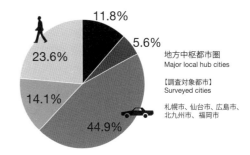

地方中枢都市圏
Major local hub cities

【調査対象都市】
Surveyed cities

札幌市、仙台市、広島市、北九州市、福岡市

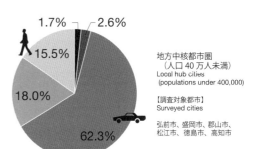

地方中核都市圏
（人口40万人未満）
Local hub cities
(populations under 400,000)

【調査対象都市】
Surveyed cities

弘前市、盛岡市、郡山市、松江市、徳島市、高知市

地方中心都市圏
その他の都市
Local central cities and
other cities

【調査対象都市】
Surveyed cities

湯沢市、伊那市、上越市、長門市、今治市、人吉市

are sidewalks, pedestrians often have to squeeze past utility poles planted by the power company (→160).

　There were a number of ways that Japanese cities with only narrow streets adapted to the introduction of automobiles. One way was to widen streets through city planning. So-called "city-planning roads" are indicated by red lines on city planning drawings prepared by local governments in Japan. However, to widen a street, land along it must be bought up; this means negotiations with landowners, and the process can consume enormous amounts of time and money.

　A second way was to create a bypass so that automobile traffic did not enter the center of the city. This was adopted as a standard city-planning technique by many local cities. This was far less costly than the first solution and intended to expand the urban area and make the city more convenient. However, this led thriving stores from both inside and outside the city to concentrate along the bypass and made people even more dependent on automobiles in their everyday lives. At the same time, it reduced the level of activity in the old central district and created difficulties for commercial enterprises in that area.

161　都市規模別の交通手段分担率
　　　Ratio of different transportation modes by city scale

を下げることである。この方法は乗用車にしか有効ではないが、東京はほぼ成功し、地方都市では富山市が果敢に挑戦をしており、長岡など大半の地方都市はほとんど策がない。

14. 立派な交通基盤と貧弱な連携

　日本の道路が貧弱で自動車時代には不都合であるという意識は、官僚や政治家の大きいトラウマになり、立派な道路建設に向けて莫大な情熱が注がれてきた。たとえば、かつては、日本では未舗装道路はありふれていたが、道路の舗装率を自治体の業績の指標として競争的努力をした結果、いまでは都市部では草の生える余地がないほどにアスファルトによる道路舗装がゆき亘っている。また、技術系官僚の本性として、少しでも仕様を上げることが正しいと頑張るので、農地のなかを過大な幅員の歩道をもった直線の道路を作り、不要で大げさな立体交差を作る。しかし、そのあいだにも人口は減り、道路需要が低下している。いまや、日本では自動車用の道路が供給過剰になっている。現代の日本に必要なのは、新たな自動車用の道路を作ることより、自転車レーンの設置など「小さい交通」のための施設の充実とともに、複数の交通網を相互に結びつけ、もっと効率的で快適なネットワークにすることである。たとえば、東京湾には、二つの川を走る観光船ルートがあるが、湾岸の埋め立て地に建つ高層住宅に専用の桟橋はなく、展示場などの公共施設でも桟橋は数えるほどである。そのうえに、鉄道、地下鉄との連絡が悪い。

　地方都市の公共交通はバスが主体であるが、バス路線どうしの連絡が悪い。バスの運賃システムは路線ごとに課金される硬直的なシステムなので、路線を乗り換えようとすると、そのたびに別の運賃を支払わなければならずバス利用の意欲を削ぐ。電子カードの読み取り機を設置しているバスが増えているのに、距離制の運賃体系や時間制の運賃体系を取り入れようというバス会社がほとんどないことは驚愕ですらある。到底、乗降客数増を真剣に考えているとは思えない。

　日本の大都市では自転車の利用率は北欧並みに高いのに、鉄道駅の駐輪場には標準的なシステムがなく、大半は鉄道駅とのつながりが悪く、また、個々の建物には駐輪場が付置されて

A third way was to shut out automobiles from the built-up area and turn it into a pedestrian zone. The shopping district of Asahikawa, Hokkaido, took this approach starting in 1972. Following this example, the development of pedestrians-only streets and the restriction of vehicular traffic on certain streets during specified hours, for example on weekends, to create so-called "pedestrian heavens" took place all over Japan, and such streets have become fully accepted in cities. Traditionally, streets in Japan have served as public spaces and occasionally as spaces for festive events; calling streets reserved for use by pedestrians "heavens" is an ironic reminder of that tradition. Cities have not shut out automobiles from the central district, the exception being the imposition of indirect control local governments in the Capital Region such as Tokyo and certain prefectures such as Osaka, Hyogo and Aichi have exercised through restrictions on vehicles with diesel engines that do not meet a NOx and PM exhaust criterion.[66] The strong voices in Japan in favor of the right to use automobiles resemble the strong voices in the United States in favor of the right to possess guns.

A fourth way was to improve public transportation and lower the dependence on automobiles. This approach, which is effective only for passenger cars, has been quite successful for Tokyo. Among local cities, Toyama has boldly taken up the challenge but most, including Nagaoka, have adopted virtually no measure of this kind.

14. Excellent Transportation Infrastructure and Poor Coordination

Japanese bureaucrats and politicians fear the perception that roads are in poor condition and not well suited to the age of automobiles. They have been ardent supporters of the construction of roads of high standards. In the past, unpaved roads were commonplace in Japan, but competitive efforts by local governments, who used the percentage of roads that are paved as an index of their achievement, have resulted in asphalt paving on streets in urban areas so complete that weeds now have little chance to grow on them. Improving specifications as much as possible is second nature to technocrats, and the consequence is the construction of straight roads of excessive width, complete with sidewalks, in the middle of

いない。また、自宅からバス停に自転車を利用しようとしても、バス停に自転車置き場があることはまずない。首都圏に自転車を裸で持ち込める鉄道路線もバス路線もほとんどない。日本の鉄道会社の一般的な規則では、自転車を袋に入れなければならないから、遠出をする自転車愛好家しか利用できない。勿論、首都圏の中心部の鉄道車両は常時乗客で溢れかえり、とても裸の自転車を載せることなど考えられないが、首都圏の周辺部や地方都市では可能なはずである[67]。日本では、総じて自動車施設に比べて自転車施設は冷遇されている。

15. 水上交通は都市の宝石

江戸幕府が鎖国したことが功を奏して、安定した政治と経済が発展し、江戸の人口は18世紀の初めには100万人を超え、世界的にも巨大都市となった。巨大都市を支える物流は水運が中心であった（→ 162）。江戸の下町地区は、低湿地を干拓と埋め立てによって形成されたので、排水も兼ねた掘割が縦横に走り水運が活用されていた。掘割の両岸には蔵が立ち並び、屋敷への物資搬入など多面的に舟運が活躍し、河岸は荷の積み降ろしで賑わっていた。また、堀は物流に使われただけではなく江戸市民に多様な娯楽を提供した。川沿いの料亭での食事や、屋形船に乗っての花火見物や船上から花見をしながらの逢い引きなど、江戸の人たちは川に親しんだ。一方、長岡は日本で一番長い川である信濃川の氾濫原にできた内陸の都市なので、明治時代に鉄道が通じるまでは、信濃川の水運が大きな役割を担っていた。

東京では工場が増え、下水道が完備しないまま人口が増えたので、掘割に雑排水と工場排水が流れ込んで悪臭を放ち、昔日の面影を失い汚染は東京湾にまで広がった。次の自動車時代になると道路の需要が高まり、汚れた水路は無用の長物視され多くの掘り割りが埋め立てられてしまった。一方、溢れ出るゴミの処理先として東京湾の埋め立てを続けたために、1965年以降だけでも2万5000 haの水面が消失した。これは東京湾の水面の約2割である[68]。埋め立て地の大半は工場用地として使われたので、市民が水際に近づける場所が減った。内陸の小河川も、水害を防ぎ周辺の土地利用を拡大するために、いわゆ

farmland and multilevel crossings that are unnecessary and exaggerated. However, even as these were being built, the population was declining and demand for roads was decreasing. Today, there is an oversupply of roads for automobiles. What is needed today in Japan is not the construction of new roads for automobiles but the improvement of facilities for Small Transportation such as the introduction of bicycle lanes as well as the development of more efficient and comfortable networks through the connection of multiple systems of transportation. For example, in Tokyo Bay there are routes for tourist boats running on two rivers. However, the high-rise housing projects standing on reclaimed land along the coast have no piers, and public facilities such as exhibition centers have only a few piers at best. Moreover, their connections to railways and subways are poorly arranged.

Buses are the main forms of public transport in local cities, but connections between different bus routes are poor. The fare system for buses is inflexible—a fare is charged for each route, and there is no discount for transferring from one route to another—and discourages the use of buses. Although buses equipped with electronic card readers are increasing, the failure of most bus companies to introduce fare systems based on distance or time is hard to fathom. They appear not to be thinking seriously about ways to increase the number of passengers.

Although the rate of use of bicycles in big Japanese cities is high and comparable to that in Scandinavia, bicycle parking facilities at railway stations do not have a standardized system, most facilities are poorly connected to railway stations, and individual buildings are not equipped with bicycle parking areas. Moreover, even if people wanted to use bicycles from their homes to bus stops, there are no bicycle parking areas at those bus stops. In the Capital Region, railway lines or bus routes that allow people to take along unbagged bicycles are the exception. Japanese railways generally require that bicycles be bagged and are therefore used only by cyclists on excursions. In central districts of the Capital Region, cars are always packed with passengers, and taking along unbagged bicycles is impracticable. However, it should be possible on the outskirts of the Capital Region and in local cities.[67] In Japan bicycle users are generally treated quite

る三面張りのコンクリート構造になった。その結果、川は放水路に成り下がった。

　日本での重工業の立地が難しくなると、東京都は埋め立て地に副都心を計画し1989年から建設を始め、専用の高架軌道を自動運転で走る軽量鉄道を敷き、首都圏随一の展示場を建設し、先端的なオフィスや商業施設、住宅、会議場などを誘致した。しかし、都市を「魅力的にすること」で投資とテナントを惹き付けるというポストモダン的都市戦略の意識に乏しく、この地の最大のポテンシャルである豊富な水面を活かすことに熱心ではなかった。水上交通が交通機能を超えた快楽を提供し、沿岸の建物の魅力を高めることは、年間1600万人の観光客を集めるベニス（イタリア）を筆頭に蘇州（中国）や日本の柳川、近江八幡、大垣、香取、栃木など都市河川の水上観光が変わらぬ人気があることからも明らかである。

　東京の近代の開発は、水面と水際の可能性を封殺する歴史であった。しかし、水面は、都市に住み移動する喜びを倍加する魔法である[69]。窓一面に水面が見えるオフィスや住宅、あるいは林立するヨットのマスト、カモメの声を聞けるカフェなどは、現代のグローバルシティの必須の魅惑的なアイテムである。しかも水際の開発が水際エリアの防災性も高めることも忘れてはならない。

16. 地方都市のダイナミズム

　全総は、大都市と小都市あるいは中央と地方の格差を、地方交付税と工業の再配置を梃にして全国を均等に産業化することで解消しようとした。しかし、実際には、工業の優位性はいつまでも続かず、サービス産業に地位を奪われ地域間格差解消の試みは失敗した。同時に、高速交通網の充実と通信技術の発達によって、東京と張り合ってきた関西経済圏が地盤沈下し東京への一極集中が進み、地方では中核的都市が支配領域を広げ周辺の中小都市を経済的支配下に組み入れる再組織化が進行した。1990年以降の、経済のグローバル化と情報化は、以前にもまして大都市を有利にし、大都市と小都市の格差をさらに際立たせることになった[70]。

　短期的な投資の効率性、人材の集約性、エネルギー消費の

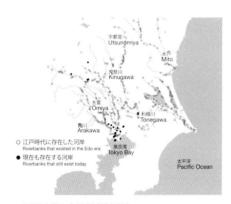

162　江戸時代の水運網（首都圏）
　　　Water network during the Edo era (Capital Region)

poorly compared with automobile users.

15. Water Transportation Ought to Be Prized by Cities

Under the Tokugawa shogunate, which adopted a policy of national seclusion, a stable political framework was created and the economy developed. By the start of the eighteenth century, Edo had become an enormous city, even by international standards, with a population exceeding one million. Water transportation was the main means of distribution in that city (→ 162). The "low-city" (shitamachi) districts had been formed by the reclamation of low-lying wetlands. Canals that doubled as drainage crisscrossed these districts and were used for water transportation. Storehouses lined the banks of canals, boats on various errands such as delivering supplies to residences navigated the waters, and riverbanks were enlivened by the constant loading and unloading of boats. Canals were not only used for distribution but provided the people of Edo with diverse forms of amusement. There were meals at riverside restaurants and flower viewing and assignations aboard roofed pleasure boats. Nagaoka is an inland city created on the flood plain of the Shinano, Japan's longest river, and until railway systems developed in the Meiji period, water transportation on the river was important to it.

The development of sewers in Tokyo failed to keep up with the increase in factories and the growth in population. Effluent from houses and factories filled canals, creating a stench, and the pollution extended to Tokyo Bay. When the age of automobiles arrived, demand for roads increased. The polluted canals came to be seen as useless, and many were filled in. Meanwhile, refuse continued to be used to reclaim land on Tokyo Bay. Since 1965, 25,000 ha of water surface, representing 20 percent of the surface area of Tokyo Bay, have been lost.[68] Most of the reclaimed land was used for factories, making it difficult for the public to approach the shore. Further inland, small rivers were turned into three-sided concrete structures to prevent flooding and to expand the use of nearby land. As a result, rivers became nothing more than drainage canals.

As conditions became less favorable for heavy industries in Japan, Tokyo developed a plan to create a subcenter on reclaimed land and began construction in 1989. Dedicated

合理性や低炭素化の観点から考えれば東京一極集中が望ましいかもしれないが、中期的にみれば対地震管理上不利であり、国防上も脆弱になる。長期的には、一極集中は文化的な多様性を減らし、創造性を弱くするだろう。たとえば、近代史初期において日本が強みを発揮したのは、識字率の高さや勤勉な国民性もあったが、地域の豊かな多様性があったからではないだろうか。個性的で自立した地域が作り出す多様なアイディアと人材こそがダイナミックな組織を育て、成長の原動力となり、明治維新前後の体制転換から近代化の爆発的発展を可能にし、戦後の驚異の経済復興を成し遂げた。

現在の多くの地方経済の最大の問題は雇用力の弱さである。理由の一つは、大企業が大都市を好むからであるが、都市間交通基盤と情報基盤が整備された今日、日本の国土のスケールであれば企業立地はどこでも成立するはずである。企業の中枢が大都市特に首都圏に無ければならないという見解も、勤労者が余暇時間に無関心なことも等しく現代日本の習慣的思考の結果に過ぎない。それどころか、企業の中枢機構が首都圏に過度に集中することは立地コストや危機管理の点から合理的ではないし、性別や年齢差に関係なく働けてしかも健康的で仕事とバランスがとれた生活を送るという点では大都市は必ずしも向いていない。経済の実態から見ても、今日の日本のGDPの7割は地方経済によって生み出され、サービス産業を中心として地域に根ざした経済活動[71]が展開されていることが十分活かされていない。

日本の地方都市の大きな弱点の一つは都市内交通である。雇用機会が増えても、自家用車以外の移動の選択肢がなければ、その都市で歳をとることに多くの人が不安をいだくだろう。

伝統的な産業の一つである第一次産業は、雇用力は現在全国で4%にすぎず、長岡でも4.3%に過ぎない。しかし、食料の安全保障の観点からも環境保全のためにも、農林水産業の健全な維持が求められる。そのために高い生産性と主体的な取り組みを促すために、世襲による事業継承から脱して、他の産業と同じようにだれもが自由に選択できる職業の一つに変わることが必要である。

elevated tracks for a light rail that operates automatically were laid, the biggest exhibition center in the Capital Region was constructed, and ultramodern office buildings, commercial facilities, condominiums and conference facilities were attracted to the area. However, Tokyo was not well-versed in the postmodern urban stratagem, namely the idea to make a city appealing and thereby attract investments and tenants, and failed to take advantage of the factor with the greatest potential in the area, the water. Water transportation is not only functional but enjoyable and enhances the appeal of coastal buildings. This is corroborated by the popularity with tourists of urban waterways in Venice (which attracts 16 million visitors a year), Suzhou, and cities in Japan such as Yanagawa, Ōmi Hachiman, Ōgaki, Katori and Tochigi.

The potential of bodies of water and watersides has been suppressed throughout the development of Tokyo in modern times, but bodies of water have a magical power to make living in and moving through cities much more joyful.[69] Office buildings and housing with windows that afford views of water, harbors full of yachts, and cafes where people can hear the cries of gulls are appealing, indispensable items for contemporary global cities. It also ought to be remembered that waterside developments strengthens the disaster-preventive capacity of coastal areas.

16. The Dynamism of Local Cities

The Comprehensive National Land Development Plan or Zensō attempted to close the gap between big cities and small cities, or between centers and localities by uniformly industrializing the entire country, using as a lever the tax allocated to local governments and the relocation of industry. However, the predominance of industry did not last forever; its place was usurped by the service industry, and the attempt to close the gap between regions failed. At the same time, the development of a high-speed transportation network and advances in communication technology led to a decline in the Kansai economic region and the continued concentration of businesses in Tokyo; in localities, there was further reorganization as core cities expanded their spheres of influence and incorporated nearby small and medium-sized cities. The globalization

17. 都市の住宅と家族

都心

第二次世界大戦前の日本の都市には町家型建築と屋敷型建築があった。町家は世界の多くの地域に存在する街区型都市建築で、主に商店との併用住宅に使われた建築タイプである。しかし、現代都市の都心を構成するには二階建ての木造建築では密度が低すぎる。そのため町家は第二次世界大戦後には新しく供給されることがなく、中高層建築に建て替わり、いまや絶滅危惧種の状況にある。一方、屋敷型の建築は、建築物の周辺に空地を取り塀で囲む形式であり、人々の高い支持を得て、住宅から公共建築まで、都心から郊外まで幅広く使われている（→163）。現在、首都圏では住民の4割弱が戸建て住宅に住み、長岡では7割強の世帯が戸建て住宅に住んでいる。

一方、歴史的にみてみると、設地型の連棟住宅（いわゆる長屋）以外に集合住宅は日本にはなかった。本格的な集合住宅が一般化するのは第二次世界大戦後のことで、当初は公共住宅を中心に普及し、1970年代から供給量が増えた[72]。この時代の集合住宅は、賃貸か分譲にかかわらず、最終的に戸建て住宅へ移行する前の仮住まいとして位置づけられてきた。そのため、長らく戸建て住宅に比べて集合住宅の戸当たりの床面積は狭かった。集合住宅が終の住処として受け入れられるようになるのは近年のことである。

郊外

首都圏で郊外開発が盛んになった1920年代では、一区画が500 m²を超えることも珍しくなかった[73]。しかし、既に述べたように、大都市で鉄道を主体に郊外化が進展したために、郊外での住宅購入時に都心に近いことと、鉄道駅に近いこと特に歩けることを重視する傾向がどの時代にも強かった。そのために、駅からの徒歩圏は限られているので住宅地の地価は上昇し、宅地は狭小化の一途を辿った。20世紀の終わりには平均的なサラリーマンが購入できる宅地は、東京の区部では75 m²程度にまで縮小してしまった[74]。東京都全体でみると一宅地の平均は135 m²、長岡では200 m²である[75]。一方、世帯収入の伸びに合わせて住宅の戸当たり床面積は増え続けたので、結果と

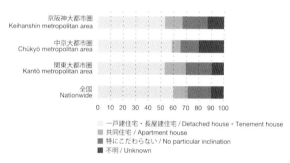

163 都市圏別好まれる住宅形式
Housing styles favored by different urban areas

and increasingly information-oriented nature of the economy since 1990 have made conditions even more favorable for big cities and further widened the gap between big cities and small cities.[70]

Extreme concentration in Tokyo may be desirable from the point of view of efficiency of short-term investments, convergence of talent, and rationality and low-carbon emission with respect to energy consumption. However, from a medium-term perspective, it is disadvantageous for earthquake risk management and weakens the capacity to defend the country militarily. In the long term, extreme concentration is likely to reduce cultural diversity and diminish creativity. In the early modern era, Japan was able to take full advantage of its strengths because in addition to a high rate of literacy and a highly industrious people, it possessed great regional diversity. The diverse ideas and talents produced by unique and autonomous regions were precisely what gave rise to dynamic organizations and became the driving force for growth; they made possible the overturning of the established order during the Meiji Restoration, the explosive development that took place in the course of modernization and the astonishing reconstruction of the economy after the war.

The biggest problem with most local economies today is scarcity of employment opportunities. One reason is that large corporations favor metropolises, but today, when inter-city transportation infrastructure and communication infrastructure are well developed, a corporation ought to be able to function successfully in any location in the country. The idea that the center of a corporation must be in a big city and preferably in the Capital Region in contemporary Japan is as much the result of old habits of thought as a lack of interest among workers in spare time. On the contrary, the overconcentration of central corporate functions in the Capital Region is not rational from the point of view of the cost of location and risk management, and metropolises are by no means suitable places for such a location if people are to be able to work whatever their gender or age and to lead healthy, well-balanced lives. If one examines the actual condition of the economy, full advantage is not being taken of the fact that 70 percent of Japan's GDP today is generated by local economies and economic activities rooted in localities and centered on the service industry are taking place there.[71]

して建蔽率が高く密集した住宅地が増えた。狭い敷地に広い住宅を建て自家用車を敷地内に駐車すると、もはや宅地と住宅の間には隙間としか呼べない空地しか残らなくなる。日本の郊外住宅は郊外居住の良さを見いだし難い状況に陥りつつある(→ 164、165)。

家族

都市の居住者を社会的側面から見ると、日本の現代の住宅地の特徴は社会階層性が弱いことにある。第二次世界大戦の敗戦後の合衆国を中心とした占領政策では、徹底的な民主化が実行され、税制でいえば高い相続税率が課せられた。これによって戦後の日本の社会は階層間の流動性を高め、同時に、高い相続税を支払うために土地を切り売りしなければ納税できなくなり、相続のたびに土地が細分化された。

一方、家族形態は核家族化が進んで、家族規模が小さくなる一方、寿命が延び、親の住宅を子供が引き継ぐ頃には、子供世帯は既にそれなりの住居を確保している。住宅市場での中古住宅の人気が極めて低いことは、結局、住宅の一世代ごとの使い捨てに繋がり、住宅が消費財化する傾向を助長している。

世帯規模が小さくなると、子育てや高齢の親の介護が家族だけでは困難になる。かつては、専業主婦の役割とされたが、高齢者人口が増えるに従って手に負えなくなっている。解決策の一つは、公共による幼児と高齢者の世話ということになるが、年々膨らむ需要に財政が追いつかない。代替策として、近隣コミュニティによる補完が必要であることが議論されているが、現状では容易ではない。それは大都市の郊外では住民の移動も多く、そのうえ都市住民の職場は住宅地の外にあるうえに日本の企業では、勤務時間が長いので、地域に対する帰属意識が薄く、住民の結束も弱いからである。

少子高齢化社会で年金制度を維持するのが財政的な大きな負担になり持続性が危ぶまれている。解決法は、高齢者全員に支給するのであれば年金支給開始年齢の引き上げと年金額の減額になる。あるいは、年金を働けなくなったときだけ支給される保険に変えることも別の選択肢である。いずれにしろ、高齢者は今より働かざるをえないが、長い人生に生きがいを作り、少子化による働き手の不足を補ううえでは有意義である[76]。

One of the major weaknesses of local cities in Japan and a major obstacle to such change is urban transportation. Even if employment opportunities increase, many people are likely to feel anxious about growing old in a city where transportation alternatives to automobiles do not exist.

Primary industries account for no more than four percent of employment at present. Even in Nagaoka, it is 4.3 percent. However, sound management of agriculture, forestry and fisheries is vital for the sake of not only those industries themselves but the security of our sources of food and the preservation of the environment. To this end, we must promote higher productivity and independent initiatives. It is necessary therefore to discontinue the custom of hereditary succession and make occupations in those industries as open to anyone who wishes to enter them as occupations in other industries.

17. Houses and Families in Cities

Center of the City

In Japanese cities before World War II, there were *machiya*-type buildings and *yashiki*-type buildings. *Machiya* were units from which urban blocks were made and used mainly as shops combined with dwellings; they represent a building type that exists in many regions of the world. However, they were two-story wooden buildings and their density was too low for the center of a contemporary city. They were not rebuilt after World War II but instead replaced by medium- and high-rise buildings. Today they are an endangered species. Meanwhile, the *yashiki*-type building, which is freestanding and surrounded by a property wall, is quite popular and used for diverse purposes from houses to public buildings and in diverse places from the center of the city to the suburbs (→ 163). At present, slightly less than 40 percent of all residents of the Capital Region live in detached houses; more than 70 percent of households live in detached houses in Nagaoka.

Nagaya or rowhouses were the only traditional type of multiunit housing in Japan. True apartment buildings were widely built only after World War II. At first, they were *mostly* public housing; the supply of apartment buildings increased from the 1970s.[72] Apartment units in that era, whether rented or purchased, were viewed as provisional accommodations because people hoped to move eventually to detached houses. For that reason, apartment units were for a long time smaller in total floor area than detached houses. Only in recent years have apartment units been accepted as ultimate places of residence.

Suburbs

Lots that exceeded 500 square meters in size were not unusual in the 1920s when suburban development took off in the Capital Region.[73] However, there has been a tendency in every period in purchasing houses in the suburbs to emphasize accessibility to the center of the city and proximity to (especially being within walking distance of) a railway station because of the major role played by railways in the development of the suburbs in metropolitan areas. Since the area within walking distance of a station was limited, the price of residential land rose and lots steadily became smaller. By the end of the twentieth century, the lot that an average office worker could purchase in the 23-ward areas of Tokyo had shrunk to about 75 square meters.[74] In Tokyo as a whole, it is about 135 square meters on average, and about 200 square meters in Nagaoka.[75] At the same time, in keeping with the growth in household income, the floor areas of houses increased. As a result, building coverage rose and densely-built residential areas increased. Once a spacious house has been constructed on a small site and an automobile has been parked, only a narrow open space is left between the property line and the house. It is becoming increasingly

都市名　地区名			name, city
■ 建物 building	■ 空地 empty area	対象街区内平均建築面積 Avg. building area	対象街区内平均敷地内空地 Avg. empty area
ニューヨーク／レビットタウン			Levittown, New York
		136m²	451m²
シドニー／オートレイ			Oatley, Sydney
		179m²	370m²
ロンドン／レッチワース			Letchworth Garden City, London
		109m²	542m²
バルセロナ／ヴィラデカンス			Viladecans, Barcelona
		121m²	356m²
ミラノ／クザーノ・ミラニーノ			Cusano Milanino, Milan
		169m²	280m²
ブリュッセル／ブリュッセル郊外			Suburb of Brussels, Brussels
		134m²	500m²
上海／西郊別荘区			West Suburb, Shanghai
		155m²	260m²
マニラ／マカティ			Makati, Manila
		332m²	399m²
シンガポール／ホーランドロード			Holand Road, Singapore
		253m²	489m²
バンコク／ラリングリーンヴィラ			Lalin Green Villa, Bangkok
		149m²	114m²
日本　Japan			
千葉ニュータウン			Chiba New Town, Chiba
		88m²	124m²
墨田区／八広			Yahiro, Tokyo
		42m²	26m²
世田谷区／成城			Seijō, Tokyo
		104m²	173m²

164 住宅地街区の国際比較
International comparison of residential districts

ここに示す事例は各都市の住宅地を代表するものでない。調査は、Googleマップから事例を採取し、それらを実測した。

The examples listed here are not representative of the residential districts in each city. This survey consisted of measuring certain examples selected from Google Maps.

同様に、女性も今以上働くことが必要であるが、それには別の障害がある。日本は女性の社会参加においては後進国であり、女性が実力を発揮できる社会とは言い難い。男女の平等ということからも社会の生産ということからも、今後は大きく変わらざるをえない。

このように日本の家族をめぐる状況は、今後も大きく変わると予想され、それは住居のあり方を大きく変える。それを列挙すれば、

・女性の有職率が高くなり、住宅地選択で通勤のしやすさがさらに高い要因になる。
・高齢者は伴侶と死別や離別することで単身者が増え家族規模が小さくなる。
・共同生活の形態が多様化し住宅の使われ方が多様化する。
・人生が長くなり子育て期間が相対的に短くなり、子供を軸に家族を考えることの意味が相対化される。
・各地で住宅が余るので中古住宅の流通が増える。
・老人の身体能力および経済力から大都市では自家用車所有世帯が減る。

165 車庫が住宅の顔になっている住宅地（長岡市）
Housing lots characterized by car garages (Nagaoka City)

difficult to see any advantage to living in the suburbs in Japan (→ 164, 165).

Family

By and large, social class is not a distinguishing characteristic of residential areas in Japan today. Under the policy adopted by Occupation forces after World War II, a thorough process of democratization was implemented and high inheritance tax rates were imposed. This increased mobility between social classes in postwar Japan, made necessary the selling of land to pay inheritance taxes and led to the gradual subdivision of land.

Nuclear families have become more commonplace while longevity has increased. By the time children inherit their parents' house, they have already acquired houses of some sort of their own. There is not a great demand for used houses in the Japanese housing market as of now, and the result is that generally a house lasts only a generation. This reinforces the tendency to perceive houses as consumer goods.

As households shrink in size, it becomes difficult for families on their own to raise children or to care for elderly

18. 死者の眠る場所

生者の数だけ死者がいる。生を受けたものはやがて死ぬ。だから、都市には赤子の誕生を助ける施設や産婆が要るように、死者を葬る墓と冥界への旅路を先導してくれる僧侶が要る。墓地は身近な人の眠る場所だから近しい場所であるが、一方近代の墓地は縁者以外の人も葬られるので忌み嫌われもする。

かつて、死は宗教的な出来事であり、墓は寺や教会で管理されたが、近代になり、死をもたらす病が病原菌のなせる業であることがわかると、死は科学的で即物的な出来事になり、衛生の問題になり、死体からの感染を恐れて墓場は住宅地から遠ざけられるようになる。同時に、国家が世俗化するに従い、墓地は自治体政府によって管理されるようになる。現代社会では、病や死は運命ではなく克服すべきものとして位置づけられている。老化から死に至る生命の原則は変えられないのに、科学の進歩により寿命が延び続けている時代にあっては、死は生における敗北と見なされさえする。

日本では、親子世帯の別居は一般的であり、ほとんどの死は病院のなかで行われ、死体も見えないように処理され、埋葬地も居住地から遠ざけられているので子供たちが祖父母の死を直接経験することは稀である。しかし、死をいくら遠ざけてもわれわれは逃げおおせない。

われわれは、親しい者の死を通じて生命の尊さに思いやり、死者に対する畏敬こそが文化の継承を動機づける。死が遠くなればなるほど、逆説的に現世は貧しくなる。

19. 多島海化する日本

日本の都市の空間構造は、大都市も中小都市も多島海化しつつある。つまり、ところどころに、独自の物的環境と管理体制をもった「島」が散在しているということである。大学のキャンパスや工場など従来からある「島」もあるが、多くの「島」は新自由主義のスローガン「選択と集中」に従って大資本による開発の結果できた。そうした「島」には超高層住宅やオフィスやホテルが入り、鉄道駅が近接し、「島」内は歩行者専用空間になる。大都市郊外や地方中小都市では、「島」にはショッピングモール

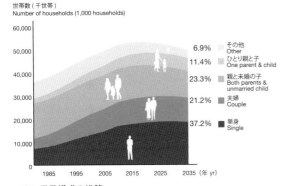

166 世帯構成の推移
Changes in houshold composition

parents. Full-time housewives were once given these responsibilities, but the progressive aging of the population has made it impossible for them to continue to manage such tasks. One possible solution is to have public services take care of infants and seniors, but public finances are unable to keep up with the ever-increasing demand. Having neighborhood communities provide help is being discussed as an alternative, but this would not be easy to achieve under present circumstances. People have little sense of belonging with respect to the places in which they live or of solidarity with one another because people move often in the suburbs, their workplaces are outside the places in which they live, and work hours are long at Japanese corporations.

As a result of the low birthrate and the aging of the population, maintaining the pension system is becoming an enormous financial burden; the sustainability of the system is in question. If all seniors are to receive pensions, the solution may be to raise the age at which payment begins and to reduce the amount paid. An alternative is to change the system into one in which pension is provided only when a person can no longer work. In any case, seniors will have to work longer than they do now, but that could be beneficial in the sense that it will make their longer-lived lives more meaningful and make up for shortages in the workforce caused by the lower birthrate.[76]

Similarly, women will have to participate more in the workforce than they do now, but that is a different problem. Japan is backward as far as the participation of women in society is concerned; it is scarcely a society that enables women to demonstrate what they can actually accomplish. Major changes will be necessary from the perspective of both gender equality and the productivity of society.

Thus the conditions surrounding the Japanese family are expected to change greatly in the future, and that is expected to lead to major changes in the nature of houses. Those changes are as follows.

* The rate of employment of women will rise, and commuting will become an even more important factor in selecting place of residence.
* Death or divorce will mean increased numbers of seniors who are single and a further shrinking of the size of families.
* Forms of communal life and ways in which houses are used will diversify.
* People will live longer, the period devoted to raising children will become relatively shorter, and the importance of children to families will become relativized.
* There will be an excess of houses everywhere and increased buying and selling of used houses.
* Households with automobiles will decrease in metropolises because of the decline in physical capacity and economic power of seniors.

18. Where the Dead Sleep

Wherever there is life, there is also death. The dead too must be accommodated. In cities graves in which to bury

やファーストフードが集められ大規模駐車場が付帯して自動車のアクセスを優先する。「島」は忽然と現れ、周辺の「海」にある商業施設の顧客を奪い、「海」への新たな開発動機を奪う。「島」への投資は原則民間企業によってなされ、自治体も政府も主に規制緩和で支援する。たとえば、日本政府は、2000年に、それまでの既存商店街保護の方針から転換して、大規模小売店舗立地法（大店立地法）を制定し秩序ある開発を条件に大規模店舗の進出を認める方針に切り替えた。

既に述べたように、日本の消費は、必要な物品を消費する段階を過ぎている。企業は戦略的に消費者を慢性的な欲求不満状態にしなければならない。それに成功したときのみ企業は成長を維持できる。そのために従来の広告宣伝に加えて、都市空間を欲望刺激装置として再定義しようとしている。第一歩は、都市空間の表層をすべて広告で覆い尽くすことである。伝統的な電柱のビラから車両の外壁、そして巨大ビデオモニターまで、あくなき追求がなされる。香港の彌敦道（ネイザンロード）に代表される香港の都市景観はもっとも古典的なものである（→167）。東京の繁華街の一つ渋谷には地下鉄2路線を含めて6路線の鉄道線が集まり、駅前には5個の巨大モニターと無数の広告看板が表層に貼り付けられている（→168）。都市の表層は消費者に向かって声高に語りかける。

次の段階は、都市の一地域を囲い込み、意図する雰囲気で覆い尽くし、街を舞台に繰り広げられる劇の出演者として消費者も取り込んでしまう戦略である。それが「島」である。「島」は、郊外でも大都市でも周辺からひときわ目立つ量塊性と要塞的な外観が特徴である。都心では金属とガラスで構成されオフィスのようであり、郊外では工場のようである。いずれにしろモダニズムの言語が幅を利かせている。ところが、ひとたび中に入ると明るく、親切で、開放的で、多くは懐古的である（→169）。構成する業種は、都心ではブランドショップと飲食店、郊外の島では、小売店舗とスーパーマーケット、そしてフードコートに加えてシネコンやスポーツジムなども含まれ、家族で出かけて休日の午後の時間を過ごせるように構成されている。「島」の内部の懐古趣味を演出する上で、歴史的遺構は欠かせない。東京の都心の

167 著者による香港の都市形態研究（1992）
Research on urban form in Hong Kong by the author

著者は、1990年と1991年の二回、イギリスから返還される前の香港の公共空間を中心に調査を行なった。

The author conducted research focusing on public spaces in Hong Kong twice, in 1990 and 1991, before it was returned by the U.K.

the dead and priests to guide them on their journeys to the other world are as needed as facilities and midwives to assist at births. We feel an attachment to graves where those dear to us are laid, but in general cemeteries are not welcomed in our neighborhoods.

In the past, death was a religious event and graves were maintained by temples and churches. In the modern era, when it became recognized that diseases that caused deaths were the work of bacteria, death became a scientific, matter-of-fact event and a problem of sanitation; fearing contamination, people located cemeteries far from residential areas. At the same time, as the state became secularized, graves came to be maintained by local governments. In contemporary society, disease and death are regarded not as matters of fate but as things to be overcome. Although the principle of life, the inevitable process of aging that ends in death, cannot be thwarted, in an era in which advances in science continue to prolong life, death is regarded as a defeat for life.

In Japan, parents and their adult children generally live in separate households. Nearly all deaths take place in hospitals, the dead are disposed of out of sight, and cemeteries are located at a distance. As a result, children are rarely exposed to the deaths of grandparents. Nonetheless, though we may keep death at a distance, we can never escape it.

We are reminded by the deaths of people with whom we are close of the preciousness of life, and respect for the dead is what motivates the preservation of culture. Paradoxically, the more remote death is from everyday experience, the more impoverished is the world we live in.

19. An Archipelago-like Structure

The spatial structure of the Japanese city, large or small, is coming to resemble an archipelago. That is, "islands" with distinctive physical environments and management systems are scattered throughout the city. There are "islands" that have existed from the past such as university campuses and factories, but most "islands" today are the results of development by big capital, produced in accordance with that slogan of neoliberalism, "concentration in core competence." Such islands contain super-tall housing, office buildings and hotels, are located

一番大きな「島」である大手町・丸の内・有楽町地区の「島」には、部分的に保存された日本工業倶楽部、東京中央郵便局、再生された東京駅や歴史的建物の復原である銀行倶楽部、三菱一号館などたくさんある。ポストモダン時代には、歴史的遺構を持たない都市は二流なのである。二流でしかありえない郊外は、代わりにテーマパーク的捏造と子供のファンタジー、ハローキティとディズニーのキャラクタなどが欠かせない要素になる。

「島」戦略が都市にもたらすものは何だろうか。プラスの面は、大資本の投資が集中し規模の経済が発揮されて、擬似的にしろ多様な商業、サービスが提供されることである。たとえば一度消滅した映画館が再びシネコンとして復活[77]している。また、全国ブランドの商品や大型書店が手近になるなどである。

一方問題も多い。第一の問題は、規模の経済が発揮される広い商圏設定をすることから、フードデザートに象徴されるように各住宅地の近隣にあった小規模な商業をほぼ壊滅させてしまうことである。第二の問題は、全国規模の大資本の遠く離れた本部が利益を吸い上げてしまい、地域への経済的還流を断ってしまうことである。第三の問題は、現代の「島」は、どんなに魅惑的なテーマでデザインされ美辞麗句で飾られても、所詮は消費空間に過ぎず、小綺麗な広場もあるが、それは消費を伴わなければ十分な満足が得られず、共同体の結束を育てる場所ではないことである（→170）。第四の問題は、都市の新陳代謝機能が低下することである。伝統的な都市では、新しい産業の芽は都市の裏通りで生まれる。たとえば若い調理人が裏通りに小さなレストランを開き、それが評判をとり店舗を拡大し、やがて表通りに店を構えるという成功物語である。ところが「島」では、利益を確実にするために評価が確立した企業しかテナントにしない。駆け出しの調理人には「島」のテナント条件は厳しすぎて出店できない。だからどこの島も同じような構成になり、見せ掛けの多様性しかできないし、「島」からは次世代のスターは生まれない。一方、島に活力を吸い上げられて「海」の中の裏通りはみすぼらしいだけの裏道に成り下がってしまう。

20. コンパクトシティは目標足りうるか

車に過度に依存しスプロールした現代都市の改造案として、

168 渋谷の駅前の風景
View of Shibuya Station

near railway stations, and offer pedestrians-only spaces. Islands in metropolitan suburbs or small or medium-sized local cities are shopping malls or concentrations of fast-food restaurants with large parking areas attached that prioritize automobile access.

Islands appear suddenly, steal customers away from commercial facilities in the nearby "sea," and rob the "sea" of the will to undertake new development. In principle, private enterprises invest in "islands," and local governments and the central government provide support mainly through deregulation. For example, the Japanese Government, which had hitherto protected existing shopping streets, enacted the Large-Scale Retail Stores Location Law in 2000. This new policy permitted the encroachment of large-scale stores as long as development was orderly.

As has already been explained, Japan has passed the stage of the consumption of necessities. As a stratagem, corporations must create in consumers a chronic sense of frustrated desire. It is only when they succeed in doing so that corporations can maintain growth. Therefore, in addition to conventional advertising, they engage in attempts to redefine urban spaces as devices to stimulate desire. The first step is to cover entirely the external surfaces of urban spaces with advertising, from the traditional posters on utility poles to the exteriors of vehicles and enormous video monitors. The townscape of Hong Kong, most notably Nathan Road, is a classic example (→ 167). The area in front of Shibuya Station, where six railway lines including two subways converge, is one of the busiest shopping streets in Tokyo and features five enormous monitors and countless billboards (→ 168). The external surfaces of the city speak loudly and insistently to consumers.

The next step is to enclose an entire area of the city, create a distinctive ambience and co-opt consumers into becoming performers in a play on the urban stage. The result is an "island." An "island" is characterized by a massive, fortress-like exterior which stands out in its immediate environment, whether that environment is a suburb or a metropolis. In the center of a city, it resembles an office complex of metal and glass, and in the suburbs it looks like a factory. In any case, it is in a modernist vocabulary. Once inside, however, it is brightly-lit, friendly,

コンパクトシティと呼ばれる都市像が1980年代から欧米で議論されるようになった。コンパクトシティの思想は欧州ではEUの都市政策に反映され[78]、米国ではニューアーバニズムとして展開されてアワニー原則(1991年)として基本的思想がまとめられた。

コンパクトシティを特徴づけるのは、用途混合を含めた多様性に満ちた密度の高い都心、都心の商業的賑わい、郊外開発の抑制、自動車利用の抑制と公共交通の重視、環境問題への積極的な取り組み、地域主義、コミュニティ指向などである[79]。これらはいずれもモダニズムの都市ビジョンが排除または関心を示さなかったことである。コンパクトシティの思想は、1990年代の初期には日本にも紹介され、中心商店街の没落問題に悩んでいた日本の都市計画の専門家や行政関係者の関心を引いた。その後、国も将来の日本の都市の向かうべき方向として打ち出すようになり(→ 171)、富山市や青森市のように具体的な施策として取り組む自治体も現れ、長岡も含めて全国の自治体の都市マスタープランのなかでコンパクトシティ化が多かれ少なかれ謳われるようになった[80]。最近は、一般のメディアにも登場するようになり、都市計画以外の人々も口にするようになっている。

コンパクトシティは確かに、スプロールした都市と比べれば、搬送エネルギーが減り、都市基盤設備の新設・更新工事が減り、エネルギー効率が改善する。また、移動交通量も減る。しかし、コンパクト化には建設工事が伴うので改造過程では逆に二酸化炭素排出量が増える。この増分で削減効果を帳消しにしないかの検証が必要である。

長岡を例に試算をしてみよう。モデルは、平均2階建ての現状の市街地を40年間かけてすべて平均9階建ての集合住宅に作り替え、市街地面積を現在の30%に縮めるという想定である。この間に長岡の人口が全国の平均と同じように30%減るという条件も織り込む。これで、現状とコンパクト化後の二酸化炭素排出量の比較をすると一人当たり年間で約14.3%の削減ができる。しかし、コンパクトにするための都市改造から排出される二酸化炭素を、この削減分で回収するにはコンパクト化を始めてから約66年を要する(fc 4J)。この試算をしてわかることの一つは、自動車が排出する二酸化炭素の削減へのコンパクトシティ化の貢献は意外に少ないことである。都市が狭くなること

169 パリ郊外の風景
View of a Parisian suburb

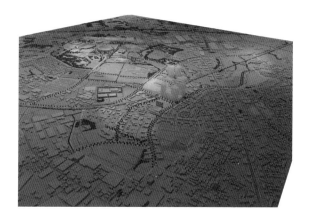

open and often evocative of the past (→ 169). In the center of a city, the tenants include brand shops and restaurants, and in the suburbs, there are retail stores, a supermarket and a food court as well as a multiplex or a fitness club. The "island" is composed in such a way that a family can spend an entire weekend afternoon there. Historic remains are indispensable inside an "island" to evoke in visitors feelings of nostalgia. The biggest "island" in the center of Tokyo, straddling the Ōtemachi, Marunouchi and Yūrakuchō districts, has a rich and varied stock of buildings, the partially preserved Industry Club of Japan and Tokyo Central Post Office, the restored Tokyo Station and reconstructions of the Bankers Club and Mitsubishi Ichigōkan, Japan's first office building. In the postmodern period, a city without historic remains is considered second-rate. The suburbs, which by that yardstick cannot help but be second-rate, must make do instead with theme park-like concoctions, children's fantasies, Hello Kitty and Disney characters.

What are the consequences of the "island" strategy for cities? The positive ones are that concentrations of large capital investments produce economies of scale, and diverse businesses and services, though oftentimes artificial in character, are provided. For example, movie theaters, once dead or dying, have revived in the guise of multiplexes.[77] In addition, people gain access to products of nationwide brands and branches of large bookstore chains.

Their negative consequences are legion. First, each "island" targets a large area and thus manages to produce economies of scale. Unable to compete, small neighborhood enterprises are driven out of business; food deserts are symbolic of this phenomenon. Second, profits are siphoned off by big capital to distant headquarters and do not contribute to local economic activity. Third, no matter how attractively they have been designed and ornamented, contemporary "islands" are ultimately nothing more than spaces for consumption. They may have neat plazas, but those are satisfactory only as long as people are engaged in consumption; they are not places for promoting communal unity (→ 170). Fourth, they weaken the metabolic function of cities. New businesses have traditionally germinated in the back streets of cities. For example, a young cook opens a small restaurant on a back

で走行距離が減る自動車交通は、出発も到着も市内となる交通である。長岡のように自律性が高い都市でも、これは全体の1/4しかない。現状で長岡の自動車由来の二酸化炭素排出量は全体の約21.7%もあるが、コンパクト化によって減らせる対象は5.6%弱しかない(→172)。現実には自動車を使わざるをえないことも多いので、バスなどに振り替えれるのはまたその何割かで微々たる量である。もう一つ効果が怪しいのは、コンパクトシティ化によって歩行圏に何でも揃うという期待である。ところが、近隣住区理論を適用した初期の団地やニュータウンでは生活に必要な諸機能を一通り取り揃えたタウンセンターが用意されたが、多くはすぐ寂れてしまった。だれしも身近に店舗があった方が良いと思うのだが、実際には多くの人は遠くても品揃えが良い店に行ってしまうので、近くの店は成り立たない。

日本でコンパクトシティを実現するとなるとさまざまな困難が待ち受けている。まず、日本の都市は既に相当高密であるが、どの程度まで高密にすればいいのだろうか。また、日本の都市は道路が狭いので高密化すると防災性が低下する恐れもある。二つ目は、都市をコンパクト化するためには現在人が住んでいる郊外を将来の非市街地に指定しなければならないが、それによって、その地域の土地の資産価値が下落する。一方、将来の市街地に指定された場所の地価は上がるので、売れば漁夫の利を得るし住み続ければ固定資産税が上がって困る。このような不公平を調整する方法を見いださない限り、将来の市街地の範囲を決めることは困難である。もっと根本的な難しさは、理念の有効期限である。都市の近代史を振り返るとわかるが、一つの都市ビジョンが半世紀以上に亘って人々に支持され続けてきたことはなかった。そのうえ低炭素化に関しても、現代の技術開発の速度は速く、元を取るまでの半世紀の間には必ず予想を超えた技術革新がありコンパクト化が無駄になるかもしれない。

コンパクトシティ政策は、長期的にわたる大掛かりなスクラップアンドビルドの言い換えである。それは環境の時代に相応しいことではない。コンパクトシティは、一見すると現代的な要請に応えているように見えて実はモダニズム的発想を乗り越えていない。それに価値があるとすれば、都市域を拡大しようとする圧力を抑え込む、わかりやすい都市イメージだということであ

170 柏の葉キャンパス駅周辺開発計画
Kashiwanoha Campus Station area development plan
つくばエクスプレス沿線は、鉄道路線敷を含めて区画整理事業で沿線開発が進められている。柏の葉キャンパス駅周辺には、東京大学の第三キャンパスや複数の政府機関、県立公園(45 ha)などが集められている。駅周辺は超高層住宅や商業施設を含む高密度で環境指向の都市開発が進められている。

Urban development along the Tsukuba Express line has been advanced through land readjustment that includes land for railway use. The Kashiwanoha Campus Station area consists of the third campus of The University of Tokyo as well as multiple government institutions and a prefectural park (45 ha). High density and eco-friendly urban planning, including high-rise residential buildings and commercial facilities, is in progress around the station.

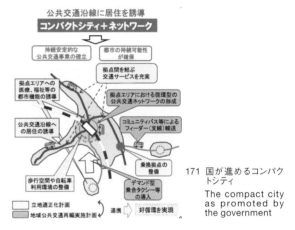

171 国が進めるコンパクトシティ
The compact city as promoted by the government

street. The restaurant becomes popular, expands in size, until eventually, the cook can open a restaurant on main street. To ensure profits, an "island" only accepts as tenants corporations with established reputations. For a cook who is just starting out, the conditions for tenants of an island are too severe; the cook would not be able to open a restaurant inside the island. Thus all islands are composed similarly; their diversity is superficial. The stars of the next generation will not be born on an "island." Meanwhile, the back streets of the "sea," drained of their vitality by the "island," go downhill.

20. Can Compact City Achieve Its Objectives?

Since the 1980s, an urban image referred to as "compact city" has been discussed in the West as an alternative to the sprawl with its excessive dependence on automobiles. The idea of a compact city has been reflected in Europe in the urban policies of the EU.[78] In the United States it gave rise to New Urbanism, and its basic philosophy was articulated in the Ahwahnee Principles (1991).

A compact city is characterized by a high-density, mixed-use city center full of commercial activity, controls on suburban development, controls on automobile use and emphasis on public transportation, active engagement in environmental problems, localism and value placed on community.[79] These represent either a repudiation of or a disinterest in the modernist vision of the city. The compact philosophy city was introduced into Japan in the early 1990s and drew the interest of city planning experts and persons involved in public administration troubled by the decline of shopping streets in central urban areas. Eventually the central government began to declare that it was the direction Japanese cities ought to take in the future (→ 171). Some local governments such as the cities of Toyama and Aomori adopted specific measures; indeed local governments throughout the country including Nagaoka more or less declare in their master plans for cities that a compact city is their goal.[80] Recently, it has begun to appear in the popular media, and people outside the field of city planning are now discussing the subject.

Certainly, compact city reduces the energy needed for transportation and construction projects to newly install or update urban infrastructure and improves energy efficiency.

る。しかし、それをそのまま都市として現実化しようとするのは賢明ではない。それどころか無用な建設投資を誘発し、地方都市の衰退に追い打ちをかけることになるだろう。

縮小する日本に必要なことは、都市全体をコンパクト化するなどと考える以前に、そこらじゅうに発生する空き地をどのように管理し、適切に再編成して都市の魅力に繋げるかを構想することである（→ 173、174）。

172 都市の活動別二酸化炭素排出量
CO$_2$ emissions per urban activity

173 長岡における中心地区の空洞化の状況(2008)
Hollowing out of the central area of Nagaoka

かつての中層ビルが立ち並び、デパート（複数）を含む大小の小売店が並んでいた長岡の中心地区は、今や空き地が目立つ地区に変わってしまった。

The central area of Nagaoka was once home to a number of mid-rise buildings and an array of large and small-scale retail shops including (multiple) department stores. It has now changed into an area pocked with many vacant lots.

However, making the city compact will require construction work, and thus the volume of CO$_2$ emission will actually increase, at least in the process of reorganization. Whether or not this increase wipes out the reduction that is ultimately achieved is something that will need to be verified. Let us investigate, using Nagaoka as an example. The model is a plan to completely replace over a 40-year period the present houses that are on average two-storied in height with apartment buildings that are on average nine-storied and to shrink the city to 30 percent of its present area. During that period, Japan's population is expected to decrease by 30 percent; for the purpose of this model, it will be assumed that Nagaoka's population will decrease at the same rate. If the volume of CO$_2$ emission after the city has been made compact is compared with the present volume, a reduction of approximately 14.3 percent annually per person will be achieved. However, 66 years will be needed from the start of the process to recover the CO$_2$ emitted through urban reorganization to make the city compact (fc 49J, fc 50J).

One of the things this calculation reveals is that the contribution made to reducing the CO$_2$ emitted by automobiles by making the city compact is unexpectedly small (→ 172). The type of automobile trip whose distance traveled will be reduced by making the city smaller is the type that starts and ends in the city. Even in a city that is highly self-sufficient such as Nagaoka, that accounts for only one-fourth of the total. At present, the volume of CO$_2$ emission originating in automobiles in Nagaoka is approximately 21.7 percent of the total; therefore, making the city compact will impact on only slightly less than 5.6 percent of the total. In reality, there are many cases where automobile use is unavoidable; those in which people can use buses or other modes of public transportation therefore represent a very small percentage. Another effect that is suspect is the expectation that making the city compact will make everything available within walking distance. Early housing projects and new towns in which the idea of the neighborhood unit was applied were provided with town centers where everyday necessities were all available, but many of them soon became deserted. Everyone is in favor of having stores nearby, but in reality many people would rather patronize stores with a wide selection of goods, however distant those may be.

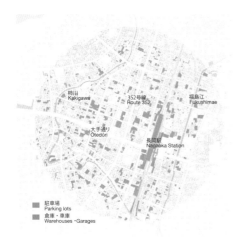

駐車場
Parking lots
倉庫・車庫
Warehouses・Garages

174 ビルが消え駐車場に変わる都心（長岡市表町）
Big buildings replaced by parking lots in the city center (Omotemachi, Nagaoka City)

There are various obstacles to making cities compact in Japan. First, Japanese cities are already of high density; to what level should density be raised? Moreover, streets in Japanese cities are narrow, and increasing density may make cities more vulnerable to disaster. Second, to make a city compact, the suburbs in which people are living at present must be designated areas for non-urban use in the future. This will lead to a drop in the value of land in those areas as assets. Meanwhile, the value of places designated as future urban areas will rise. Selling land in such places will yield unfair profits, while retaining land will mean higher fixed property taxes. Unless ways to rectify such inequalities are found, it will be difficult to determine the extent of the future urban area. A more basic problem is the period of validity of any idea. It is evident from the modern history of the city that no single urban vision has continued to be supported for more than half a century. Moreover, the present pace of technological development in lowering carbon emission is such that in the half century it would take to recover the cost of making cities compact, unexpected technological revolutions are likely to take place and may well make the process pointless.

The compact city policy might be described as a strategy for long-term scrap-and-build on a prodigious scale. it is not appropriate for the age of the environment. The compact city may appear at first glance to meet today's demands but does not in fact transcend the modernist way of thinking. If it does have value, it is that it suppresses the pressure to expand the urban area and presents an easily understood urban image. However, it would not be wise to realize it as it is. Far from it. It would induce useless investment in construction and aggravate the decline of local cities.

Before considering ideas such as making entire cities compact, we must conceive ways to manage the empty lots that are being generated everywhere as Japan shrinks and reorganize them appropriately so that cities become more appealing(→ 173, 174).

第二部：理論とデザイン

Part 2. Theory and Design

縮小の時代の都市理論

　現代都市特に日本の都市計画に関わる最重要課題の一つは、これから本格化する縮小の時代への対応の仕方を示す理論の構築である。なぜ、理論かと言えば、今起こりつつある変化が根本的な変化だからに他ならない。この世界を成り立たせてきた原理が音を立てて崩れ始めている。具体的に言えば、日本のような先進諸国ではさまざまな面で「縮小」が避けられない事態に陥り、これまで成長を前提に組み立てられて来た都市計画は、「縮小」とその先の定常社会を見据えて根本から組み直さなければならない。何世紀も続いた成長の時代に慣れすぎたわれわれは、残念ながら、都市が縮小することの具体的イメージもなければ、それを乗り切り、定常社会を運営する方法も持ち合わせていない。われわれに必要なのは、縮小する都市に包容力のある「場所」と活気に満ちた「流れ」を作り出すための実現可能な都市戦略、そしてそれらに一貫性を与える理論を構築することである。「ファイバーシティ」はそのための計画理論の素描であり、実践的都市戦略である。ファイバーシティは世界中に何万と存在する既成の都市を相手にして、それらの都市が秘め

Chapter 1. The Planning of Flow and Place

An Urban Theory for the Age of Shrinkage

　When it comes to the planning of contemporary cities, particularly contemporary Japanese cities, one of the most important tasks we face is constructing a theory that suggests ways to deal with the age of shrinkage. The reason a theory is needed is because the changes occurring now are fundamental changes. The framework on which this world has been based—its ruling principle—is definitely beginning to collapse. Specifically, the state of affairs in developed countries is now such that shrinkage in various aspects of society is unavoidable and the approach premised on growth hitherto taken to city planning must be fundamentally altered with shrinkage—and the steady-state society that shrinkage will lead to if all goes well—in mind. The age of growth lasted for centuries, and we became accustomed to it. Unfortunately, as a consequence we have neither specific images for the shrinkage of cities nor a method for weathering that process and managing a steady-state society. We must formulate workable urban strategies for creating inclusive places and vibrant flows in shrinking cities and establish a theory underpinning those strategies. "Fibercity"[81] is an outline of such a planning

持つ潜在力を引き出し、都市の体質を変え、縮小の時代に耐える体力をつける内科治療的な処方箋を提示しなければならない。近代都市計画を西洋医学の外科手術に比するなら、ファイバーシティは東洋医学的な代替医療と言ってもよいだろう。よく知られているように、後者の最大の関心は患者がもつ自然治癒力を最大限引き出すことにある。そのためには、線的要素を操作するのがもっとも効果的だというのがわれわれの見解である。

ファイバーシティは、既存の都市を対象に、比較的小さな線状要素（ファイバー）[81]を操作することで、都市のなかの「場所」と「流れ」を同時に制御しようとする内科治療的な計画理論である。

1. 流れも場所も

場所の危機

人間の存在は「場所」を得ることによって全うされ、この世界に錨を下ろすことができる。「場所」を確定することは、人が環境の中に生きることそのものであり、混沌とした空間の中に人間が意味ある領域を張ることである。建築や都市を作るということは空間に秩序を打ち立て「場所」として定着する行為である（→ 201）。

20世紀の後半になると、このような意味での「場所」の危機が多く語られてきた。曰く風景の均質化、標準化・規格化、ディズニーランド化、清潔な廃墟[82]、雑然、などである。それに対して近代以前には「個人および共同社会の一員として内側にいて自分自身の場所に所属すること、そしてこのことを特に考えることなしに知っているという感覚」があったとエドワード・レルフは述べている[83]。これらの主張は、現代都市に対する批判的視座に立っている。著者も、これらの意見には大いに頷くところである。ただ、あまりに「場所」に関心が集中し過ぎ、「移動」の意

201 近世日本人の「場所観」
Early modern Japanese "sense of place"

ヘチマの棚と茣蓙で規定された開放的な場所のあり方は時代を超えて日本人の共感を呼ぶ。久隅守景画「夕顔棚納涼図屏風」（国宝、17世紀後半）

Sponge gourd (*hechima*) trellises and open spaces set off by mats have evoked a sense of comfort for the Japanese across the ages. *Kusumi Morikage, Staying cool under the arbour of evening glory* (National Treasure, late 17th century).

theory and a set of practical urban strategies.

Fibercity is compelled to offer prescriptions for the treatment of the tens of thousands of preexisting cities in the world that are intended to summon their hidden strengths, change their constitutions and enable them to build up the power and vitality needed to endure shrinkage. Fibercity might be likened to Eastern alternative medicine as opposed to modern city planning which is comparable to surgery, a branch of Western medicine. Alternative medicine is focused on summoning the natural capacity for healing that patients possess. Our view is that manipulating linear elements is the most effective method for achieving that objective.

Fibercity is a planning theory dealing with existing cities that attempts to control simultaneously *place* and *flow* inside cities through the manipulation of relatively small linear elements (fibers),[81] and as such might be likened to internal medicine.

1. Both Flow and Place

Crisis of Place

Obtaining a place completes human existence; it enables one to put down roots in the world. To settle on a place is to live in the environment and to define a meaningful domain within a chaotic space. To create buildings and cities is to give order to and establish places within space (→ 201).

In the second half of the twentieth century, there was much talk of a crisis of place in such a sense; for example, the homogenization and standardization of the landscape; Disneylandification; what sociologist Matsubara Ryūichirō has referred to as "clean ruins;" [82] and disorder. The geographer Edward Relph explains that, by contrast, in premodern times people instinctively knew they belonged to a certain place as both individuals and members of a community.[83] Assertions such as his are made from a point of view that is critical of contemporary cities and can be said to express the voice of intellectual conscience. The author is sympathetic to such views. However, they seem to focus excessively on place and to neglect the significance

義を無視していることが気になる。レルフは、「本物の場所のセンス……は、技術的発展を遂げた社会における多くの人間にとって、それを育む可能性は、空間内の移動の増大と場所の象徴性の衰退によって蝕まれてきている」として、現代の高速移動が本物の場所のセンスを蝕んでいるとしている[84]。レルフにとっては、「本物の場所」はあっても「本物の移動」はなさそうである。しかし、これは本当だろうか。われわれがどこかに移動するということは、単なる目的的な行動なのだろうか。

日本の多くの大都市や中都市では、ほとんどの居住者は郊外に住んでいる。首都圏で言えば郊外居住者は7割を超える[85]。彼らの生活時間のなかで、通勤、通学の時間や、買物で外出する時間が大きな割合を占める。人々は移動するあいだ何もしていないわけではない（→202）。スマホでゲームする人もいれば、本を読む人もいる。それは動く居間である。自動車で移動する人は、CDをかけて音楽を聞いているかもしれない。あるいは窓外を流れる風景に見入っているかもしれない。多くの現代の大都市の住人は自宅と勤め先や学校、そして乗り物のなかの三カ所に生きている。近年のパソコンの軽量化と高速通信網と高速処理技術の進展はめざましく、かつては雑誌の記事の中にしか存在しなかったSOHOやリモートオフィスやモバイルオフィスなどの現実感が増している。鉄道車内でも飛行場でもパソコンを開けば仕事ができるようになり、オフィスで働くことと移動していることの区別が曖昧になりつつある。

生まれてから死ぬまで同じ場所に住む人を定住者と呼び、それ以外の人を都市の遊牧民（ノマド）と呼ぶとすると、東京のような現代の大都市で働いている人や道を歩いている人のうちかなりの人たちは都市の遊牧民ではないだろうか。在学期間だけ滞在する学生もいれば、出張で日帰りの人もいる。収入に応じて住まいを変えることもごく普通の行動である。実は、都市が遊牧民によって構成されるということは、参勤交代があった江戸時代からの東京（江戸）の一つの特徴であった。おびただしい種類の江戸案内本が出版されていたのは江戸にはオノボリさんが溢れていたからであった。だから流動性は決して現代都市だけの特徴ではない。グローバル化はさらに移民や出稼ぎ労働者という大量の都市の遊牧民を生んだ。高度成長期の日本では、東北や九州の中学生が金の卵とありがたがられて大都市に

202 スマホに見いる乗客
Passengers engrossed in their smart phones

of *movement*. Relph states that for many people in a technologically advanced society, the possibility of developing an authentic sense of place is being undermined by the increase in movement in space and the decline in the symbolic character of place today, that is, that high-speed travel today is having an adverse effect on an authentic sense of place.[84] Relph implies that *authentic place* may exist but not *authentic movement*. Is that truly the case? Is moving somewhere for us simply a means to an end? Practically all residents of large and medium-sized cities in Japan live in the suburbs. In the Capital Region, more than 70 percent of the population are suburbanites.[85] For them, time spent going to and from work or school or time spent going shopping constitutes a large percentage of daily life. People must do something while moving (→ 202). There are those who play games on their smartphones and others who read books. They are essentially in a moving living room. People traveling by automobile may listen to music on CDs or look at the moving landscape outside windows. Many residents of metropolises today live in three places: home, place of work or study, and means of transport. In recent years, personal computers have become extremely lightweight and great advances have been made in high-speed telecommunication networks and high-speed processing technology, and the small office/home office (SOHO), "remote office" and "mobile office" that once existed only in magazine articles have increasingly become real. One can do work inside a train or at an airport simply by switching on a computer, and the distinction between working at an office and traveling is becoming less clearcut.

If we call those who live in the same place all their lives "permanent residents" and everyone else "urban nomads," then a considerable number of those who are working in, or walking on the streets of, contemporary metropolises such as Tokyo are probably urban nomads. There are students who reside in the city only while they are enrolled in schools there, and others who are in the city for the day on business. Changing one's residence in response to a change in income is also quite commonplace. In fact, Tokyo has been a city made up of urban nomads since it was still called Edo, when the rule of alternate attendance (*sankin kōtai*) was in effect. Many types of guidebooks to Edo were published precisely because the city was full of

集団就職した。彼らが上野駅に到着する風景は毎年春の恒例のニュースであった。現代では、労働者は海外に求められ始めている。未だに鎖国的と言われる日本ですら、コンビニのレジや居酒屋の給仕、そして建設現場に外国人労働者が混じるようになった。シンガポールは積極的に移民労働者を受け入れ、特に高級な技術者は歓迎される。有能な経営者や学者、技術者やデザイナーも、最高の待遇とやりがいのある仕事を求めて世界を渡り歩いている。

移動は快楽

これらの現代的な移動の経験は本物の場所の感覚を「蝕む」存在だけなのだろうか。どうして場所だけがそれほど特権的なのだろうか。移動は「生きられた世界」の一部を構成できないのだろうか。

ところが、現実には、移動は、それ自体が快楽ですらあるのだ。江戸時代の日本は封建制であった。封建制は一種の国家連合なので、移動の自由は厳しく制限されていた。各「国」の間は尚更であった。それでも、人々は限られた機会を活かして富士登山や伊勢神宮を始めとする寺社仏閣巡りを楽しんだ。その後、市民社会に移ると観光への関心はますます高まり、今日では、世界各地の風景や生活のイメージが、テレビやインターネットを通じて居間のテレビや手元のスマホに飛び込み、人々を旅に駆り立てる。海外旅行の費用は劇的に下がり、海外旅行は異郷への冒険というよりショッピングモールに出かけるようなものである。1年間に約3億人の日本人が国内旅行をし、1700万人が国の外を旅行している。

アムステルダムで旅行客にもっとも高い人気を誇るのは運河遊覧である。人々は動くこと自体を楽しむ。それは、遊園地のライドも似ていて、すべての経験は「移動」とともにある（→ 203）。

地中海に面する欧州の都市には散歩が都市生活の一部として定着し、都市のなかに散歩道がきちんと位置づけられている。人々は夕方になると少しめかしこんで遊歩道に出て来てぶらぶら歩く。かつて東京にも銀ブラという文化があったし、夏には夕涼みの習慣があった。筆者の故郷は岐阜市である。鵜飼で有名な長良川が近くを流れていた。河畔には旅館が並び、川岸は二段になり下の道は車が通れず、鵜匠が鵜舟を係留して

203 都心部を流れる運河（ユトレヒト）
Canal that flows through the city center (Utrecht)

ユトレヒトでは、運河の水面は地下一階のレベルにある。運河に沿う町家の地階の倉庫は水際にテラスを持つ。

The waters in Utrecht's canals flow at the basement level. Storage spaces on the basement floors of houses along the canal have terraces by the waterside.

people newly arrived from the provinces. Mobility therefore is by no means a characteristic unique to the contemporary city. Globalization has produced large numbers of additional urban nomads in the form of immigrants and migrant workers. In the period of intensive economic growth in Japan, junior high school students from Tōhoku and Kyūshū were welcomed with open arms in metropolises. The arrival of trainloads of students at Ueno Station always made the news in spring each year. Today, workers are increasingly sought overseas. Even in a country as insular as Japan, foreign workers are to be found among those operating cash registers at convenience stores, waiting tables at *izakaya* and engaging in manual labor on construction sites. Singapore actively accepts migrant workers and courts highly trained engineers. Capable managers, scholars, engineers and designers are wandering the earth in search of good pay and work worth doing.

Movement Is Pleasurable

Are those contemporary experiences of movement unworthy of consideration? Are they merely things that undermine an authentic sense of place? Why is place alone so valued? Is not movement also a part of the lived world?

In fact movement itself is pleasurable. Edo-period Japan was ruled by a feudal system. That system was founded on an alliance among domains, and movement between those domains was strictly controlled. Despite those controls, people took advantage of the limited opportunities they had to go on pilgrimages to places such as Mt. Fuji and Ise Shrine. Interest in tourism greatly increased once Japan became a civil society, and images of landscapes and people's lives in different parts of the world are transmitted by broadcasting companies or the internet to television sets and smartphones and stimulate the urge to travel. The cost of overseas travel has decreased dramatically. Going abroad is no longer regarded as an adventure to strange lands but as something as casual as visiting a shopping mall. Every year approximately 300 million Japanese travel within the country and 17 million Japanese go overseas.

The most popular attractions in Amsterdam among tourists are canal cruises. People enjoy the sheer joy of movement; a cruise in that sense is not unlike an amusement park ride (→ 203). In European cities on the

鵜飼の準備をし、その間を子供たちが遊んでいた。夏になると夕涼みの市民も多数集まってきた。高齢社会に生きる現代人にとっては健康維持が最大の関心事である。多くの人がウォーキングやジョギングを楽しんでいる。いつでもランニングシューズを持ち歩き、走ることに中毒になっている人たちが増えている。ランニングは純粋に移動自体が目的であり、快楽である。

速度に身を任せることはそれだけで目眩を覚える快楽である。速度は富の源泉なので、かつては権力によって独占されていたが、近代社会は、回転木馬、自転車、鉄道、ジェットコースター、カーレース、アルペンスキー、オートバイ、飛行機など速度の快楽を普通の市民に開放した(→ 204)。

場所と流れの相互性

ここまで、主に現代都市における移動の問題を考えて来たが、現代都市では人や物の移動だけではなく、何度も述べて来たように、エネルギーや情報や資本の移動そして移住・移民も拡大している。それらをまとめて「流れ」という概念で捉えて、都市のあり方を再考してはどうだろうか。そうすると、近代都市計画は、都市の活動を場所の占拠と流れに分離し、しかも場所を優位に置くという発想に基礎をおいていたということに気づくのである。果たしてこのような分離は適切なのであろうか。そんな疑問をもつのは、たとえば、身体を分析していかに正確な解剖図を作って、それを理解しても、それだけでは生きている生物の実相には迫れないからである。生物が生きると言うことはとりもなおさず運動を続けるということである。身体の付随として運動があるのではなく、運動は身体の本質であり、運動は一連の「流れ」である。流れのなかで身体の各部は相互に働きあい、変形し、滑らかな全体を形作る。

生物は、一般的に外皮と内皮を通して外界にあるエネルギーの吸収と老廃物の排出の二つの「流れ」を自ら作ることによって生命を維持しているので、「流れ」が作れなくなると器官は停滞し死を迎える。人間の集落も同じように、交通や水、エネルギー、廃棄などの「流れ」によって存在している。電流が磁場を作ることからもわかるように、「流れ」は「場所」の原因である。都市でも「流れ」が集中するところには力が生まれる。たとえば「文明の十字路」という表現がある。また、鉄道網が発達した

204 ツール・ド・フランス最終日
　　Last day of the Tour de France
23日間フランス中をスピードに酔わせた自転車レースはパリで終わる。
This 23-day long bicycle race that takes France by storm ends in Paris.

205 ドン・ルイス1世橋 (ポルト、1886)
　　Dom Luis I Bridge (Porto, 1886)
ポルトガルの第二の都市ポルトはドウロ河沿いにできた街である。アーチ橋の上段をメトロと歩行者、下段を自動車道、川面をポートワインを積んだ船が行き交う。設計はテオフィロ・セイリグ

Porto, the second largest city in Portugal, is a city founded along the Douro River. The metro and pedestrians come and go above the arched bridge, while automobile traffic flows below it and ships carrying port wine travel on the water's surface. The bridge was designed by Théophile Seyrig.

Mediterranean, going for a stroll is an integral part of urban life, and paths are provided for that purpose. In the evening, people dress up a bit and go out for a walk. In the past, strolling in the Ginza district was so popular there was a word (*Ginbura*) for it. It was also a custom in Japan to go out on a summer evening to enjoy the cool air. The author comes from Gifu City. Nagaragawa, a river famous for cormorant fishing, flowed nearby. Inns were lined along the river. The riverbank was two-tiered, and the road on the lower tier was closed to vehicular traffic. There, cormorant fishing masters moored their boats and made preparations, while children played around them. In summer, many local residents gathered to enjoy the cool evening air. For people today living in an aging society, maintaining health is a major concern. Many people walk or jog. Increasing numbers of people have become enthusiasts of running and always carry their running shoes with them. Movement itself is the objective and the pleasure of running.

To yield to the sensation of speed is dizzying and pleasurable. Speed was once monopolized by the powerful, but in modern society merry-go-rounds, bicycles, trains, roller coasters, car races, Alpine skiing, motorcycles, and airplanes provide ordinary people with the pleasure of speed (→ 204).

Interrelationship of Place and Flow

We have mainly considered the issue of movement in contemporary cities up to now, but as has already been explained several times, this is not limited to the movement of people and things. The movement of energy, information and capital, change of residence, immigration and emigration are also increasing. Let us think of these as manifestations of the concept of *flow*, and reconsider the nature of cities from the perspective of that concept. It then becomes obvious that modern city planning has been based on the idea of separating urban activities into the occupation of place and flow and, moreover, on putting priority on place. Is such a separation appropriate? For example, one can analyze bodies, draw accurate anatomical charts and understand them, but one will not understand the true nature of a living organism by those actions alone. For an organism, to live is to continue to move. Motion is not incidental but essential to the body, and motion is a series of flows. The parts of a body interact

東京では「場所」の価値は、鉄道駅と集まる路線の性格と本数が決める。

このように「流れ」は「場所」に生気を吹き込む(→205)。しかし、これは常に「場所」が「流れ」の結果だということではない。逆に「流れ」は「場所」の結果でもあるという当たり前のことも明記しておかなければならない。「場所」は多数の「場所」の群として存在し、相互に関連しあっている。「場所」はそれぞれ個性をもち、強弱があり、高低がある。それゆえ「場所」のあいだに「流れ」が生まれる。「場所」の様態が「流れ」を呼び起こすのであり、「場所」と「流れ」は相互に依存し条件づける関係にあり、決して一方が他方の目的や手段ではない。

都市空間に介入して、それを再組織化しようとするなら、都市における「流れ」と「場所」の相互作用についての理解を深めなければならない。

近代建築と都市計画における流れ

20世紀は「流れ」の分野で目覚ましい技術革新が起こった時代である。建築家や都市計画家や都市計画技術者たちは、それをどのように捉えたのだろうか。事例でみてみよう。

近代的な「流れ」が強く意識された最初の都市計画はオースマン[86]のパリとフレデリック・ロー・オルムステッドによるシカゴの郊外の住宅地リバーサイド(1868)だろう(→206)。パリの街路構成はバロック式の幾何学をもつが、長いビスタをもつブルバールはパリコンミューンでの経験から暴徒を鎮圧するためには軍隊を素早く動かし街路の見通すことが必要だという認識ができたと言われる[87]。新古典主義的な美学に支配されたパリに対して、リバーサイドの区画街路はペイズリー模様のように滑らかに弧を描いてダイナミックであり馬車の走行に呼応している。

産業革命は「流れ」の技術革新によって駆動され、新しい技術成果は世紀の転換点の芸術家たちにも霊感を与えた。1909年にイタリアの詩人フィリッポ・トンマーゾ・マリネッティによって起草された「未来主義創立宣言」では、「吾等は世界に一の美なるものの加はりたることを主張す。而してその美なるものの速なることを主張す。廣き噴出管の蛇のごとく毒氣を吐き行く競爭自動車、銃口を出でし弾丸の如くはためきつつ飛び行く自働

206 リバーサイド(1868)
Riverside (1868)

and change within flow and form a seamless whole.

An organism generally maintains life by creating two flows through its epidermis and endodermis that absorb energy from the outside world and eliminate waste matter. When it can no longer create flow, its organs stagnate and it dies. A human community similarly exists through the flow of such things as traffic, water, energy and garbage. Place is caused by flow, as is evident from the creation of a magnetic field by an electric current. In cities too, the concentration of flow produces power. For example, there is the expression "crossroads of civilization." In Tokyo, a city with a highly developed rail network, the value of any place within it is determined by how many and which lines serve the nearest railway station.

Thus flow breathes life into place(→205). However, this does not meant that place is always the result of flow. On the contrary, flow can just as often be the result of place. Place exists as a cluster of multiple interrelated places. Each place has its own individuality, strength and altitude. That is why flow occurs between places. The mode of place triggers flow. Place and flow are in an interdependent, interactive relationship. One is by no means the objective or the means to an end of the other.

If we are to intervene in and to reorganize urban spaces, we must first gain a deeper understanding of the interaction between flow and place in cities.

Flow in Modern Architecture and City Planning

The twentieth century was a time when remarkable technological innovations were achieved in flow. How did architects and city planners perceive those innovations? Let us consider some examples.

The first instances of city planning that reflect a strong awareness of modern flow were probably Haussmann's Paris[86] and Riverside (1868), the Chicago suburb planned by Fredrick Law Olmstead (→206). The streets of Paris are organized in a static, Baroque geometry, but the experience of the Paris Commune is said to have brought recognition of the advantages of boulevards with long vistas that enable swift movement of troops and provide unobstructed lines of sight in quelling mobs.[87] By contrast to the neoclassicist aesthetic of Paris, the layout of Riverside, with its Paisley-like pattern of smooth, dynamic curves is more in tune with horse-drawn carriages.

車はsamothrakoの勝利女神より美なり。[88] 未来派の芸術家たちは機械の速度に圧倒され興奮しデザインに取り入れようとした。同じ頃、ロシア構成主義者たちも積極的に動きをテーマにして絵画、彫刻、建築などの造形を考えた(→ 207)。

近代都市の理念を表現したエベネザー・ハワードの「田園都市」(1898)もル・コルビュジエの「現代都市」(1922)も新しい交通を組み込んでいる。田園都市は母都市(ロンドン)と鉄道で結ばれている。「現代都市」の中心部には飛行場と高速道路と地下鉄、郊外線の駅が上下に重なっているという説明がついている。ただ、両方とも街路構成は古典的装いである。田園都市では、空間構成を示すダイアグラムは円形の組み合わせで表されている。円形は求心性をもち自立完結した「場所」を示すのに適した図形である。「現代都市」の街路パターンは完全な矩形で左右対称にレース編みのテーブルクロスの模様のように区画されている。コルビュジエが都市プランに「流れ」の形を本格的に導入するのは、1929年のサンパウロと1930年のアルジェの計画まで待たなければならない(→ 208)。

建築家たちも、近代的な「流れ」の力を実感し、これを建築に取り込もうと考え始める。建築家としてのル・コルビュジエは自らがデザインしたサボワ邸(1931年竣工)の斜路を指して「建築的プロムナード」と呼んだ。ただし、この名称に真に値するのはフランクロイドライトが設計したソロモン・R・グッゲンハイム美術館(1959年竣工)であろう。もっと複雑な機能をもった施設に適用した事例として、筆者の設計になる「東京大学カブリ数物連携宇宙研究機構棟」(2009年竣工)がある。螺旋的な「流れ」が研究者たちが集まり議論する「場所」を、生み出している(→ 209)。

都市計画思想を「流れ」から見たときの大きな変化は、CIAM[89]の内部造反として生まれたチーム・テンによって始まった(1953)。これは、日本のメタボリズム[90](1959)、イギリスのアーキグラム[91](1961)などの後続の建築運動に大きな影響を与えた。チーム・テンのキャンディリスとウッズのベルリン自由大学のデッキシステムや、メタボリズムの丹下健三による「東京計画1960」などはいずれも、「流路」が都市の骨格となっている。ただ、彼らは都市の「流路」を建築の手法と語彙を拡張して作ろうとした。それが彼らの提案の特徴であり建築家としての野心で

The Industrial Revolution was driven by technological innovations in flow, and new technological achievements inspired artists at the turn of the century. The following is from The *Futurist Manifesto* drafted by the Italian poet Filippo Tommaso Marinetti in 1909. "We believe that this wonderful world has been further enriched by a new beauty, the beauty of speed. A racing car, its bonnet decked out with exhaust pipes like serpents with galvanic breath...a roaring motorcar, which seems to race on like machine-gun fire, is more beautiful than the Winged Victory of Samothrace."[88] The artists of Futurism were overcome with excitement by the speed of machines and attempted to express that speed in designs. Around the same time, the Russian Constructivists conceived paintings, sculptures and buildings that had speed as their theme (→ 207).

Both *Garden Cities of To-morrow* (1902) (originally published as *To-Morrow: A Peaceful Path to Real Reform* (1898)) by Ebenezer Howard and *Ville Contemporaine* (1922) by Le Corbusier incorporated new systems of transportation. Garden cities are linked to a mother city (London) by railway. It is explained that in the center of the Ville Contemporaine an airport, expressway, subway and suburban railway stations are layered one on top of the next. However, in both schemes, the street pattern is classical. The spatial organization of a garden city is drawn as a combination of circles. The circle is a figure that is centripetal in character and appropriate for indicating a self-sufficient place. The street arrangement for the Ville Contemporaine is a rectangle divided into an axially symmetrical pattern resembling a lace tablecloth. It was only with his plans for São Paolo (1929) and Algiers (1930) that Corbusier introduced forms of flow into his city plans in earnest (→ 208).

Architects too sensed the power of modern flow and began to consider ways to introduce it into their designs. Corbusier referred to the ramp in his Villa Savoye (1931) as an "architectural promenade." However, the Solomon R. Guggenheim Museum (1959) designed by Frank Lloyd Wright is probably more deserving of that description. Kavli Institute for the Physics and Mathematics of the Universe, the University of Tokyo (2009), designed by the author, is an example of a facility with a more complex function in which a spiral flow generates places where researchers may gather and engage in discussions (→ 209).

The schism in CIAM[89] that led to the birth of Team X (1953) initiated what was from the perspective of flow a major change in the approach taken to city planning. This had an enormous influence on subsequent architectural movements such as Metabolism[90] (1959) in Japan and Archigram[91] (1961) in England. In both the deck system for the Free University of Berlin by Team X's Candilis and Woods and *Tokyo Plan 1960* by Tange Kenzō, channels of flow formed the urban framework. However, they attempted to create urban channels of flow by expanding the method and vocabulary of architecture. This was what distinguished their proposals and demonstrated their ambition as architects, but it was also what limited their schemes. An excessive integration of place and flow is inappropriate because their technologies have different lifespans. The schemes were also limited by the fact that they were concerned only with transportation. The architects were unable to envision a mature consumer society in which the flow of things is combined with the flow of information.

The importance of flow continued to grow in the real world, but to this day the ideas of flow and place remain unintegrated in architectural design and city planning theory and in the actual practice of planning. The planning of flow is dealt with in architectural planning as the planning of mechanical systems and in city planning as the planning of transportation and water supply and sewage systems; in both cases flow is treated as a technical problem separate from spatial planning. Both architects

あったが、同時に計画の限界でもあった。なぜなら、「場所」と「流れ」の過度な統合は、両者の技術の寿命が違うので不都合だからである。もう一つの限界は、彼らの「流れ」に対する関心は交通に集中し、モノの「流れ」と情報の「流れ」の合体という成熟した消費社会を具体的に想像することが叶わなかったことである。

現実世界では、その後も、「流れ」の重要性は増し続けたのだが、現代の建築設計や都市計画の理論においても計画実務においても、「流れ」と「場所」の概念は統合されないままである。建築計画における「流れ」の計画は設備計画として、都市計画では交通計画と上下水道計画として、いずれも技術的な問題として括られ、空間計画から分離されている。空間の設計者を自認する設計者も彼らの後方を固める理論家も、地区や街区や建築あるいは広場つまり「場所」をどうデザインするかに注力するばかりである。これは、建築家や都市計画家が「流れ」の技術の急速な発展に追いつけないということもあるのだが、おそらくそれだけではないだろう。流れを敵視する場所論者と共通して、伝統的な都市における安定した空間概念の解体への抵

207 旧秩父セメント第二工場 (1956)
Former Chichibu Cement Plant No.2 (1956)

建築家谷口吉郎が設計監修をした工場は近代建築の息づかいを感じさせる。

This plant, designed and supervised by architect Taniguchi Yoshiro exudes the spirit of modern architecture.

208 ル・コルビュジエによる「アルジェ都市計画-計画A」(1930)
Le Corbusier's "Urbanisation de la ville d'Alger Projet A"

海岸沿いの長い壁状の構築物は、18万人の住人を想定した集合住宅で屋上は高速道路である。ル・コルビュジエとピエール・ジャンヌレによる提案。

In this proposal by Le Corbusier and Pierre Jeanneret, the long, wall-like structure along the coast is a mass-housing unit for 1.8 million residents. Its roof acts as a highway.

who see themselves as designers of spaces and theorists who provide those designers with their rationale are concerned only with how to design areas, blocks, buildings and plazas, that is, places. This is accounted for in part by the inability of architects and city planners to keep up with the rapid development of the technologies of flow, but that is probably not the only reason. A deep-seated intellectual tradition of resistance against the dissolution of stable spatial concepts in traditional cities and of skepticism toward the modern era appears to be at work in them as in theorists of place who are hostile to flow. That is, they are so conscious of the crisis of place that they totally ignore flow. Or perhaps the professional image that architects have of themselves as designers of place is making it difficult for them to deal with flow.

It is evident from a survey of history that there has been little progress or development in the technology of constructing place, and moreover, though there have been developments and changes in the spatial principles behind the formation of place, the concept of progress basically does not exist. Therefore, a brick-construction house from ancient Rome or a wood-construction *shinden*-style

209 東京大学国際高等研究所カブリ数物連携宇宙研究機構 (2009)
University of Tokyo Kavli Institute for the Physics and Mathematics of the Universe (2009)

宇宙の研究者にとって議論は重要な研究活動である。議論の場を研究個室が螺旋状に配され、囲んでいる。設計は大野秀敏。

For those researching the universe, debate and discussion are a core part of research activities. Personal offices encircle this space for debate in a spiral fashion. Designed by Ohno Hidetoshi.

抗、あるいは近代に対する懐疑という根強い知的伝統も作用しているように思える。つまり「場所」の危機の意識が彼らを支配し「流れ」について無関心を許しているのかもしれない。あるいは場所の設計者としての職能の意識が「流れ」を遠ざけているのかもしれない。

歴史を概観すればわかることだが、「場所」を構築する技術の進歩発展は遅く、しかも、「場所」を形成する空間原理には発展変化はあっても基本的に進歩という概念がない。だから、古代ローマの煉瓦造の住宅も平安時代の木造の寝殿作りの御殿も、住宅設備を入れ替えさえすれば現在でも十分住める。それに対して「流れ」に関わる技術は確実に進歩し続け、特に近代に入ってからの「流れ」の技術革新にはめざましいものがある。だから、ローマ時代や平安時代の交通手段や上下水道のままで現代都市を運営することは不可能なのである。

膨張する大きな流れ

流れの技術は休むことなく発展し、現代社会には過剰で強力な「流れ」が溢れている。それは社会の構造すら変える。世界のあらゆる地域が、「流れ」の技術革新と「場所」の「境界」の開放により、相互依存関係で結ばれた今日、物と情報の移動なくしては、どんな地域もやっていけない。つまり、われわれの生活に占める「流れ」の存在感は昔に比べてはるかに大きくなっている。それだけに「流れ」を使える技術を持つ人とそうでない人の間の格差、「流れ」の恩恵を受ける都市とそうでない都市の間の格差が決定的になりつつある。「流れ」から遠い人や都市は取り残されてしまう。たとえば、都市が自家用車を前提とする構造に変化すると、自家用車をもてない人は決定的に不便になる。まさに手足をもぎ取られた状態である。日本でいう「買い物難民」もイギリスでいう「フードデザート」も取り残された地区や人の問題である。

大きな交通とは「遠く・速く・大量に」を特徴とするが、その代表の弾丸列車網が世界中に広まっている。日本でも、徐々に在来線が新幹線に置き替えられている。そうすると、在来線時代には特急停車都市だったのに新幹線は停まらない不便をかこつ都市が生まれる。そういう都市にとっては、まさに「格下げ」の事態であり、経済的にも心理的にも大きな打撃を受ける。大き

residence from the Heian period would make a quite adequate dwelling today if proper mechanical systems were installed. By contrast, progress has continued to be made in the technologies related to flow, and especially since the advent of the modern era, technological innovations in flow have been remarkable. That is why it would be impossible to operate a contemporary city with means of transportation or water supply and sewage systems from ancient Rome or the Heian period.

Expanding Big Flow

The technology of flow develops ceaselessly, and contemporary society is filled with excessive, powerful flow. It is changing even the structure of society. Today, when all regions of the world are linked in interdependent relationships by technological innovations in flow and the opening of the *boundaries* of place, no region can long survive without the movement of things and information. That is, flow has a far greater presence in our everyday lives today than it ever did in the past. The gap between those people who have the skills to use flow and those people who do not, and between those cities that receive the benefits of flow and those cities that do not is growing and reaching a critical level precisely for that reason. People and cities that are distant from flow will be left behind. For example, if the structure of a city were to change into one premised on people having their own automobiles, then those who do not have automobiles would be significantly inconvenienced. It would be as catastrophic as the loss of the use of their limbs. The Japanese who have difficulty purchasing everyday goods because nearby stores have all closed (referred to as "shopping refugees" *kaimono nanmin*) and the so-called "food deserts" in England are problems of people and areas that have been left behind. Big Transportation is characterized by the need to carry people and things "further, faster and on a larger scale," and good examples of it are the networks of bullet trains that are proliferating all over the world. Old railway lines are gradually being replaced by Shinkansen lines in Japan as well. When that happens, some cities that were stops for express trains on the old lines are inevitably bypassed by the Shinkansen. Those cities have essentially been "demoted," and for them it comes as a major economic and psychological blow. The

な「流れ」の皺寄せは、最終的には弱者が受けることになる。これは、超高齢社会の先頭を走る日本にとって大きな課題である。「流れ」の拡大はエネルギー過剰消費をもたらす。目指すべき低炭素社会では、過剰な「流れ」の抑制こそ課題になり、それは即座に「場所の」のあり方に影響を与える。

現代都市に意識的な都市社会学者や環境論者は「流れ」の拡大に困惑し苛立っている。エドワード・レルフもセルジュ・ラトゥーシュも大衆ツーリズムやエコツーリズムを唾棄している。物流や消費に関する技術や社会の仕組みの変化があまりに大きく、「流れ」の加速が交通に関するエネルギー消費を増大させ、多国籍企業の支配を強め、「場所」の安定を損ない「場所」のアイデンティティを侵し始めていると危惧している。というのは、これらの困惑は、深いところで、近代の自由の理念の根本的な矛盾に繋がるからである。近代市民社会はあらゆる拘束を取り払い、人々が自由に活動できる滑らかな空間を理想としているが、「流れ」が過ぎると「場所」に対する脅威となる。近代人が掲げる理念と技術が「場所」を根無し草にするという脅威である。この自己撞着の解消法には三通りあるだろう。第一の途は自由をさらに推し進める途である。第二の途は、自由という市民社会の理念そのものを問い直し、過剰な技術を封印する途である。第三の途は市民的社会の理念である自由を保持しつつも、それの利用に適度な制限を加えて中間的な解決を見つけることである。第一の途は、本書の第一部でみたように環境問題が突きつけられている現在の状況からすれば、問題の先送りにすぎず、その場しのぎの無責任な途でしかない。第二の途は魅力的であるが次のような二つの疑問が残る。第一の疑問は、封印すべき技術を誰が選択するかということ、そして、どのような権力が封印を実行するかという疑問である。現在は問題の多い技術が将来改善される余地も封印するべきかという疑問も残る。第二の疑問は、人は自由な通信や自由な移動の拡大の可能性を知った後に、それ自体への興味を封じることが可能だろうかというものである。これらの難問は、原子力、遺伝子組換えなど先端技術が共通して直面する問題であり、容易には解答は得られないだろう。しかし、当面は、技術の封印を迫ったり近代の理念を捨て去ったりしないで、第三の途を選ぶほかはないだろう。つまり、空間の専門家が「場所」だけではなく「流れ」に

burden of Big Flow is ultimately borne by the weak. This is a major issue for a country like Japan that is on the leading edge in the evolution into a society that is highly advanced in age. The expansion of flow results in an excessive consumption of energy. To achieve a low-carbon society, we must restrict excessive flow, and that will have an immediate effect on the nature of place.

Urban sociologists and environmental theorists aware of the nature of the contemporary city are perplexed and exasperated by the expansion of flow. Wakabayashi Mikio declares that the connection between place on the one hand and nature or the past on the other is being lost as the result of high-speed travel and the flood of images to which we are exposed. Both Edward Relph and Serge Latouche abhor mass tourism and eco-tourism. The changes in the technologies and the social mechanisms related to distribution and consumption are so great that the acceleration of flow is increasing transportation-related consumption of energy, reinforcing the dominance of multinational corporations, disturbing the stability of places and beginning to obliterate the identity of places. On a deep level, these anxieties reveal a fundamental contradiction in the modern idea of freedom. Modern civil society has thrown off all constraints and made spaces in which people can freely engage in activities an ideal.

When flow becomes excessive, it becomes a threat to place. The danger is that the very ideal and technology people of the modern era embrace may uproot place. There are three ways to resolve this self-contradiction. The first is to further promote freedom. The second is to reconsider freedom, that is, the very idea of a civil society, and to contain or suppress excessive technology. The third is to impose moderate limits on freedom and to search for an intermediate solution. Given our environmental problems and the circumstances in which we find ourselves today as the result of those problems, the first way is an irresponsible act of temporization. The second way is attractive but dubious for the following two reasons. First of all, it begs the question of who would select the technologies that ought to be contained and what sort of power would implement that containment. It also leaves unanswered the question of whether even the opportunity or scope for improvement in the future of technologies that are problematic today ought to be suppressed. It is also

も取り組み、難しい均衡点を見いだすようなトータルな都市像を提示すべきである。どちらかに肩入れし、閉じた王国を作っても、それは激流の中では脆い存在でしかない。

2. 見取り図

合理的な線による介入

　私たちは、「場所」だけに関わるのではなく「流れ」をも同時に制御しようと考えている。それは、とりもなおさず都市計画の戦線を拡大しようとしていることになる。その一方、第一部では、21世紀の都市にさまざまな厳しい条件を課したばかりでもある。私たちは、戦線拡大の進軍ラッパを吹きつつ、後ろを振り返って兵站線は絞れと叫ばなければならない状況にあるということである。この無理難題にたいする解答があるとすれば、それは、縮小の時代の都市には、線状要素（ファイバー）による介入が一番適っているというのが、ファイバーシティの基本的な主張なのである。

　20世紀の英雄的な改革者たちは都市の「面」全体に関わろうとした。目の前にある古い建築と都市を壊して全面的に作り替えるか、あるいは、まったく新天地に理想都市を作るかという発想である。第二次世界大戦前には理論と小規模な実験しかできなかったが、戦後になると、戦勝国でも敗戦国でも戦災復

210 表参道（東京都渋谷区）
Omotesandō (Shibuya Ward, Tokyo)

明治神宮への参道である表参道には世界的な有名ブランドの店が軒を並べている。立派な欅並木と谷地形の双方がこの道を独特にしている。

Omotesandō, the street forming the approach to Meiji Shrine, is filled with world-renowned brands. The Japanese zelkova trees that line the street together with the area's valley formation give the street a unique character.

211　公園の二つの形
　　　Two forms of a park

二つの図形は公園の平面図である。二つの公園の面積は同じであるが、周長が違う。線的な公園は矩形の公園の5.1倍の周長を持つ。

These two figures depict plans for parks. While both measure the same area, their perimeters differ. The perimeter of the linear park is 5.1 times that of the rectangular one.

2. A Sketch

Rational, Linear Interventions

　We are thinking of not only involving ourselves with place but simultaneously controlling flow. In other words, we are attempting to expand the front lines of city planning. Meanwhile, in Part 1, we just imposed various severe conditions on twenty-first century cities. This is like ordering troops to expand the lines of engagement despite limited supplies. The basic assertion of Fibercity is that linear elements (fibers) are the most appropriate means of solving this difficult problem and intervening in cities in the age of shrinkage.

　The heroic reformers of the twentieth century tried to deal with the entire *plane* of the city. Their idea was to destroy the existing city and completely remake it or to create an ideal city elsewhere. Before World War II, they were only able to develop theories or undertake small-scale experiments. After the war, however, reconstruction was an urgent problem for all countries. Moreover, the increase in population as the result of the baby boom had to be accommodated. The opportunity arose to realize visions of

doubtful that once they are aware of the possibility of the expansion of free telecommunication or free movement, people will be able to contain their interest in such expansion. We are confronting these difficult problems in all advanced technologies including nuclear power and genetic recombination, and there seem to be no easy solutions. However, for the moment, our only viable solution is not to attempt to contain technology or abandon the idea of freedom but to select the third way. That is, specialists in space ought to grapple with flow as well as place and present a total urban image in which a difficult equilibrium is attempted. If we throw our weight behind the first or second approach, the result is likely to be too fragile to long survive in our quickly changing environment.

興が緊急課題となり、しかもベビーブームによる人口増に応えるために、近代都市計画のビジョンを大規模に実施する機会が訪れた。各国に団地が作られ都心に真新しいビルができた。ところができてみると、彼らの鼻息の荒さとは裏腹に評判は芳しくなかった。ジェイン・ジェイコブス[92]やクリストファー・アレグザンダー[93]が指摘したように、都市は近代主義者たちが考えるような単純なものではなかった。どれほどの天才であっても所詮限られた経験しか持ち合わせない少数の専門家が、非常に短い時間に、部分的とはいえ面に関わるには、都市はあまりに複雑で巨大な組織体なのである。

反対に、点的な介入はどうだろうか。都市の点、すなわち建築を変えることで町全体を変えようという筋書きは政治家や、職能的欲望から建築家が好むところである。ところが、期待するような影響力を持たせようとすると、相当強烈なインパクトの強い建築を投入しなければならないから、下手をすると環境を混乱させるだけの騒々しい介入になる可能性が大きい。われわれは、そのような例をたくさん目撃してきている。

最後に残るのが、面と点の中間である線的介入である。遊歩道を新しく作るとか、並木を植えるとか、壁を築くといった介入である(→ 210)。これは、街の全体を作り直すよりは遥かに安くでき、しかも、都市に与える影響は大きく、その割には周辺の環境を乱すことを最小に抑えることができる。歴史的に大事な構造物や守りたい自然の保全もしやすい。線的介入は穏やかでありながら、それなりの影響を都市に与え、かつ実現がしやすいという特徴をもっている。

線的介入がいかに有効かを2種類の公園の形態を比較して説明しよう(→ 211)。図に示す矩形の公園とヒトデ形の公園は、面積が等しくなるように描かれているので、地区の地価が均一だとするなら公園用の土地の総価格は同じである。この二種類の形態の公園の周辺の市街地への影響を、公園が見える宅地の数と地価の変化で評価すると、ヒトデ型の公園の方が矩形の公園より周長が5.1倍長いので、公園に面した良好な宅地を5.1倍作ることができる。つまり、線的介入が同じ介入(買収費用)でも、大きな効果(良好な宅地と地価の上昇)を挙げることができる方法だということである。線的介入が縮小する都市に向いている方法だと納得できるだろう。

modern city planning on a large scale. Housing projects were created and gleaming new office buildings were constructed in central urban districts in every country. However, for all the arrogant self-regard of modernist planners, the cities, once built, were not well received. As Jane Jacobs[92] and Christopher Alexander[93] have pointed out, a city is not as simple as modernists thought. A city is too complex and too enormous a system to be dealt with as a plane, even in part, by a small number of experts of limited experience, no matter how brilliant, in the extremely short time they allotted themselves.

What about the opposite approach, that is, to intervene at certain points? Politicians and ambitious architects favor schemes to change an entire city by changing points in the city, that is, by introducing new buildings. However, to produce such a result requires the introduction of buildings of considerable impact. If it is done badly or carelessly, the intervention is likely to be disruptive to the environment. We have witnessed many examples of such projects.

What is left then is halfway between intervention at points and planar intervention, namely linear intervention, for example, creating a new promenade, planting trees along a street, or building a wall (→ 210). This is far less costly than remaking the entire city, and the disruption it causes to the environment is small relative to its impact on the city. It makes it easy to protect historically important structures and preserve nature. Linear intervention is restrained but influences the city nonetheless; moreover, it is easy to realize.

Comparing two types of shapes for parks can help explain the effectiveness of linear intervention (→ 211). The illustration shows two parks of the same area, one rectangular and the other star-shaped. If land prices in the area are uniform, then the two parks would cost the same. If we were to evaluate the effect of these two differently-shaped parks on their surrounding areas by the number of lots from which a park is visible and the rise in land prices, then the star-shaped park, having a perimeter that is 5.1 times longer than that of the rectangular park, increases the number of favorable lots facing the park more than five-fold. That is, the two interventions may cost the same, but the linear intervention can have a greater effect (as measured by the increase in the number of favorable lots and in land prices). This makes it clear that linear

流路と境界

このような線状の形態の特質を活かして、現実の都市の「場所」や「流れ」を制御しようとするのが、ファイバーシティの基本的な方法である。次に、もう少し方法の理論としての精度を高めるためには、具体的な線状の要素（ファイバー）が都市空間のなかでとりうる二つの形態である「流路」と「境界」に着目する必要がある。

「流路」は、都市の線状要素（ファイバー）の重要な働きである。伝統的な都市から今日の都市まで都市の基幹的な施設である。具体的には、川や運河や鉄道の線路や、鉄道や道路の高架構造物、電話線や電力線、地下の配水管などとして都市のなかに存在する。これがなければ水流や電流や自動車の走行、あるいは鉄道の運行などの「流れ」が具体化しない。

「境界」は、都市の線状要素（ファイバー）のもう一つの重要な働きである。それは「場所」と「場所」が接するところに現れる。建築的スケールでは、住宅団地の縁にある樹林帯や塀、住宅の門などがあり、地理的なスケールでは河岸段丘や山裾、海岸などは皆「境界」である。一つの線状要素（ファイバー）が「流路」と「境界」を兼ねることも往々にしてある。その代表は川や幹線道路などである。建築計画や都市計画で「場所」を定義し、「場所」の質を一定に保つために最初にすることは「境界」で「場所」を囲繞することである。家も国も囲い込むことから始まる。国境線はもちろん、税関や検疫なども「境界」である。常に侵略の危機にさらされた大陸の都市では、近代までは都市を高い壁で物理的に囲い込むのが普通であった。

近代都市計画も囲い込みを採用している。それはゾーニング制である。それは都市内の諸活動を類型化し、類型ごとに配置をまとめて相互に隔離する計画手法である。このシステムが考案された背景には、産業革命によって工場から容赦なく煤煙と工場排水が吐き出され、街には貧困と不道徳と疫病がはびこったことがあった。中・上流階級の住民の健康な生活と倫理を守るために、工場地帯や商業地から住宅地を隔離しようと考えたのである。だからゾーンの「境界」は閉鎖的で「流れ」を断つように作られる。多くの場合緑地帯がその役割を果たすことになる。ハワードもコルビュジエも田園のなかに、隔離された静的な都市、自律した「場所」を提示したことも、この発想とは無縁

intervention is an approach suited for a shrinking city.

Channel and Boundary

The basic method of Fibercity is to try to control place and flow in an actual city, taking advantage of such characteristics of linear forms. To get a slightly more precise idea of what this method as a theory entails, it is necessary to focus on two types of linear elements (fibers) in urban space: *channel* and *boundary*.

First, functioning as channels is an important task performed by linear elements (fibers) in a city. Channels have been basic facilities for cities, both traditional and contemporary. They take such forms as rivers, canals, railway tracks, elevated structures for railways and roads, telephone and power lines, and underground water pipes. Without them, flows such as water and electrical currents, the movement of automobiles and the operation of railways would be impossible.

Functioning as boundaries is another important task performed by linear elements (fibers) in cities. A boundary appears where two places come into contact. At an architectural scale, a boundary can take the form of a zone of trees or a wall at the edge of a housing project or the gate to a house; at a geographical scale, fluvial terraces, foothills and seashores are all boundaries. A linear element (fiber) can often be both channel and boundary; examples are rivers and arterial roads. In architectural or city planning, the first thing one does to define a place and preserve a certain quality of place is to surround the place by means of a boundary. To build a house or a country, one begins by surrounding a place. Boundaries include not only national borders but customs and quarantines. Cities on the continent that were constantly threatened by hostile forces were usually enclosed physically by high walls until the modern era.

Modern city planning also used enclosure in the form of zoning. This was a method of planning in which urban activities were categorized into a number of different types; activities belonging to the same type were consolidated and isolated from activities of a different type. This system was devised at a time when, as the result of the Industrial Revolution, factories were spewing smoke and waste water, and poverty, dissipation and epidemics were rampant in cities. The idea was to isolate factory and commercial zones from residential zones in order to protect the health and moral well-being of middle- and upper-class citizens. The boundaries of zones were meant to close off areas and stop flow. In many cases green belts served that function. Both Howard and Corbusier were influenced by that thinking and suggested isolated, static cities and autonomous places in rural settings.

However, boundary is not something that merely closes off and protects a place. The moment a boundary creates an inside and an outside, exchange begins across the boundary. From ancient times, that exchange has in fact been the source of a city's vitality. Whether it is a difference in wages and costs that creates profits, a difference in culture that creates curiosity and admiration, or a difference in climate that creates distinct vegetation, difference in potential is the driving force of urban activity. Boundary controls flow between inside and outside, and at times that means stopping exchanges. A prison is a system designed to isolate prisoners from the outside world; what this means is that stoppage of flow is painful and the heaviest form of punishment.

Texture

Boundary is not the only way in which a line (fiber) and place relate to one another in a city. Multiple lines can intertwine and define place.[94] For example, street pattern is an example of texture on an urban scale, but from the point of view of the reorganization of urban space, changing the

ではないだろう。

しかし、本来「境界」は場所を閉ざし守るためだけのものではない。「境界」が内と外を作った時点から「境界」を挟んで交流が始まる。この交流こそ古来より都市の活力の源であった。労賃、物価の差が作り出す経済差益、文化の差が作り出す好奇心と憧れ、気候が作り出す異なる植生など、いずれにおいてもポテンシャルの差こそ都市活動の原動力である。「境界」は、交流の遮断も含めて内外の間の「流れ」を制御する。監獄は囚人から外部との関係を絶つ仕組みであるが、これが意味するところは、「流れ」の遮断は苦痛であり最大の罰だということだ。

織目

都市の線(ファイバー)と「場所」の関わり方は「境界」だけではない。複数の線どうしが絡み合うことで「織目(テクスチャ)」を作り、一定のテクスチャが一定の範囲を覆うことで「場所」を定義することができる[94]。たとえば街路パターンは都市スケールでの織目の例である。しかし、都市空間の再組織化ということで考えれば面的な介入になるので、ファイバーシティの戦略としては限定的になろう。

都市的スケールをもった線状要素(ファイバー)は「流路」として「流れ」を具体化し、「境界」や〈織目〉として「場所」の定義に関わる。「境界」はまた「流れ」の制御もする。

流路と場所

では、どのようにすれば、「流路」と「境界」を道具にして都市空間の制御ができるのだろうか。「流れ」と「場所」の関係を示すいくつかの事例からヒントを得よう。

まず、「流れ」が強烈な「場所」を作る例をみよう。最初は、ターミナル駅や飛行場である(→212)。こうした場所は、定住地でも目的地でもない。かといって単純な通過点でもない。「流れ」の速度が変わるところであり「流れ」の様相が変わるところである。それによって明確なアイデンティティをもつ「場所」が生み出されている。飛行機や列車の出発を待ち旅立つ人、到着して街に出る人、乗り継ぐ人、見送りの人や出迎えの人など、そして、そんな人たちを見に来る石川啄木[95]のような人、食事をしたり土産物を買ったりホテルに宿泊をしたりといろいろな「流れ」が交

street pattern is a planar form of intervention. As a stratagem for Fibercity, therefore, it is likely to be of limited use.

As channels, linear elements (fibers) on an urban scale give shape to flows and, as boundaries and textures, they are involved in the defining of places. Boundary can also control flow.

Channel and Place

Then in what way can channel and boundary be used as tools to control urban space? Some examples indicating the relationship between flow and place provide hints.

To begin with, let us look at instances of flow creating a place of intense activity (→ 212). First, there is the railway terminal or airport. A place of this kind is neither a home nor a destination, but it is not a simple point of transit either. It is a place where the speed and mode of flow change. This gives rise to a place with a clear identity. People about to go on a journey waiting for an airplane or a train, people who have arrived and are now going into town, people transferring from one means of transport to another, people seeing off or meeting others, people who come to watch such people, an example being Ishikawa Takuboku,[95] people dining, buying souvenirs and making hotel reservations—many different flows intersect. The facility accommodates a bundle of different channels that expand and contract.

In the other example, pedestrians (flow) endow a place with character. There is a small park on 53rd Street in Manhattan called Paley Park, which is narrow and deep. Like a pool of still water along a riverbank, it offers to people coming and going on the avenues a place to get away from the hustle and bustle of the city and relax for a moment. It is precisely the flow of the avenues that makes this park meaningful and attractive. This is a good example of a channel and a place generated alongside it.

There are instances where flow is highly symbolic. In Japan's premodern cities, temples and shrines were arranged nearby, and this was conceivably meant to ward off malevolent deities and spirits of the dead as the scholar of religion Mircea Eliade maintained.[96] Toward the end of the Heian period, 33 temples dedicated to Kannon (Skt. Avalokiteśvara) were established around Kyoto, and

わる。「流路」の束が施設そのものとなり、「流路」には膨らみや澱みがある。

　もう一つの例は、行き来する歩行者（流れ）が場所に性格を与えている例である。ニューヨークマンハッタンの53番街に面して、奥に向かって細長く伸びる小さなペイリーパークという名の公園がある。外部空間の設計で知られる芦原義信が日本に紹介した。川の岸辺にできる水流の澱みと似て、大通りを足早に行き交う人々に都市の喧噪から少し離れてゆったりとした時間を過ごせる「場所」を提供している。この公園の存在も魅力も、大通りの「流れ」があってのことである。「流路」とその傍らに発生する場所の好例である。

　流れに高い象徴性が伴う場合もある。日本の近代以前の都市では周辺に寺社を配置するが、それは、宗教学者ミルチャ・エリアーデが言うように「魔神や死者の霊を防ぐ」魔術的目的のためだと考えられる[96]。平安末期に京都の周辺に三十三所観音が置かれたが、後にこれを巡る洛陽三十三所観音巡礼が成立し、この写しが全国に多数生まれる。日本の巡礼の特徴は、明確な目的や到達点がなく、ただ巡ること自体に価値が認められることである。このような円環状の周遊は、室町時代から江戸時代に作られた回遊式庭園にも共通する。その特徴は、いずれも、巡ること、つまり環状の流れが、それによって囲われた「場所」に力を与え、人は「場所」を感得することである。ここでは「流路」は巡礼路であり、「場所」は京都である。

境界と場所

　次に「境界」と場所の関係についても事例をみてみよう。日本の大学は都心に立地していてもキャンパス型が多い。つまり校地がひとまとまりになって塀を巡らすタイプである。一方、東京の駿河台地区には、明治大学や日本大学や中央大学など歴史ある大学が集中している。各大学では、かつて校地が複数の街区に分散し、それぞれに中層の校舎が建てられていた。学生も教師も講義のたびに道路を渡らなければ目的の教室に行けない。欧州の都心に建つ古い大学には多いが、日本では珍しいタイプであった。大学の構内も街の一部のようになり、学生と近辺で働く人との区別がつかない。

　一方、東京大学の本郷キャンパスは江戸時代は封建領主前

subsequently the custom of making a pilgrimage of those temples developed. Pilgrimages modeled on this one came into being all over the country. Many Japanese pilgrimages such as the one around the Kyoto temples are characterized by the absence of any clear objective or destination; simply going from place to place is seen as having value. A circular movement also characterized the tour-style gardens of Japan built from the Muromachi period through the Edo period. In each case, circular flow endowed the encircled place with power and enabled people to understand place. In the case of the temples dedicated to Kannon, the channel was the pilgrimage route, and the place was Kyoto.

Boundary and Place

Next, let us examine an example of the relationship between boundary and place. Many Japanese universities, even those located in the middle of a city, are of the campus-type. That is, the university grounds are consolidated and surrounded by a wall. However, a number of universities located in the Surugadai district of Tokyo, including Meiji University, Nihon University and Chūō University, were once each composed of medium-rise buildings scattered over a number of city blocks. Students and teachers alike had to cross streets to get from one classroom to the next. This type of university, though common in older universities located in the middle of cities in Europe, was unusual in Japan. The university campus was a part of the city, and there was no way to distinguish students from people working in the neighborhood. Meanwhile, the Hongō campus of the University of Tokyo occupies what was once the residence of the Maeda clan who were *daimyō* in the Edo period. The spatial organization of the campus reflects this origin. A fence surrounds the grounds, and the campus has a distinctive environment quite different from the city around it (→ 213). Even though these institutions are all universities and located in the middle of the city, differences in boundary conditions generate entirely different environments.

An entirely closed boundary and an open situation with no boundary at all both create a uniform, expressionless landscape. A city gains vitality when boundaries exist to an appropriate degree, modulation is introduced, and interchange between places is promoted.

田家の屋敷であったが、その由来が、キャンパスの空間構成に反映して塀で囲われ、構内は周辺の街とはまったく異なる独自の環境をもつ（→ 213）。同じ大学という機能で同じように街中にありながら、「境界」の性情の違いがまったく異なる環境を生みだしている例である。

完全に閉じた「境界」も、「境界」が無く開かれた状態もともに、均質でのっぺりとした風景を作る。都市の活力は、適切な「境界」性を強化して都市に抑揚を持ち込み、都市の「場所」間の交流を促進することによって得られる。

近世までの多くの都市は、現在の国家と同じように領域の縁に「境界」を持ち、「境界」を制御して内部の秩序を維持してきた。具体的には、関税や入国管理, 施設としては城門、港、関所、橋、軍の基地などが設けられていた。EUは国民国家の概念に挑戦する壮大な実験であり、域内の国を移動するときに相互に特別な入国手続きを要求しない。つまり「境界」を消したのである。新しい状況はわれわれ部外者の意識のありようにも働きかけ、国のまとまりより欧州連合内の同質性に目が向かうことになる。

212 香港国際空港 (1998)
Hong Kong International Airport (1998)

国際空港では滞留と移動がひっきりなしに発生する。ノーマン・フォスター設計。

Congestion and movement constantly occur at an international airport. Designed by Norman Foster.

213 東京大学本郷キャンパス
University of Tokyo Hongo Campus

本郷キャンパスがある場所には、かつては江戸時代の有力大名前田藩の藩邸があった。境界を形成する樹林は、大学と街の両方に景観と環境の両面で貢献をしている。

The site of the present-day Hongo campus was once the residence of the Maeda clan, who were powerful daimyo in the Edo era. The trees that form the site's boundaries contribute both to the landscape and environment of the university and the city.

Until the modern era, many cities had boundaries just as countries today do, and maintained order inside by controlling them. Boundaries were equipped with fortified gates, harbors, checkpoints, bridges and military bases for the purpose of imposing tariffs and controlling immigration. EU is a grand experiment that challenges the concept of the nation-state, and special immigration procedures are not required when moving from one member country to another. That is, it has erased boundaries. The new conditions also affect the consciousness of outsiders such as the Japanese; we focus more on commonality within the EU and less on national distinctions.

The nature of a boundary reflects the spatial culture of that region (→ 214). The Imperial Palace illustrates quite well the interrelationship between the shape of a boundary and the characteristics of a place in Japan. The Palace is the official residence of the emperor, the symbolic head of state of Japan, but few people are familiar with its facade. The best-known exterior feature of the Palace is Nijūbashi, a bridge over the moat that leads to the main gate of the Palace (→ 215). This is one of the boundaries of the Palace. The Imperial Palace has multiple, layered

214 一遍上人絵伝巻一
Ippen Shōnin Eden Scroll 1

モンスーン地帯の建物の大きな特徴は、縁側である。縁側は部屋の前についた境界装置であると同時に、それ自身が場所である。

One of the greatest features of buildings in monsoon climates are their verandas. These verandas, attached to the rooms, are both boundary makers as well as places in and of themselves.

「境界」のあり方は地域の空間文化を反映するものである (→214)。皇居は日本の「境界」の形態と場所の性格の相互関係を知る好例である。皇居は日本の象徴的元首である天皇の公邸であるのに、宮殿の正面なるものはほとんど知られていない。もっとも代表的な「外観」は宮殿正門に向かう途上にある堀を渡る二重橋である (→215)。これは宮殿領域の「境界」の一つである。皇居は、日本の空間構成の典型である重畳[97]する「境界」構成を踏襲している。幾重にも重なる「境界」こそ皇居の壮大さであって、単体としての宮殿の建物やその正面は、天皇の威光を示すうえでそれほど重要ではないという日本の建築文化が反映している。

「境界」の開口部は取引や交渉の場になる。市場もその一つであり城塞都市では多かれ少なかれ公的に管理される。日本語の「都市」に当たる中国語は「城市」というが、いずれも漢字の一字が「市 (イチ)」を含むことは意味深い。「境界」は、都市活動というドラマの舞台である。

仲の悪いキャピュレット家とモンタギュー家の娘と息子ロミオとジュリエットは、両家の間に築かれた憎悪の「境界」に開かれた小さなバルコニーで繋がる。「境界」は光の舞台でもある。黒田清輝の油彩の佳品『読書』は、窓辺で鎧戸を通して入る柔らかな光を受けて読書に耽る西洋夫人の姿を描いている。窓辺は常に画家が好む題材である。

線状要素 (ファイバー) は、「流路」の形や「境界」の形を通して、「流れ」や「場」を性格づけることができる。

断片的な線

このような線はどのように扱われるべきなのだろうか？ ひと言「線」といってもその様態は多種多様である。

実は、それの解答は冒頭で「比較的小さな線状要素 (ファイバー) を操作する」と述べている。なぜ、小さな線でなければならないのか (→216)。それは、都市の再組織化に柔軟性を保つためである。最初に目指すべき都市の姿を描いて、後はそれに向けて完成に向けて粛々と作業を進めるような、都市の再組織化の方式では、変化の激しい現代社会に対応できないからである。状況が変化し、ときには目標像すら描き直せるような柔軟さが必要であり、そのためには介入の単位を小さくするほう

215 皇居宮殿へのアプローチ
Approach to the Imperial Palace

天皇主催の晩餐会や謁見が行なわれる鳳明殿に行くためには、手前に見える「正門石橋」を左方向に渡り、向きを変えて、奥に見える「正門鉄橋 (二重橋)」を右方向に渡って進むことになる。二つの橋が重なって見える。

In order to access the Houmeiden, where the Emperor hosts his dinner parties and receives visitors, you must first cross the Seimon Ishibashi bridge (to the front in the photo) to the left, then change directions and cross the Seimon Tetsubashi (Nijubashi) to the right. The two bridges look as though they overlap.

boundaries which are typical of spatial organizations in Japan.[97] The grandeur of the Palace and the authority of the emperor are expressed, not by any individual palace building or the facade of such a building, but by the layering of boundaries. This reflects the nature of Japan's architectural culture.

An opening in a boundary is a place for transactions and negotiations. The market is one such place and administered more or less publicly in a walled city. The word for city is *toshi* in Japanese and *chéng shì* in Chinese; it is interesting that they both contain the character meaning "market." Boundaries are dramatic stages for urban activities. Romeo and Juliet meet on a small balcony that represents an opening in the wall of hatred that has been built between their respective families, the Montagues and Capulets. A boundary can also be a stage of light, as witness the well-known painting by Kuroda Seiki entitled "Reading," which shows a Caucasian woman sitting by a window and reading a book by the light entering through louvers. A subject positioned near a window has always been a favorite subject matter for painters.

A linear element (fiber) in the guise of a channel or a boundary can give character to a flow or a place.

Line Segments

How should such lines be dealt with? We may call them all lines, but they exist in many different modes.

The answer was in fact given at the outset of Part 2, where I explained that "relatively small linear elements (fibers)" are to be manipulated. Why do the lines have to be short? (→ 216) It is to preserve flexibility in the reorganization of the city. If we were to first draw the image of the city we intended to realize and then assiduously attempt to make that image real, we would not be able to adapt to the rapid changes that the contemporary city constantly undergoes. When the situation is changing, flexibility—that is, the ability to redraw the image we are aiming for—is needed, and for that, it is convenient to make the unit of intervention small. Moreover, each unit needs to be autonomous to a certain extent so that it can function even if the intervention is interrupted. This conforms to the characteristic of "group form" as defined by Maki Fumihiko (→ 217).[98]

が好都合である。しかも、いつ中断しても、部分が自立して機能するように、単位ごとに一定の自律性を保持していなければならない。これは槇文彦の「集合体へのアプローチ」として示されたグループフォーム(群像型)[98]の特質に通ずる(→ 217)。

少ない投資で小さな空間的介入を行ない、大きな効果を目指すことは、成熟した都市の空間の再組織化に必要なばかりか、現代の環境デザインに対する倫理的要請でもある。

小さい線でなければならない二つめの理由は、市民が制御できるスケールを保つためである。成熟社会の市民は、居住地の環境、自然、健康な食、地域文化、移動の自由などを今まで以上に高い水準で求める。こうした価値は、市場機能だけで達成することができない。都市空間の再組織化を市民の手に残しておくためには、介入のスケールを小さくすることが重要なのである[99]。

小さい線でなければならない三つめの理由は、「大きい流れ」に対抗するためである。「大きい流れ」は領土的野心の表現でもある。書籍通販アマゾンのおかげで、田舎にいても誰もが世界一大きな書店を使えるのだから夢のようなことである。しか

し、地域経済の問題としてみると悪夢に近い。地元の本屋で買えば、本屋の販売の利益、それにかかる所得税、本屋の店員の給料、運送費、本屋の建物の維持費、地代と固定資産税などの多面的に地元経済に関わるが、アマゾンから買うと、地元雇用された宅配便の配達員の給料以外は、消費税も含めて日本国はおろか、すべてを飛ばして、シアトルにある本部機構に吸い上げられてしまう[100]。大きな「流れ」は地域の経済力と活力を吸い上げ、地域を脆弱にする。大きな「流れ」にだけ頼っていると、やがて、足下が崩れ始める。

大きな流れが卓越する現代都市では、市民の能力で把握と制御ができ、時間的変化に対応できるように、介入単位は小さく、地域に小さい流れが生まれるように、一定の自律性を持たせることが必要である。

大都市の魅力は、大都市が提供する選択の豊かさにあるのだが、いかに多様なモノや情報が用意されたところで、それらを手に入れて使えなければ、実際の選択肢とはならない。過去の階級的な社会では権力者が高速移動手段を独占していたから

216 ピエト・モンドリアンによる「線の構成/第二段階」(1916-1917)
Piet Mondrian's *Composition in Line, Second State* (1916-1917)

217 槇文彦による「集合体への三つのアプローチ」(1964)
Maki Fumihiko's "Approaches to Collective Form" (1964)

槇文彦が提示した集合体の三つの形式(パラダイム)。左から構成的形態、メガストラクチュア、群像型(グループフォーム)。

Three types (paradigms) of collective form as proposed by Maki Fumihiko. From the left: Compositional Form, Mega Form, Group Form.

Small but highly effective spatial interventions carried out by means of small investments are not only necessary for the spatial reorganization of mature cities but are demanded of contemporary environmental design from an ethical point of view.

The second reason for making the lines short is to keep interventions to a scale that enables the public to control them. Members of a mature society demand higher standards for all things including the environment of their residential areas, nature, food, local culture and freedom of movement. Those standards cannot be met simply through the functioning of markets. Keeping the scale of interventions small is important if the reorganization of urban spaces is to be managed by the public.[99]

The third reason for keeping the lines short is to compete against Big Flow. Big Flow is an expression of territorial ambition. Thanks to the online retailer Amazon, anyone, even someone living in a rural area, can purchase books from the world's biggest bookstore. Though this may seem like a dream come true, it is a nightmare for local economies. When a purchase is made at a local bookstore, that impacts on the sales profits of the store, the wages of the store staff, the transportation costs borne by the store, the income tax levied on the store owner, and the maintenance costs and the local fixed assets tax for the building occupied by the store. When a purchase is made from Amazon, everything except for the wages of local home delivery service employees is swallowed up by the company headquartered in Seattle including the consumption tax.[100] Big Flow sucks dry local economic vitality and weakens localities. When we rely solely on Big Flow, we eventually undermine our own economic environment.

In the contemporary city where Big Flow excels, the unit of intervention must be made small to enable the public to comprehend and control it and to adapt to changes in time, and the intervention must be autonomous to a certain extent so as to give rise to small local flows.

A metropolis is attractive because it offers many choices, but even when diverse things and bits of information are on offer, they are not real choices if people do not acquire and use them. In the class society of the past, those with power

一般市民は限られた選択肢しかなかった。現代社会では誰でも平等に高速移動手段と情報伝達手段を利用できるようになった。これこそが近代都市がもたらした最大の福音であろう(→ 218)。問題は、新幹線の例やアマゾンの例で見た通り、大きい「流れ」だけを追求すると、その恩恵に浴することができない地域が増えることである。見た目の派手さに目を奪われて、たまにしか利用しない「大きい流れ」だけを受け入れると、実はそれと引き換えに毎日使っていた便利な「小さい流れ」を失うのである。

「大きい交通」に対して、「近く・遅く・少量」の交通を「小さい交通」と呼ぶと、小さい交通は毛細血管に比せられるであろう。それが無ければ組織を構成している細胞にまで栄養が届かない。「大きい交通」だけでは都市は機能しない。都市には、歩くには少し遠い距離を移動する需要がある(→ 219)。それは身体能力に依存するので需要は多様であり、一つ一つの需要は小さい。「小さい交通」は、求めに応じて細い道にも入り都合のよいところで停められなければならない。「大きい交通」と違って「小さい交通」はわれわれに密度の濃いきめ細かな体験を提供する(→ 220)。「小さい交通」には自転車、車椅子、タクシー、そしてコミュニティバスなどがある。車種のバリエーションが近年急激に増えている。そのなかで、自転車はもっとも使いやすい「小さい交通」の代表である。快適な利用のためには自転車レーンの充実が必要であるが、日本では、自転車交通は自動車交通のお余り的な扱いしか受けていない。

グラミン銀行のマイクロクレジット[101]や非貨幣的な価値の交換も可能にする地域通貨などが注目されている。「大きい流れ」では解決できない難しい問題を小さいお金の「流れ」を作ることで解決しようとしているのである[102]。「流れ」を小さくすることで、それに関わる人の顔が見えるようになり、再び「信頼」[103]という、社会を成立させる核にある原理を復活させることに繋がる。

「小さい交通」の究極には徒歩があり、「歩いて暮らせる」都市は褒め言葉である。しかし、「コンパクトにして歩いて暮らせる都市にしよう」というスローガンには注意が必要である。すべての住人が歩ける距離のなかに便益施設が配置できるのは小さい村だけである。都市というからには、歩いて暮らすことができない。だから都市には「小さい交通」が要るのである(fc 3J)。

had exclusive use of the means of high-speed travel, and the general public had limited choice. In contemporary society, everyone has equal right to use high-speed means of travel and means of communication. That is probably the greatest blessing of modern society (→ 218). The problem is that, as we have seen from the examples of the Shinkansen and Amazon, pursuit solely of Big Flow results in increasing numbers of localities that do not receive its benefits. If, captivated by its immediate appeal, we make use of only Big Flow, we are going to lose the small flows we have made use of every day up to now.

Small transportation as opposed to Big Transportation is transportation that is "near, slow and on a small scale." It might be compared to capillaries, without which nutrients would not reach the cells that make up tissue. A city would not function with Big Transportation alone (→ 219). In a city, there are transport demands for distances that are slightly too far for walking. The demands are diverse because they depend on physical capacity, and each demand is small. In response to a request, small transportation must enter narrow streets and stop where it is convenient. Unlike Big Transportation, small transportation provides us with high-density, small-scale experiences (→ 220). Means of small transportation include bicycles, wheelchairs, taxis and community buses. Vehicular types have increased sharply in recent years. Among these types, bicycles are the easiest to use. Development of bicycle lanes is necessary if they are to be used comfortably, but in Japan such facilities are mostly an afterthought.

The micro-credit loans made by the Grameen Bank[101] and community currencies that make possible transactions outside the national monetary system have drawn great interest. They attempt to solve, through the creation of small flows of money, problems that cannot be dealt with by means of Big Flow. Making flows small personalizes transactions[102] and helps revive that core principle of society called trust.[103]

Walking is the ultimate means of small transportation, and to describe a city as one in which people's everyday needs can be met within walking distance of their homes is to praise it. However, we need to be skeptical of any suggestion to make a city compact so as to achieve such an objective. Only in a small village can all facilities for

小さい線でなければならない理由の説明に戻ろう。第四の理由は、ファイバーシティは都市空間の再組織化のための手段として既存の構造物の再利用や改修を中心に置くからである。これまでのように、何でも新規に建設するのではなく、自然も人工物も、いまあるものをできる限り継承し、改修して使い続けることが求められる。建築や都市施設における改修は小さい工事の集積である。新築は規模のメリットが受けられるが、改修や修復は現場を見ながら一つ一つ個別の対応を考えなければできないので工事の単位は小さくなる。改修の優れた点は、現場に即した設計と工事にならざるをえないので地元の職人と企業が中心になり、お金が地元で循環することである。

既にあるものと新しいもの、そして自然と人工物が一緒になって緻密で豊かな織物を織り上げることこそが都市デザインの目標となる。

庭師のように

以上、ファイバーシティの方法の特徴を述べたが、それだけでは完全ではない。なぜなら、あらゆる都市的介入には計画者や設計者などの専門家が介在するからである。問われるべきは、専門家の姿勢である。何に向けて、どのようにファイバーシティの方法を用いるかである。

この問は、同時に、断片的な線的要素の介入を柱とするファイバーシティは、都市の全体性に対してどのような責任を負うのかという疑問に答えることでもある。

われわれは、更地に理想都市を計画するというモダニズム的都市計画に慣れすぎているので、計画者が都市の「全体」を作ることができると考えてしまう。この「全体」は建築が持つ「全体」と同じである。建築は設計図を描き、それに合わせて施工すれば、当初に構想した通りにできる。都市も、新都市であればそのように作ることができるだろう。しかし、ひとたびそこに人が住み始め、樹々が育ち、物が動き、建物が古びて新しい建物に建て替えられ、新しい住人が移り住み、新しい生命が生まれ、古い住人が去る。つまり、いったん歩み出した都市の再組織化は、最初に新都市を生み出すこととはまったく違った行為である。それは計画するというより「介入」するというべきであ

218 シンガポール都心の高速道路（イースト・コースト・パークウエー）
Expressway in central Singapore (East Coast Parkway)

219 テント張りの仮設店舗と群れるリキシャー（インド、ハイデラバード、チャル・ミナール通り）
Makeshift stores under tents and crowded rickshaws (Char Minar Road, Hyderabad, India)

220 ムラーノ島の運河（フォンダメンテ・デイ・ヴェトライ、ベニス）
Canal on Murano Island (Fondamenta dei Vetrai, Venice)

people's everyday needs be arranged within walking distance of their homes. That is not possible in a city, and that is why small transportation and medium-sized transportation are needed in an urban environment (fc3J).

Let us return now to the explanation of why the lines must be short. The fourth reason is that in reorganizing urban spaces, Fibercity will focus on recycling and repairing existing structures. There is a need to preserve, repair and continue to use as much as possible existing things, both natural and man-made, instead of constantly constructing new things as we have previously. Buildings and urban facilities are repaired through the cumulative effect of small construction projects. With new construction, one has the advantage of working on a large scale, but a repair or restoration project is necessarily small in scale because it requires scrutiny of the site and consideration of each response to an existing building. The advantage of a repair project is that it is done primarily by local workmen and businesses since design and construction work has to be done at the site; thus money is circulated locally.

The objective of urban design will be to weave a rich, closely-knit fabric made up of the existing and the new, the natural and the man-made.

Like a Gardener

The distinctive characteristics of the method adopted by Fibercity have now been explained, but that still does not give a complete picture because a planner or architect is involved in every urban intervention. The stance taken by the architect needs to be examined. How will he or she use that Fibercity method and to what end?

In answering this question, we will also attempt to settle any uncertainty the reader may have concerning what sort of responsibility Fibercity, an approach based on fragmentary intervention by means of linear elements, assumes for a city's overall character.

Accustomed to the modernist approach to city planning, which is to start with a clean slate and plan an ideal city, we are apt to believe that a planner can create an entire city. A building can be created as initially conceived if drawings are prepared and construction is carried out according to those drawings. A city, if it is new, can probably be created in this way as well. However, the reorganization of a city

る。日本の都市計画や建築設計ではあまり馴染みがない「介入」という用語を、ここまで断りなしに使ってきたが、それはこのような理由からであった。

いったん歩み出した都市には、もはや完成という最終状態は無く、ただ変化し続けるだけなので、建築のような意味での「全体」を想定することはできない。そうすると、当初の問いかけである断片的な線的要素の介入の「都市の全体性に対する責任」は、断片的な線的要素による介入が「何に向けて、どのようになされるか」という問いに変わらなければならない。つまり、全体性ではなく、歩み続ける都市に度々なされる「介入」を貫く姿勢あるいは思想の一貫性である(→ 221)。

比喩で言えば、人生全体をデザインして完成像を描くことができないのと似ている。ここに一人の若者がいるとしよう。彼の先にあるのは壮年の彼であり、老年の彼であって、決して固定的な彼ではない。彼が行なうことは、その時々で異なる状況に対して、心に描く「幸せ」や「徳」や「正義」と照らし合わせて、毎時毎日大小の決断をすることである。「彼」に明瞭な輪郭があるとしたら、個々の決断の背後に潜む姿勢の一貫性あるいは、態度の型であろう。

このように見ると、20世紀型の都市への「介入」の姿勢の型は、「対象地を更地にして、そこに道具としての都市を構想する」というものだった。「道具として」というのは、都市をわれわれの外にある対象として見るということである。道具は交換可能である。それは技術者的な発想だと言ってもよいだろう。自然現象を要素に還元し、それらの間に合理的な関連を見いだし、それを基に新たな道具あるいは機械を創り出そうとする。それは普遍性を目指す。コルビュジエの「住宅は住むための機械である」という有名なアフォリズムはこのような姿勢の簡潔で力強い表明であった。しかし、環境は外にある対象ではなく、環境はわれわれを取り囲み、われわれの存在に色濃く影響を与えている。人間は環境の中に存在しているが、同時に、中にいる人間がそれを包む環境に影響を与えてもいる。

そのような認識を踏まえて、縮小の時代の都市への介入における姿勢を考えてみよう。ファイバーシティV.1.0に添えた拙文「縮小する都市のデザイン戦略」では、縮小する都市のモデルとしてモダニズムの「機械」モデルから「布」モデルに変わるべきだ

221 ファイバーシティのイメージ図
　　 Fibercity concept image
切れ切れの短い線的介入によって都市を活性化するのが、ファイバーシティの都市計画である。

The urban plan proposed by Fibercity vitalizes the city through small, fragmented linear elements.

once that city has taken on a life of its own—that is, once people have begun to live there, trees have grown, things have been moved, buildings have grown old and been rebuilt, new residents have moved in, children have been born and old residents have left—is an undertaking completely different from its initial creation. It is not so much an act of planning as it is an act of intervention. "Intervention" is not a word much used in city planning or architectural design in Japan, but the above reason is why we have been using it.

Once it takes on life, a city never reaches completion, that is, an ultimate condition; it simply continues to change and evolve. Though an entire building can be realized as designed, one can never assume what the entirety of a city will be like. That being the case, the initial question concerning "the responsibility that fragmentary interventions by means of linear elements have toward a city's overall character" need to be changed to "how and to achieve what objective are fragmentary interventions by means of linear elements to be carried out?" That is, the question concerns the consistency of stance or philosophy that connects the repeated "interventions" in a city that has a life of its own (→ 221).

This is similar to the impossibility of designing one's life in its entirety and depicting what the final, complete image of that life will be like. Let us say there is a young man. Ahead of him still are the man in his prime and the man in old age—he is by no means settled into any one self. What he does is to check the ever changing conditions of his life against his own ideas of happiness, virtue, and justice, and make decisions, large and small, every hour and every day. If "he" has a clear profile, it is a consistency of stance that forms the context in which he makes his decisions—an attitude.

The stance taken in urban interventions in the twentieth century was to start with a clean slate and to create a city conceived as a tool. "As a tool" because we regard the city as something that exists outside ourselves. Tools are interchangeable. This could be said to be the mindset of an engineer. One reduces natural phenomena into elements, discovers rational connections between them, and based on those connections tries to create new tools or machines. These efforts are in pursuit of universality. Corbusier's well-known aphorism that "a house is a

と述べた。「機械」は精巧に設計、製作された要素が計画された通り相互に絡み合って所期の目的を達成する。これに対して「布」は柔らかで一部が破断しても全体が壊れない。部分の性格はテクスチャーとして現れる。目を寄せれば同じような構造が繰り返されるフラクタル性があると主張した(fc 25JE)。しかし、その後、私たちは「布」でも不十分だと考えるようになった。布は服となって身を包むが未だ道具的である(→ 222)。一方、都市で繰り広げられる活動は、時間のなかで活動が繰り広げられるのだから、もっと環境的な比喩が必要である。われわれが思い至ったのは、生物を素材として作られ、しかも空間的である「庭」であった。そして、庭を作り管理するのは庭師である。庭師は環境の計画者であり管理者である。これこそ、縮小する社会の環境のデザイナーのモデルではないだろうか(fc 55E)。

1)庭は、人工物と自然物の混成系である。
2)庭は、計画地の気候や地形に強く支配される。
3)庭は植物の成長に伴って日々変化する。当初の構想を立てた庭師がいなくても自律的に変化し続ける。すべての庭はそれぞれが個性的である。
4)庭の将来の姿は初期のデザインだけではなく実現後の管理のやり方に大きな影響を受ける。

造園の素材である樹木は自然物であり、人間には作り出せない。だから庭師は、無から有を作り出す「発明」をするのではなく、自然物、人工物を問わず、要素を関係づけようとする。縮小の時代の都市のデザインも、既にあるものと新しいもの、そして自然と人工物を等しく位置づけ、両者が共同して大きな環境システムを形成することを目指す。このようなデザインは「発明」とは異なる種類の創造行為である「編集」でなければならない。発明から編集への移行は、現実と歴史を白紙に戻さず、それらの両方に敬意を払い、都市空間の再組織化を既存の文脈への介入あるいは要素の組み替えとして位置づけ直すことである。進歩と成長を前提として革命主義である近代主義者は、既存の環境は改善されるべき劣った環境としかみなかったから計画された都市の姿は世界中で似てしまったのである。

庭師的態度で介入を受け、変化を続ける都市の姿は、当然一

	Modern city				Fibercity
面	Surfaces	計画における操作対象	Manipulated object	断片的な線	Line segments
機械	Machine	組織体の性格	Organizational nature	庭	Garden
分離	Separation	境界の役割	Role of the boundary	交換	Exchange
発明 白い画布に理想を描く	Invention Drawing an ideal vision on a tabula rasa	デザインの目標	Design goals	編集 既存のものへの介入	Editing Intervening in existing things

222 近代都市とファイバーシティの比較
Comparison of Fibercity and the modern city

machine for living" is a concise and powerful expression of such a stance. However, the environment is not something that exists outside. The environment surrounds us and exerts a deep influence on us. Humans exist in the environment, but at the same time, humans also exert an influence on the environment that surrounds them.

Keeping those things in mind, let us consider the stance to be assumed for urban interventions in an age of shrinkage. In an essay entitled "Design Strategy for a Shrinking City" attached to Fibercity Volume Version 1.0, I wrote that a city undergoing shrinkage ought to be modeled, not on a machine as had been the case with modernism, but on a fabric. In a machine, elements that have been precisely designed and manufactured interrelate with one another exactly as planned and achieve the desired objective. By contrast, a fabric is soft and continues to serve its purpose even when a part of it is torn or frayed. The characteristic of a part of a fabric appears as texture. There is a fractal quality to the way the same structure is repeated (fc 24JE). Since then, however, we have come to believe that even a fabric is insufficient. A fabric is still tool-like in that it can be made into clothing that is wrapped around one's body (→ 222). Meanwhile, the activities that unfold in the city require a more environmental metaphor because they unfold in time. The image we eventually arrived at was that of a garden, a spatial entity made of organic materials. A person who creates and manages a garden is a gardener. A gardener is a planner and manager of an environment. A gardener is the model for a designer of the environment in a shrinking city (fc 56E).

1. A garden is a composite system made from man-made and natural materials.
2. A garden is highly influenced by the climate and topography of the area of the site.
3. A garden changes constantly with the growth of plants. It continues to change on its own, even in the absence of the gardener who first conceived it.
4. The future appearance of a garden is heavily influenced by, not just its original design, but the way it is managed after it has been realized.

A gardener works with trees and other natural materials

つとして同じにはならないので、原理的には、完成形として示すことはできないが、ただ、目標を広く共有するためには、大凡の姿を示すことはしなければならないだろう。「断片的な線」でできた都市の姿を想像する上で示唆に富む先行事例を求めるのなら日本の近代以前に成立した都市や建築空間が最適である(→223)。もう一つの先行事例は、アメリカの都市研究者ケビン・リンチが1960年に著した『都市のイメージ』のなかで示したイメージマップである(→224)。彼は、都市のパブリックイメージを構成する要素としてパス、ランドマーク、エッジ、ノード、ディストリクトと名付けられた5つを取り出した。これらの要素は、現実の都市の物的要素を反映しているが、一人一人の選択と関心によって選ばれ、再構成されたものである。われわれが提案している線状要素(ファイバー)は、5つの要素のうちのパスとエッジに相当する。そして、われわれが小さな「流れ」や断片的な線状要素(ファイバー)と同じように、イメージマップに描かれたパスとエッジは断片的で離散的である。これらのイメージを構成する要素の統合は、都市を使う一人一人が行なっている、というのがリンチの論である。

中世都市のように市壁で囲まれて「場所」が優位な単純な都市とは異なり、「場所」と「流れ」が拮抗し、時には「流れ」が勝つ現代都市では、空間構造も過去の都市とは異なるべきであろう。現代都市の空間構造は市民に開かれ、各市民が異なった風に組み立てられるような自由と多様さを備えるべきである。

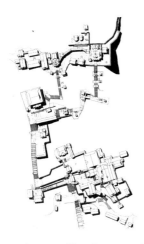

223 金刀比羅宮参道の空間構成
Spatial composition of the approach to Kotohira-gū

山の高い場所にある社殿にむかって参道が屈曲しながら登って行く。折れ曲がる度に、ありがたみが増してゆく。

The approach winds up the mountain to the shrine located high above. Each bend in the path heightens your sense of appreciation.

that humans cannot create. Therefore a gardener does not "invent" something from nothing but instead attempts to establish relationships between elements both natural and man-made. The design of cities in an age of shrinkage also places equal value on both existing things and new things and on both the natural and the man-made, so that these things might work together and form a large environmental system. This sort of design must be a different type of creative activity from *inventing*; it must be a form of *editing*. To shift from inventing to editing is to respect reality and history instead of wiping the slate clean, and to regard the reorganization of urban spaces as intervention in existing contexts or as the rearrangement of elements. Modernism, an ideology of revolutionary character premised on progress and growth, regarded the existing environment as something inferior that needed improvement. Consequently planned cities throughout the world have all come to resemble one another.

Cities that receive interventions made with a gardener-like stance and continue to change will naturally differ in appearance. In principle, cities will not converge on some ultimate, complete form. However, they will have certain shared objectives. It is necessary to suggest what they will look like in general. In imagining the appearance of cities resulting from line segments, the most suggestive precedents are the architectural and urban spaces of premodern Japan (→223). Still another precedent is the image maps indicated in *The Image of the City*, the 1960 work by the American urban planner Kevin Lynch (→224). Lynch isolated five elements from which the public's image of a city is composed: path, landmark, edge, node and district. These elements reflect actual physical elements of the city but are selected and reconstructed by each individual according to his or her interests or concerns. The linear elements (fibers) we are proposing correspond to the path and edge among these five elements. Like the small flows and fragmentary linear elements (fibers) we are describing, the path and edge drawn in the image maps are fragmentary and dispersed. It is Lynch's contention that the integration of the elements from which images are composed is something that each person using the city does in his or her mind.

Unlike simple cities in which priority is given to place (such as a medieval city surrounded by a wall),

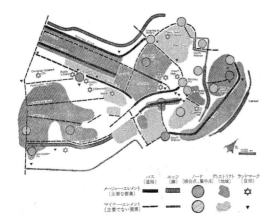

224 ケビン・リンチによる「イメージマップ」(1960)
Kevin Lynch's "image map" (1960)

ケビン・リンチは、人々がどのように都市空間を認識するかを解明し、人々は「イメージ地図」を持っているという仮説を提示した。

Kevin Lynch clarified how people understand urban spaces and proposed the hypothesis that people have their own image maps of the city.

contemporary cities are cities in which place and flow compete against each other, and at times flow wins out. They probably ought to be different from past cities in spatial organization. Contemporary cities ought to be open to the public and sufficiently free and diverse so that each person can assemble his or her own spatial organization.

以上現代の都市が直面する課題群とそれを枠づけてきた都市計画思想を概観し、これまでの都市計画とは大きく変わる縮小する都市のための新たな都市計画思想ファイバーシティを構築するための思考を進めてきた。現時点で明言できることを次の8項目にまとめてみよう。

1.［縮小の時代の既成都市のための内科治療的な計画理論］
　ファイバーシティは、既存の都市を対象に、比較的小さい線状要素（ファイバー）を操作することで、都市のなかの「場所」と「流れ」を同時に制御しようとする内科治療な計画理論である。

2.［都市を「流れ」と「場所」で考える］
　都市の「流れ」と「場所」は相互に依存し条件づける。「流れ」は「場所」に生気を吹き込み、「場所」の様態は「流れ」を呼び起こす。

3.［線状要素によって「場所」を定義する］
　線状要素（ファイバー）は「境界」として、あるいは編まれて〈織目〉として「場所」の定義に関わることができる。

4.［線状要素によって「流れ」を制御する］
　線状要素（ファイバー）は「流路」として「流れ」を具体化し、「境界」として「流れ」を制御する。

5.［自律性をもった小さい単位による介入］
　大きい流れが卓越する現代都市では、市民の能力で把握と制御ができ、時間的変化に対応できるように、介入単位は小さく、地域に小さい流れが生まれるように、一定の自律性を持たせることが必要である。

6.［最小介入の原則］
　少ない投資で小さい空間的介入を行ない、大きい効果を目指すことは、成熟した都市の空間の再組織化に必要なばかりか、現代の環境デザインに対する倫理的要請でもある。

3. ファイバーシティは何をめざすのか
3. What Are the Objectives of Fibercity?

　The above has been an overview of the issues confronting contemporary cities and the city-planning ideas in which those issues have been framed, and a discussion of the thinking behind the construction of Fibercity, a new approach to city planning for shrinking cities that will differ in major ways from cities of the past. What can be said definitely about this approach at this point might be summarized as follows.

1. [Internal medicine-like approach to planning for existing cities in an age of shrinkage]
　Fibercity is an internal medicine-like approach to planning intended for existing cities that attempts to control both place and flow in a city through the manipulation of relatively small linear elements (fibers).

2. [Consider the city as both flow and place]
　In a city, flow and place are interdependent and condition each other. Flow breathes life into place, and the mode of place gives rise to flow.

3. [Define places by means of linear elements]
　Acting as boundaries or as woven textures, linear elements (fibers) can help to define places.

4. [Control flow by means of linear elements]
　Linear elements give shape to flow as channels and control flow as boundaries.

5. [Intervention by means of small, autonomous units]
　In the contemporary city where Big Flow excels, units of intervention need to be small so that people can comprehend and control them. The units also need to be autonomous to a certain extent so that small flows are generated in the community.

6. [Principle of minimum intervention]
　Small spatial interventions made with small investments intended to have big impacts are not only necessary for the spatial reorganization of a mature city but demanded of environmental design today from an ethical point of view.

7.［編集による計画］
　既にあるものと新しいもの、そして自然と人工物が一緒になって緻密で豊かな織物を織り上げることこそが都市デザインの目標となる。

8.［都市空間への介入者は庭師のように振る舞うべきである］
　ファイバーシティのモデルは庭であり、庭を作り管理するのは庭師である。庭師は、以下の点を考慮して物と自然物の混成系の形と関係をデザインする。

1) 庭の計画の内容は、計画地の気候や地形に強く支配されること
2) 庭は樹木の成長に伴って日々変化する。当初の構想を立てた庭師が世を去った後も自律的に変化し続ける。
3) 庭の将来の姿は初期のデザインだけではなく実現後の管理のやり方に大きな影響を受ける。

7. [Planning through editing]
　The objective of the design of cities will be to weave together existing things and new things, and the natural and the man-made, into a rich, closely-knit fabric.

8. [Those who intervene in urban spaces ought to conduct themselves in the manner of a gardener]
　Fibercity is modeled on a garden, and a person who creates and manages a garden is a gardener. A gardener designs the form and relationship of a composite system made up of man-made things and nature, taking into consideration the following points.

1. The substance of any plan for a garden is controlled by the climate and topography of the site.
2. A garden continually changes as its trees grow. It will continue to change on its own even after the gardener who first conceived it passes away.
3. How a garden will appear in the future is determined as much by the way it is managed as by its initial design.

理論的な組み立ての素描を基に、この章では、東京首都圏と長岡を対象にして具体的な都市戦略を描く(→ 226、277)。

これらの都市戦略は、前章で述べたファイバーシティの8項目の実践であるが、より普遍的に使える計画の道具となることをめざして、モデル化をしようと思う。そのために、ここでは、現実の都市を再組織化するための線的介入のパターンを地形的な形態に比喩してモデル化する。それは、ファイバーシティが都市計画家に庭師のように都市に取り組むことを求めるからである。ここでは、汀、川、運河、乱流、庭が選ばれている。いずれも「流れ」と「場所」に同時に関わる地形概念である。5つのモデルの展開としてそれぞれに1～3のデザイン・プロジェクトを示した。各デザイン・プロジェクトには、計画行為における操作に加えてデザイン・プロジェクトの特徴を表す名前が与えられスラッシュ(／)で結ばれている。

ここで示すデザイン・プロジェクトは、ファイバーシティの方法論を示すための単なる事例ではなく、現実の東京や長岡にとって、忍び寄る縮小の時代を乗り切り、衰退に打ちひしがれる市民を元気づけ、真に活力のある都市にするためのヒントとなるような提案をめざしている。なお、第一部の第四章で検討した20項目の都市的課題と現状はこれから述べる提案の計画条件であり、同時に注釈でもある(→ 225)。

第二章:
デザイン・プロジェクト

Chapter 2.
Design Projects

Having outlined the theoretical structure of Fibercity, we will attempt in this chapter to formulate specific urban strategies for the Tokyo Capital Region and Nagaoka (→ 226, 277). These urban strategies are practical applications of the eight items of Fibercity explained in the previous chapter, but an attempt has also been made to model them so as to make them more universally usable tools of planning. To that end, patterns of linear intervention for reorganizing actual cities are likened to five topographical forms and modeled. Topography is referenced because Fibercity requires the city planner to be engaged with the city much like a gardener. Shore, river, canal, turbulence and garden are topographical concepts, each related simultaneously to flow and place. Design projects are developed in one to three places that suit each topographical model. Each design project is designated by a planning operation, slash, and a name expressing a characteristic of the design project.

These design projects are not simply examples indicating the methodology of Fibercity but intended to be proposals that provide Tokyo and Nagaoka with actual hints to overcome the approaching age of shrinkage, encourage citizens disheartened by decline and give cities true vitality. The 20 urban issues and present conditions examined in Part 1, Chapter 4, are the planning conditions as well as annotations for the proposals that are about to be explained. (→ 225).

汀 Shore		襞をつくる Folding	
		めぐらせる Circulating	
川 River		置き換える Replacing	
		数珠つなぎにする Stringing Together	
		組み合わせる Combining	
運河 Canal		結びつける Connecting	
乱流 Turbulence		縁飾りを付ける Edging	
		囲う Enclosing	
庭 Garden		灌漑 Irrigation	
		小さい庭 Small Gardens	

225 ファイバーシティの都市戦略群
Fibercity's urban strategies

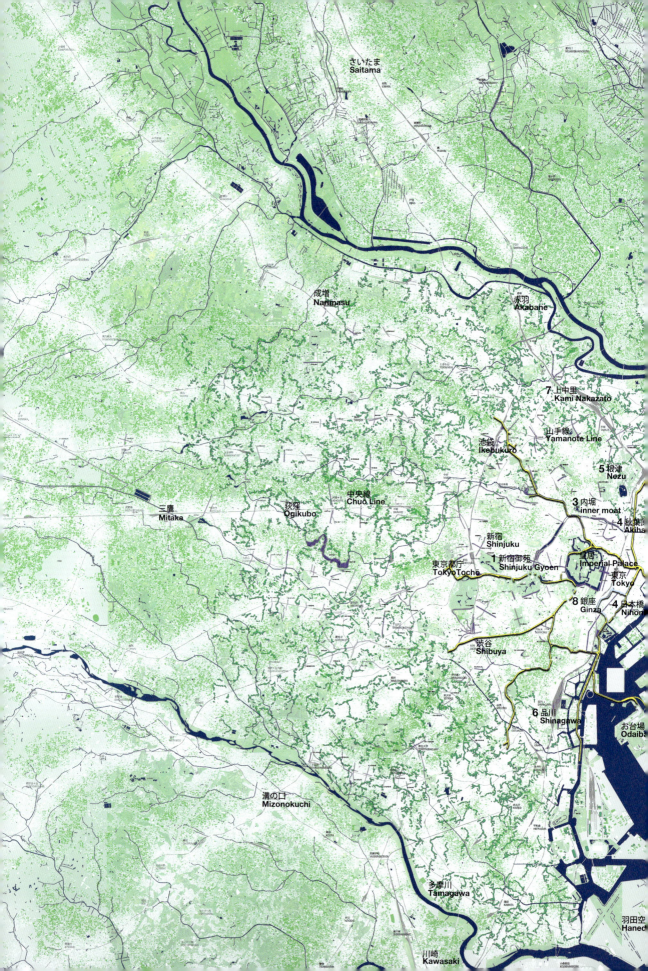

		リアス	Rias
1		新宿御苑	Shinjuku Gyoen

		緑の網	Green Web
2		隅田川	Sumidagawa
3		内堀	inner moat
4		日本橋	Nihonbashi

		緑の間仕切り	Green Partition
5		根津	Nezu

		青いネックレス	Blue Necklace
6		品川	Shinagawa

		緑の花輪	Green Garland
7		上中里	Kami nakazato

		生命の回廊	Cloister of Life
8		銀座	Ginza

三つの東京
Three meanings of TOKYO

- 鉄道駅の東京 / Name of a railway station
- 行政区としての東京都 / Name of an administrative territory
- 首都圏 / Another name for the Capital Region

226　ファイバーシティ東京2050
　　　Fibercity Tokyo 2050

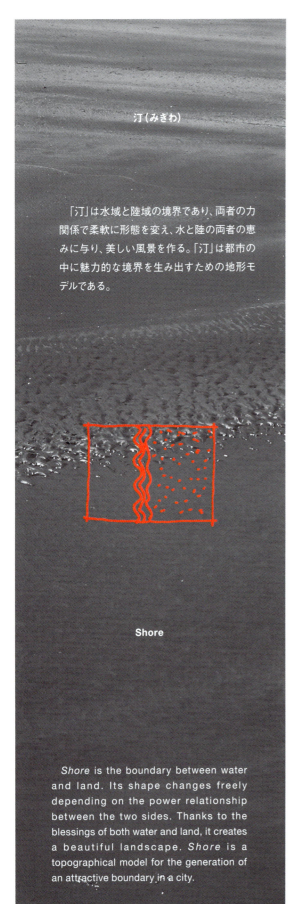

汀（みぎわ）

「汀」は水域と陸域の境界であり、両者の力関係で柔軟に形態を変え、水と陸の両者の恵みに与り、美しい風景を作る。「汀」は都市の中に魅力的な境界を生み出すための地形モデルである。

Shore

Shore is the boundary between water and land. Its shape changes freely depending on the power relationship between the two sides. Thanks to the blessings of both water and land, it creates a beautiful landscape. *Shore* is a topographical model for the generation of an attractive boundary in a city.

襞をつくる／リアス（新宿御苑）

「襞をつくる」は、線状要素（ファイバー）をウネウネと曲げることで、境界面を長くする操作である。それは、ちょうど、消化器官が効率よい吸収を可能にし、ラジエーターのフィンが効率よく放熱できることと同じである。

〈リアス〉を適用するのは新宿御苑の北側の境界である。新宿御苑は、東京のいくつかの業務商業的中心の一つである新宿にある（→ 227）。日本の都市公園のなかには近世までの武家屋敷の庭園を引き継いだものがある。新宿御苑も江戸時代には高遠藩内藤家の下屋敷であった。1879年に皇室の植物園になり、第二次世界大戦後は一般に公開され、現在は、皇居外苑、京都御苑、千鳥ケ淵戦没者墓苑とともに環境省が直接管理する国民公園の一つである。日本の近代の造園家は公園の空間構成をするにあたって、大名庭園を範とし、周囲に森を配して園を閉鎖的に構成することが多い。この構成法は園の空間的独立性を高めるが、周辺の市街地に対して閉鎖的になる。新宿御苑もその例に漏れない。これに加えて、新宿御苑の場合

Folding / Rias (Shinjuku Gyoen)

"Folding" is an operation that lengthens a boundary surface by twisting a linear element (fiber) into a serpentine form. The result is similar to the way digestive organs are arranged to make efficient absorption possible or a radiator is equipped with fins to radiate heat efficiently.

"Rias" is applied to the north boundary of Shinjuku Gyoen. Shinjuku Gyoen is located in Shinjuku, one of the districts of Tokyo (→ 227). A number of city parks in Japan can be traced back to gardens on feudal *daimyō* estates. Shinjuku Gyoen was a villa of the Naitō clan, lords of Takatō domain in the Edo period. In 1879 it became a botanical garden for the imperial family. After World War II, it was opened to the public. Today it is one of the National Gardens administered directly by the Ministry of the Environment, the others being the Outer Garden of the Imperial Palace, Kyoto Imperial Garden and Chidorigafuchi National Cemetery. Modern Japanese gardeners tend to model the spatial organizations of parks on *daimyō* gardens, surrounding them with woods to create enclosed spaces. This compositional method enhances the spatial

は地区の目抜き通り（新宿通り）との間に一街区が挟まっているため、新宿を訪れる人々からは公園の存在が目に入らない。一方、公園の中からは、樹冠越しに周囲の高層ビルが見えるものの都市的雰囲気は乏しい。われわれの興味は、公園と市街地のいずれにとっても他方の存在が有意義であるような関係を創り出すことができないだろうかということである（→ 228）。

〈リアス（新宿御苑）〉は、御苑と市街地との境界を襞状に折り曲げることで両者の関係を深めることを意図している。新宿御苑の園地の面積を変えないで、現在の園地の一部を市街地にして建物を建て、現在の市街地を公園にして樹々を植える。

こうして、公園に接する街区が増え、一帯を優雅な地区に変えることができる。同時に公園の一部が新宿通りに顔を出し、道行く人に公園の存在を強く訴える（→ 229）。さらに、新宿通りの反対側の街区にある寺の境内や小さな公園にも、緑地が視覚的に繋がり、辺り一帯に緑地の網の目を作る。

227 新宿御苑と新宿副都心
Shinjuku Gyoen and the Shinjuku subcenter of Tokyo

江戸に起源をもつ皇居の東側の古い都心に対して、新宿は近代の郊外住宅地を背景とした都心である。新宿駅の西側は副都心として位置づけられ、東京都庁舎も含めて、超高層オフィスビルが集まる地区である。新宿御苑がある駅の東側は比較的古い地区で商業地区として繁栄している。

In comparison to the old city center to the east of the Imperial Palace, which dates back to the Edo era, Shinjuku, to the west of the Palace, is a city center based on the context of modern suburban housing. The western side of Shinjuku Station is considered a subcenter and is a district alive with high-rise office buildings such as the Tokyo Metropolitan Government building. The eastern side of the station, where Shinjuku Gyoen is located, is a relatively old neighborhood currently thriving as a commercial district.

228 〈リアス〉新宿御苑
Rias（Shinjuku Gyoen）

市街地に進入する公園、公園に進入する市街地。襞をなす境界
A park that penetrates into the urban fabric and an urban fabric that penetrates into a park. A boundary that creates folds.

229 新宿通りに顔を出す御苑
Shinjuku Gyoen making an appearance onto Shinjukudōri

autonomy of a park but also means the park is closed to the surrounding urban area. Shinjuku Gyoen is no exception. People visiting Shinjuku do not see the park because, in addition to this closed-off quality, the park is a block removed from the main street of the district (i.e. Shinjuku Avenue). Nearby high-rise buildings can be seen over the treetops from inside the park, but otherwise there is not much of an urban atmosphere in the space. Can a relationship be created between the park and the surrounding urban area? Can the presence of the park be made meaningful to the urban area, and vice versa? (→ 228)

"Rias (Shinjuku Gyoen)" is intended to deepen the relationship between the park and the urban area by bending into folds the boundary between them. Parts of the present park area are built on and integrated into the urban area, and parts of the present urban area are planted with trees and integrated into the park; thus the total area of Shinjuku Gyoen remains the same.

In this way, more city blocks border the park, and the entire zone is transformed into an elegant district. At the same time, parts of the park face Shinjukudōri Avenue and make the presence of the park known to passersby (→ 229).

Furthermore, the greenery of the park is visually linked to a temple precinct and a small park in the block on the other side of the avenue, creating a green network stretching over the entire district.

めぐらせる／暖かい巡回

「めぐらせる」は、往復運動する小さな流れを作り出すことで、「流れ」の痕跡が「場所」を生み出す操作である。

〈暖かい巡回〉は、公共サービスを自動車に載せて複数の地域にサービスを届けることで、「買物難民」を救うことを目指す。現代社会においては、郵便、保健、医療、福祉、教育など多くの公共サービスが人々の生活を支えているが、その経済的基礎は税収である。縮小社会では税収が減る。財政難のもとでサービスを維持する方法として、公共サービスの提供形態まで遡って考え直すのは一つの方法である。高度成長を続けて来た日本では、公共サービスが公共建築と分ち難く結びつけられてきたために、公共サービスのコストが高くなり都市が縮小すると、公共サービスの維持がままならぬ事態が発生する。公共サービスを車に載せて巡回すれば、当然、サービス水準は下がるが、まるでなくなってしまうのとは大違いである。医療で言えば、歯科のように予約診療が可能な診療科とか人工透析（血液透析療法）のように慢性症状の定期的施療に向いている。生活の基幹的サービスだけではなく、文化的な催し（劇、音楽、展覧会、図書、相談など）にも巡回サービスを拡大することが可能である。このようなサービスは現在でも行なわれているが、建物施設で行なうサービスの補助的な扱いである。しかし、巡回サービスの機動性の高さを考えると、移動サービスはもっと表舞台に出るべきであり、本格的な技術的開発と運営のノウハウの蓄積を図るべきである。もし、各自治体がこのような移動公共サービス方式を備えれば、災害時にはサービス車を被災地に派遣することができる。到着直後から救援活動に入れる。〈暖かい巡回〉は分散システムであり、これまでの階層的システムの公共サービスとは異なるサービス提供システムなのである（→ 230）。

今日の地方都市では、自家用車を使えなくても生活ができる環境整備が急務である。私たちは、このような取り組みを総称して「暖かい（＝オレンジ）インフラストラクチャー」と呼んでいる。それは縮小期においても、日本国憲法ですべての国民に保障された健康で文化的な生活をおくるための物的仕組みである。縮小する都市のための戦略に課せられた課題の一つは、「大きい流れ」から取り残され、孤立した「寂しい個人」のために「小さ

230 地域間公共サービス共有システムとしての〈暖かい巡回〉
Orange Rounds as an interzonal public sharing system

コンパクトシティ政策は、住人をサービスの近くに移住させようとする。〈暖かい巡回〉は、人の住む場所にサービスを届ける。

The compact city policy attempts to move residents closer to services. In contrast, Orange Rounds brings services to areas where people live.

Circulating / Orange Rounds

"Circulating" is an operation in which a small, circulating flow is created, and the route followed by that flow generates place.

"Orange Rounds" is aimed at providing relief to "shopping refugees" by delivering mobile public services to multiple communities. In contemporary society, many public services such as mail, public health, medical care, welfare and education are essential to people's lives. Tax revenues finance these services. In a shrinking society, tax revenues decline. One way to maintain public services despite financial constraints is to rethink their format. In Japan, a country that long experienced intensive growth, public services have become closely linked to public buildings. As a result, when the cost of public services rises and cities shrink, maintaining those services becomes next to impossible. If public services were loaded on vehicles that make the rounds of communities, the standard of service naturally suffers, but the result is much better than having no access to services at all. From the point of view of medical care, this best suits types of treatment that can be done by appointment such as dentistry and periodic free treatment of chronic conditions such as artificial dialysis (hemodialysis). Mobile public services could be expanded to include not only basic services essential to everyday life but cultural events and functions (such as dramatic and musical performances, exhibitions, libraries and consultations). Such services exist today but are treated as supplemental to services provided by facilities housed in buildings. However, mobile public services should assume a more prominent role given their great flexibility. Real technological development needs to be made in the provision of such services, and know-how regarding their administration amassed. If each local government equipped itself to provide such services, it could dispatch service vehicles to stricken areas in times of disaster, and they could begin relief activities immediately upon arrival (→ 230).

Developing an environment in which people can live without having their own cars is an urgent task in local cities today. We call these measures "orange (i.e. warm in the sense of friendly) infrastructure." Orange Infrastructure is a physical mechanism that enables everyone to lead

い流れ」を提供し、暖かい関係を取り結ぶことができる都市環境を作ることである。

　これまでの日本の行政的発想であれば、〈暖かい巡回〉を始めるとなると、すぐさま、そのための拠点施設を新たに作ろうという話しになるのだが、そこはありあまる既存の施設を使うべきであろう（→ 234）（→ 235）。そのなかで一番可能性があるのは社寺である[104]。建物だけではなく、広い境内を持ち、地理的にコミュニティの中心に位置することも好都合である。社寺は宗教施設を超えて地域コミュニティの場であった。特に寺は、本堂に大きな広間をもち、コミュニティ施設への転用に大きな可能性を秘めている。また、寺の大きな屋根と境内の森は一般の家並みの上に盛り上がり、視覚的にも集落の中心として親しまれてきた（→ 232）（→ 233）。ところが、近年、人口減少、脱宗教化などにより檀家や氏子の繋がりが薄れ、社寺の経済基盤が弱くなっている。やがて廃社、廃寺ということになれば、日本の集落景観の視覚的中心をなし、それを近世と結びつける唯一と言ってもよい要素、いわば臍が失われることになる。われわれは寺院の境内と建物の活用を考えてもいい時期ではないだろうか[105]。

従来のシステムは、
［建築＝空間＋サービスコンテンツ］
提案するシステムは、
［建築＝空間］＋［乗り物＝サービスコンテンツ］

231　長岡市栃尾地区に適用した〈暖かい巡回〉
Orange Rounds adapted to the Tochio district in Nagaoka City
長岡市栃尾地区は山間地にあり、小さい集落が深い谷川沿いに散在している。
The Tochio district in Nagaoka is located in a mountainous region with small villages scattered along a deep mountain river.

healthy, culturally rewarding lives as guaranteed by the Japanese Constitution, even in an age of shrinkage. It provides small flows to "lonely individuals" who have been left behind and isolated by Big Flow and creates an urban environment in which warm relationships can be established.

The mindset of most administrative authorities in Japan, if faced with the task of initiating Orange Rounds, would be to immediately create a new facility to serve as a base of operations, but there are more than enough existing facilities that can be used (→ 234) (→ 235). Among them, the facilities that have the most potential are temples and shrines.[104] Conveniently, they possess not only buildings but spacious precincts and are in central locations in their respective communities. Temples in particular have main halls, each with a large space, and can be readily converted into community facilities. The large roof of a main hall and the woods of the precinct soar above most houses, and the temple has long been regarded as the focal point of a community (→ 232). However, in recent years, the decline in population and the secularization of society have weakened the ties between members of communities and undermined the economic foundation of temples and shrines. If temples and shrines were to be abandoned, Japanese communities would lose their focal points (and arguably their centers of gravity) and the only elements linking those communities to the premodern period. Perhaps the time has come to consider the utilization of temple precincts and buildings.[105]

The conventional system is (building = space + services).
The proposed system is (building = space) + (vehicle = services).

232 寺の屋根が支配する日本の田園風景
A temple roof soaring above a rural landscape in Japan

日本の田園地帯の風景の核をなすのは、お寺の大きな屋根と境内の森である。

Massive temple roofs and forests in temple precincts form the heart of rural landscapes in Japan.

233 廃寺を転用した〈暖かい巡回〉の拠点
Orange Rounds base utilizing an abandoned temple

234 〈暖かい巡回〉の拠点で展開するサービス
Services offered at one of the bases in Orange Rounds

人口減少は膨大な空き家を生み出す。自動車が運んで来るモジュールによって、空き家は歯科診療所になったり図書館になる。

The shrinking population will lead to a massive rise in vacant houses. These vacant homes can turn into dental facilities or libraries based on modules carried there by automobiles.

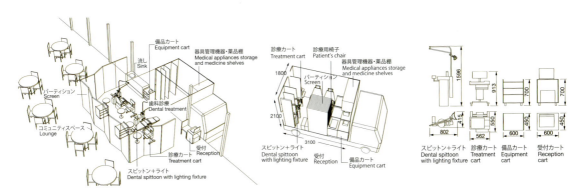

235 移動サービスの事例検討。歯科診療
Study of a mobile dental clinic

川

「川」は水だけではなく土も石も流し、山を削って段丘を作り、沃野を作り、洲を作る。「川」は濁流となることも淀むことも枯れることもある。ときには逆流することすらある。下流域では流路を変え中州をつくり蛇行して三日月湖を残す。「川」は流れが場所を作る地形モデルである。

River

River is not just water; it carries soil and stones, carves mountains and creates terraces, plains and sandbanks. It can muddy, stagnate and dry up. At times it can even reverse its flow. It can change course downstream, create sandbars, meander and form an oxbow lake. River is a topographical model for the creation of place by flow.

置き変える／緑の網

「置き換える」は、既存の流路をそのままにして、流すものだけを現代的な流体に置き換え、「流れ」に新しい生命を吹き込む操作である。流路の建設や改変には普通は費用がかかるので、流すものだけを変えることで場所や流れの性格を変えられれば効率的である。

〈緑の網〉は、東京の首都高速道路の一部の道路機能を廃して、その構造体を転用して、未来の東京のための複合機能を持った新たな社会基盤として再生することを目指す。新たな機

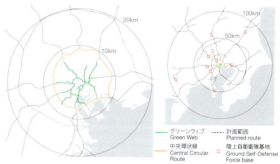

237 首都圏の自衛隊駐屯地
Japanese Self-Defense Force bases in the Capital Region

自衛隊の駐屯地は首都圏を囲むように、放射幹線道路近くに配置されている。
Japanese Self-Defense Force bases are located near major radial roads and encircle the metropolis.

Replacing / Green Web

"Replacing" is an operation that leaves an existing course as is, changing only what flows to a contemporary "fluid" and thereby breathing new life into the flow. Constructing or altering a course is ordinarily costly; it is therefore more efficient if the character of a place can be changed by altering only the thing that flows.

The idea of Green Web is to convert parts of the Tokyo Metropolitan Expressway to new uses and transform them into a new social infrastructure with composite functions for a future Tokyo (→ 238). They will serve three new functions. One is to serve as roads for relief activities in times of disaster, a second is to serve as raised parks and promenades or as recreation roads for light vehicles, and a third is to serve as conduits for a heat network.

The Metropolitan Expressway is a system of expressways for cars and trucks only, constructed as part of the reorganization of Tokyo for the 1964 Olympics. Routes over rivers and existing streets that made expropriation of land unnecessary were chosen because of the limited time available for construction, and much of the system was raised above grade. It was extended after the Olympics and now reaches a total length of 301.3 kilometers. It plays

能は三つある。一つは災害時の救援路、もう一つは平時の空中公園と遊歩道および軽車両のための遊歩道、そして最後に熱のネットワークのための管路である（→ 238）。

首都高速道路は、1964年の東京オリンピックのために行なわれた都市改造の一環として建設された自動車専用道である。短期間に行なわれたために、土地収用の必要のない河川や道路の上空を選んで路線が決定され、大半の区間は高架構造で作られた。オリンピック後も延伸され、今日までに総延長301.3kmに達し、首都圏の道路網の中での役割は大きく、現在は1日に約111万台、東京区部の自動車の走行台キロの約30%

を担っている[106]。

災害時の救援

日本では地震、津波、火山爆発、山崩れ、台風、竜巻、大雪など地球上で起こる自然災害のほとんどが起こり、その強度も大きい。防災は、日本の都市づくりの最大の優先事項である。そのなかでも地震は最大の脅威である。建築・土木の構造物には、地震とそれに付随する火災に関して過去100年間の経験と研究が蓄積されている。大きな地震を経験するたびに、日本の耐震工学は精度を上げてきた。建築物では伝統的な耐震設計

236 都市の空中庭園〈緑の網〉
Urban hanging garden in Green Web
首都高速1号線
Metropolitan Expressway Route 1

a major role in the road network in the Capital Region and today carries approximately 1,110,000 vehicles a day, or approximately 30 percent of all traffic flow as measured in vehicle-kilometer, in the 23-ward areas of Tokyo.[106]

Relief Activities in Times of Disaster

Nearly all natural disasters known on earth including earthquakes, tsunami, volcanic eruptions, landslides, typhoons, tornados and blizzards visit Japan and are moreover quite destructive. Disaster prevention is of the highest priority in urban development in Japan. Of these disasters, earthquakes are the greatest threats. Japan's buildings and civil engineering works are the results of a century of experience in and research into earthquakes and the fires that accompany earthquakes. In addition to traditional seismic design, rapid advances have been made in recent years in the technology for mitigating seismic motion in buildings and base-isolation technology for preventing the transfer of seismic forces onto buildings. The Japanese custom of using buildings for only 30 to 40 years before reconstructing them has ironically helped to disseminate these technologies. As a result, damage to buildings from earthquakes has undoubtedly been reduced.

	緑の網 Green Web
	大規模緑地 Large green tract of land
Ⓜ	主な美術館 Major art museum
	高速機能を残す首都高速道路 Segment of Metropolitan Expressway that continues to maintain expressway functions
··········	JR 山手線 JR Yamanote line
	大規模再開発地域 Large-scale urban redevelopment area
■	地域冷暖房施設 District heating and cooling plant
◯ 500,000 ◯ 1,000,000	地域冷暖房施設と清掃工場の能力 Plant's capacity for heating and cooling
	地域冷暖房供給区域 District heating and cooling service area
□	清掃工場 Waste incineration plant
□	下水処理場 Sewage disposal plant

地域冷暖房供給区域　District heating and cooling service area
1. 大手町 (410 / 800,177), 2. 丸の内一丁目 (120 / 311,359), 3. 八重洲・日本橋 (86 / 63,482),
4. 箱崎 (254 / 100,916), 5. 丸の内二丁目 (146 / 270,565), 6. 新川 (68 / 141,427), 7. 有楽町 (113 / 120,867),
8. 東京国際フォーラム (212 / 136,311), 9. 銀座2・3丁目 (26 / 42,796),10. 銀座四丁目 (43 / -),
11. 明石町 (143 / 187,489), 12. リバーシティ 21 イーストタワーズ (154 / -),13. 日比谷 (53 / 84,596),
14. 銀座5・6丁目 (74 / 39,250), 15. 東銀座 (18 / 42,040),16. 豊洲三丁目 (48 / -), 17. 内幸町 (265 / 255,065),
18. 汐留北 (94 / 388,839), 19. 竹芝 (85 / 118,533), 20. 芝浦 (95 / 234,248),
21. 東京臨海副都心 (3,050 / 1,045,426), 22. 天王州 (200 / 248,195), 23. 品川東口南 (119 / -),
24. 品川駅東口 (110 / 263,406), 25. 芝浦4丁目 (196 / 101,847), 26. 田町駅東口 (40 / -),
27. 虎ノ門四丁目城山 (47 / 61,367), 28. 虎ノ門二丁目 (60 / 122,168), 29. 霞が関三丁目 (55 / 84,817),
30. 赤坂・六本木アークヒルズ (72 / 224,558), 31. 永田町二丁目 (70 / -), 32. 赤坂五丁目 (102 / 268,080),
33. 赤坂 (52 / 60,435), 34. 六本木ヒルズ (127 / 552,864), 35. 広尾一丁目 (50 / 41,418),
36. 恵比寿 (97 / 290,923), 37. 紀尾井町 (47 / 56,033), 38. 青山 (60 / 96,736), 39. 北青山二丁目 (42 / 57,773),
40. 渋谷道玄坂 (36 / 72,911), 41. 新宿南口東 (50 / -), 42. 新宿南口西 (94 / 246,564),
43. 西新宿一丁目 (144 / 149,287), 44. 新宿副都心 (243 / 1,409,703), 45. 初台淀橋 (105 / 193,242),
46. 西新宿 (118 / 371,038), 47. 西新宿六丁目西部 (40 / 22,109), 48. 新宿歌舞伎町 (10 / -),
49. 西池袋 (126 / 317,664), 50. 東池袋 (124 /360,628), 51. 後楽一丁目 (216 / 83,638),
52. 神田駿河台 (107 / 45,334), 53. リバーサイド墨田 (23 / -), 54. 東京スカイツリー (102 / -),
55. 錦糸町駅北口 (44 / 142,558)

数値は、各施設の熱供給対象面積（単位 ha）と 2008 年度の供給実績を示す（単位 GJ）
Figures indicate the area serviced by each plant's heat supply (ha) and the annual amount of heat supplied by the plant in 2008 (GJ).
出典：資源エネルギー庁電力・ガス事業部政策課熱供給産業室監修、日本熱供給事業協会編『熱供給事業便覧平成 22 年度版』日本熱供給事業協会 (2010)
Source : Data adapted from Agency for Natural Resources and Energy Electricity and Gas Industry Department and Japan Heat Service Utilities Association, eds., Report of the Heat Supply Business in 2010 (2010)

清掃工場　Waste incineration plants □
1. 墨田 (136,730), 2. 豊島 (50,908), 3. 新江東 (331,128), 4. 中央 (82,441), 5. 有明 (234,169), 6. 港 (239,756),
7. 渋谷 (35,798), 8. 目黒 (102,914)

数値は、2014 年度の各施設の熱供給と売電の実績の合計値である（単位 GJ）
Figures indicates annual supply of heat and sales of electricity by each plant in 2014 (GJ)
出典：各清掃工場 2015 年環境報告書、東京二十三区清掃一部事務組合「東京 23 区のごみ処理」
http://www.union.tokyo23-seisou.lg.jp/index.html
Source：Data adapted from "Environmental Report of Waste incineration plants in 2015," Clean Authority of TOKYO

238　〈緑の網〉の全体図
General diagram of Green Web

に加えて、建物に加わる地震力を打ち消して弱める制震技術や建物に加わる地震力を減らす免震技術も近年急速に進歩した。こうした技術の普及には、建物を30〜40年ほどしか使わず建て替えるという日本の消費型の建設文化が皮肉にも幸いしている。その成果があって、地震による建物被害は確実に減っている。

安全な避難を考えると、問題はその先の都市空間にある。特に大都市都心の商住混合地域は、火災が広がらないようにする対策、安全な避難場所の確保、そしてそこに至る避難経路の確保で不安が残る地域が多い。一方、都心の業務地区が昼間に被災すると、経済と政治と行政の中枢機能が長期間停止するなど甚大な被害が予想される。そこで働く大量の人が避難する場所が十分確保されておらず、さらに交通遮断によって、東京北部に正午にマグニチュード7.3の直下型地震が起きた場合、帰宅困難者は1都3県で650万人に及ぶと予測されている[107]。実際、東日本大震災のときには、東京では、震源地から離れていたにもかかわらず、地震発生時の外出者の約28%、首都圏で合計515万人が当日自宅に帰れなかったという[108]。こうしたリスク回避を目指して首都機能移転なども議論されたこともあるのだが、1980年代末からの土地バブル経済がはじけて地価が下がった途端に忘れ去られてしまい、東京は事実上無防備のままである。

道路は災害時の避難経路、救援路になるだけではなく、街区間の延焼防止の緩衝帯として働くので、防災的観点からも極めて重要であるが、東京中の狭い道路を拡幅する見通しは立っていない。東京都は災害時には、首都高も、指定された地上の幹線道路も一般車両の通行を禁止して消防、警察、救急車などの活動が展開できるようにしている[109]。しかし、昼間に地震が発生すれば、道路はたちどころに渋滞を起こし、沿道の古い建物のなかには倒れて道を塞ぐものもあるだろう[110]。そうなれば人々は車を乗り捨てるしかない。一方、都心が被災したときに救援に動員される陸上自衛隊の駐屯地は、首都圏の外縁部にあり都心から放射状に伸びる幹線沿いに配されているが（→237）、混乱のなかを救援隊が被災地にたどり着くには長い時間を要する。現状でただちに利用できるのはヘリコプターであるが、輸送力が低く焼け石に水である。

239 東京港を渡る〈緑の網〉
Green Web stretching across Tokyo Bay

From the perspective of safe evacuation, the problem lies instead in urban spaces. There are many districts with mixed commercial and residential uses in the center of a metropolis in which uncertainties exist concerning measures to prevent the spread of fires and the securing of safe places of evacuation and evacuation routes to such places. Meanwhile, if the business district of a metropolis were to suffer a disaster in the daytime, enormous damage is anticipated including long-term cessation of central economic, political and administrative functions. Places where the large numbers of people working there can evacuate have not been adequately secured, and transportation systems are likely to be interrupted. If an earthquake of a magnitude of 7.3 were to occur directly above the epicenter at noon in northern Tokyo, people in the metropolitan region who would have difficulty going home are expected to number 6.5 million.[107] In fact, when the Great East Japan Earthquake struck, approximately 28 percent of those people who were out at the time or a total of 5.15 million persons in the Capital Region were unable to get home that day, even though the epicenter was at some distance from Tokyo.[108] The idea of moving the capital elsewhere to avoid such risks has been discussed, but it was forgotten as soon as the real estate bubble of the late 1980s burst and land prices plummeted. Tokyo remains to all intents and purposes defenseless.

Roads are extremely important from the perspective of disaster prevention because they function not only as evacuation routes and roads for relief activities but as buffer zones to prevent the spread of fires from one block

〈緑の網〉は都心の被災地に救援部隊が進む太く確実な救援ルートを提供する。そのために、首都高の1レーンを災害救援道路として常時空けておき、残りを緑化して空中線状公園に転用する。首都高は阪神淡路大震災のあと補強工事をしているので、ある程度の地震には耐えられるはずである[111]。

日本の国土を貫くいくつかの高速道路網は東京に集まり、首都高は巨大なインターチェンジの役割も担っているので、都心に無用な交通まで入ってきてしまう。これが首都高の慢性的な渋滞の一因となっている。この問題を解決するために、首都高速道路株式会社と東京都は、JR山手線を取り巻く位置に都心を迂回する中央環状線を建設して2015年に全通した。これで、中央環状線のなかの都心部の首都高の交通ネットワーク上の役割はずいぶんと軽減された。それでも、〈緑の網〉を実行すれば、かなりの交通を一般道路で捌かなければならなくなることは否定できない。しかし、これは都市の危機管理の問題であり、今後の人口減少と適切な自動車利用抑制策を組み合わせれば、都心部の首都高の転用は十分見通しが立つ。

空中公園と遊歩道

〈緑の網〉の第二の機能は、平時には、空中歩道として歩行者や軽車両に開放することであり、第三の機能は、世界に例を見ない一大空中緑地網を作ることである。〈緑の網〉の全面積は約120 haあり、これは新宿御苑の面積50 haの2.4倍に相当する。沿道ビルの中間階に空中庭園を作って首都高と橋で結べば一続きになる（→ 236）。沿道ビルは二つのレベルに入り口をもつことができ資産価値が高まる。首都高からみる東京の風景は、空が広く壮快であるが、今は自動車に乗らなければ見られない（→ 240）。〈緑の網〉は、この楽しみをすべての市民に開放する。首都高が地下に潜るお堀端は都内でも第一級の景勝地と言われているが、今は狭い歩道をお堀端ランナーたちがひしめき、散歩者は景勝をゆっくり楽しむことができない。地表面の自動車交通のうち沿道サービス用に二車線だけを残して、残りの二車線を地下に移せば、お堀端に広い緑道を作ることができる（→ 241）。

空中に浮かぶ線状緑地とお堀端の広い緑道はともに脱自動車社会建設の象徴として、また東京の英断として世界に誇るこ

240 墨田川沿いの〈緑の網〉
Green Web along Sumidagawa

隅田川河岸は下町の中心にあり、かつては遊行と放蕩と宗教の場所であった。花見や花火の時には、多数の江戸・東京の市民を集めた。

The banks of Sumidagawa are located at the heart of the downtown area and were once sites for aimless wandering, debauchery and religion. These places gathered crowds of Edo/Tokyo residents during fireworks displays.

to the next. However, it is impossible to predict when Tokyo's narrow streets will be widened. In a time of disaster, the Tokyo Metropolitan Government prohibits the use of the Metropolitan Expressway and designated arterial roads at ground level by ordinary vehicles so that fire trucks, police vehicles and ambulances can go about their business.[109] However, if an earthquake struck during the day, traffic jams would occur and old buildings would collapse and block roads.[110] In that case, people would have no alternative but to abandon their cars. Meanwhile, the Ground Self-Defense Force posts from which units would be mobilized for relief activities if the center of Tokyo were to suffer a disaster are arranged along arterial roads extending radially from the central districts toward the periphery of the Capital Region (→ 237), but in the confusion, it would take considerable time for relief units to reach stricken areas. Under present circumstances, helicopters can be used immediately, but their transport capability is low and inadequate.

Green Web offers routes that are reliable and of ample proportions for relief units to make their way to stricken areas in central districts. One lane on the Metropolitan Expressway will be always kept open for possible use as a relief route, and the rest will be greened and converted into a linear park-in-the-air. Reinforced after the Hanshin-Awaji Earthquake, the Metropolitan Expressway ought to be able to withstand earthquakes of considerable intensity.[111]

The expressway systems that run through Japan converge on Tokyo, and the Metropolitan Expressway

とができる。それだけではなく、昨今、外貨を稼げる新たな産業として注目され、国も力を入れている東京の都市観光の重要な目玉になる（→239、240、241）。古い高架構造物を遊歩道にするプロジェクトは既に他の大都市で実現しているのだから夢物語ではない。これらの先行例はいずれも遊歩道利用なので、次に述べる防災対策とエネルギーネットワーク利用を兼ねることは真に革新的である。

熱のインターネット

都心にいくつもある地域冷暖房施設やゴミ焼却場など大規模な熱源[112]を繋げて熱の相互利用を可能にするネットワーク、すなわち熱のインターネットを作るのが〈緑の網〉の第四の機能である（→242）。

地域冷暖房の長所は、大規模で高効率の熱源機器から熱を供給できることに加えて、タイプの異なった熱需要を組み合わせて熱供給を最適化できることである。たとえば、暖房を必要とするホテルと、冬でも冷房を必要とする電子計算機を多数使う事務所の間の熱のやりとりをすればエネルギーの無駄がなくなる。現在の東京都内の地域冷暖房システムの問題は、供給範囲が狭く長所を活かしきれないことと、それぞれの熱源センターが一カ所なのでエネルギーセンターが被災すれば一帯へのエネルギー供給が滞ってしまうことである。そこで、「緑の網」を使って複数の地域冷暖房センターやゴミ焼却場を繋げば、余剰熱を融通しあうことができるので、それぞれのエネルギーセンターの余裕を小さく見込める。緊急時には、他のセンターから熱の供給を受ければ、被災後の迅速な復旧が可能になり、帰宅難民の一時逗留もできる。2011年の東日本大震災のようなプレート型の大地震より1995年の阪神・淡路大震災のような直下型

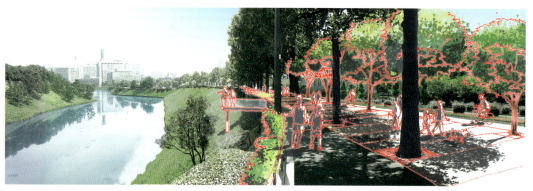

241 お堀端の〈緑の網〉
Green Web near the Imperial Palace moat

皇居の半蔵濠から三宅坂にかけては東京随一の景勝地である。ここでは首都高は地下に潜っている。地上交通の半分を地下トンネルに振り替えることで、地上の歩行者空間を拡大することができる。

From the Imperial Palace's Hanzō Bori (moat) to Miyakezaka stretches one of the most scenic areas of Tokyo. The Metropolitan Expressway runs underground in this area. By transferring half of the vehicle traffic above ground into an underground tunnel, pedestrian spaces above ground can be largely expanded.

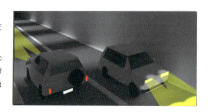

serves as an enormous interchange. As a result, extraneous traffic is introduced into the center of Tokyo. That is one reason for chronic congestion on the Metropolitan Expressway. Metropolitan Expressway Company Limited and the Tokyo Metropolitan Government constructed the Central Circular Route outside the JR Yamanote line to permit traffic to go around the center of Tokyo, and this opened in 2015. This has greatly reduced the role played by the parts of the Metropolitan Expressway within the Central Circular Route in the transportation network. Nevertheless, implementing the Green Web proposal will lead to increased traffic on ordinary city streets. This, however, is a problem of risk management for the city. Combined with the future decline in population and suitable measures to control automobile use, the conversion of the Metropolitan Expressway in the center of Tokyo ought to be quite feasible.

Parks-in-the-Air and Promenades

The second function of Green Web is to serve in ordinary times as elevated sidewalks, open to pedestrians and light vehicles. Its third function is to create a system of parks-in-the-air of a scale that is without precedent anywhere in the world. The total Green Web area is 120 ha, which is 2.4 times the size of Shinjuku Gyoen (50 ha)(→ 236). If gardens-in-the-air were created at intermediate levels of buildings along roads and connected by bridges to the Metropolitan Expressway, they would be continuous with the network. The buildings along roads would have entrances on two levels, which would raise their value. The Metropolitan Expressway offers exciting views of Tokyo's townscape and the sky above it, but at present one can only enjoy them from a vehicle (→ 240). Green Web will make those views available to all members of the public. The embankment of the moat around the Imperial Palace, where the Metropolitan Expressway goes underground, is said to be one of the most scenic areas in Tokyo, but the narrow sidewalk there is filled with joggers, making the leisurely enjoyment of the view impossible for pedestrians. If two of the four lanes at ground level were transferred belowground

地震の方が被害範囲は狭いが頻度は高い。そのようなときにこそ、熱のインターネットは威力を発揮し首都機能の継続を可能にする。さらに、使い道の少ない首都高の高架下にエネルギー工場を設ければ、より高い性能の熱システムを構築することができる[113]。沿道の建物は、このシステムから熱を受けることができるので、冷却塔などで埋められた建物の屋上が活用でき、しかも機械室を小さくできる。

このようなシステムを東京で構築しようとすれば、新規に深い地下トンネルを作って配管することが考えられるが[114]、費用が高く実現性が低くなる。しかし、〈緑の網〉を利用して、高速道路の路面上や桁側面に管を敷設すれば、非常に安価で効率的な広域の地域冷暖房網を実現することができる（→ 243）。

沿道地域の高度利用

人口減や日本を取り巻く経済環境を考えたとき、東京都心で今後どの程度の床需要があるかについては、過大に期待できないが、建て替え需要も考えれば今後も一定の新規のオフィスや住宅の供給が続くと考えてよい。その際、合理的な土地利用

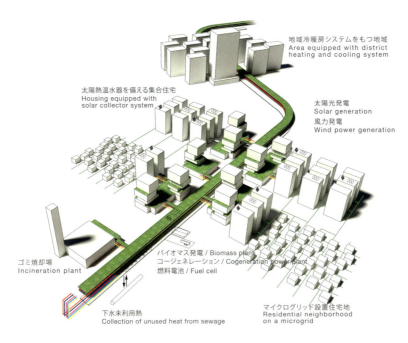

242　熱のインターネットとしての〈緑の網〉
　　 Green Web as an internet of heat
既存の地域熱供給施設と清掃工場、そして将来の住宅地のマイクログリッドを〈緑の網〉に接続する。

Existing district heating and cooling facilities and waste incineration plants are connected to Green Web together with future residential districts on a microgrid.

leaving only lanes to be used for servicing places along the street, a broad green road could be created along the embankment (→ 241).

The elevated linear green areas and the broad green road on the moat embankment would be symbols of the development of a post-automobile society in which Japan could take pride (→ 239, 240, 241). Not only that, they would be centerpieces of urban tourism in Tokyo, a new industry that has great money-earning potential and that the government is emphasizing. This is not just a pipe dream—other metropolises have already realized projects to convert old elevated structures into promenades. These precedents are all promenades. To create promenades that also can be used for disaster-prevention measures and an energy network would be highly innovative.

Internet of Heat

The fourth function of Green Web is to link together large-scale heat sources[112] such as district cooling and heating facilities and waste incineration plants to create a network that enables those facilities and plants to use each other's heat, that is, to create an internet of heat (→ 242). The advantage of district cooling and heating is that, in addition to being able to supply heat from a large-scale and highly efficient heat source device, it can combine different types of demand for heat in order to optimize the supply of heat. For example, with proper give and take between a hotel, which requires heating, and an office that uses multiple computers that require cooling even in winter, waste of energy is eliminated. The problem with the present district cooling and heating systems in Tokyo is that the area over which each system supplies cooling and heating is too small—full advantage is not taken of those systems. Each has one heat source center, so that if it suffers from a disaster, then the supply of energy over the entire zone will dry up. If multiple district cooling and heating centers and incineration plants were linked together using Green Web, then surplus heat from one facility could be passed on to another facility and the margin for each energy center could be smaller. Furthermore, if a supply of heat were

形態に誘導することが必要である。

　沿道の建物を〈緑の網〉に接続すれば、空中交通網兼災害時救援ルートが利用でき、しかも、経済的にも環境的にも効率的なエネルギー利用ができる。東京の更なる高度利用を促進するためには、後述の〈緑の間仕切り〉と連携して、当該地域への容積移転を行なえば、防災性と環境性を備えた都市を目指すことができる（→ 244）。

新しい名所／日本橋

　首都高は、1911年に建造された日本橋の上を通り、あろうことか2つの高架橋の桁が、欄干を飾るブロンズの麒麟の像を挟む位置にあり、古い橋に対してまるで敬意が感じられない（→ 245）。日本橋は江戸時代から下町の晴れの場所であり、現在は日本の道路元標が埋め込まれている。これは東京にとって重要な記念碑である[115]。このようなことから、首都高を移設して日本橋の上空の高架橋を撤去する提案がなされている[116]。しかし、明治の建築家が作った西欧様式の橋と、昭和の土木の技

244 街区型都市と輝ける都市の合体：著者による上海市楊浦区への提案
Union of an urban system based on the city block and La Ville Radieuse
伝統的な街区型の都市システムの上に「輝く都市」を載せる。
La Ville Radieuse is overlaid atop a traditional urban system based on the city block.

243 〈緑の網〉で実現するプラグインシティ
Plug-in city as realized by Green Web
〈緑の網〉はメタボリズムやアーキグラムの夢を実現するだけではなく、低炭素化に貢献する。
Green Web not only realizes the dreams of the Metabolists and Archigram, but it also contributes to a low-carbon society.

received from another center, speedy restoration of service would be possible after a disaster and people who have difficulty returning home could stay where they are for the time being. Earthquakes where the epicenter is directly below the stricken area (i.e. "inland shallow earthquakes" occurring along active faults) such as the 1995 Great Hanshin-Awaji Earthquake inflict damage over a smaller area but are more frequent than earthquakes associated with convergent plate boundaries such as the 2011 Great East Japan Earthquake. At such times, the heat internet will demonstrate its power and enable Tokyo to continue to function as the capital. Moreover, if energy factories were installed below the elevated structures of the Metropolitan Expressway, a higher-performance heat system could be constructed[113]. Buildings along the expressway would be able to receive heat from this system, freeing up rooftops currently occupied by cooling towers for other uses and making it possible to reduce the sizes of machine rooms. If such a system were to be constructed in Tokyo, deep underground tunnels would need to be newly constructed and installed with pipes;[114] the cost would be so high as to reduce the feasibility of such a project. However, if Green Web were used and pipes were installed either on the surfaces of expressways or underneath them, the cost of creating an efficient network of district cooling and heating systems covering a wide area would be extremely low (→ 243).

Intensive Use of Areas Along Expressways

　In considering the decline in population and the economic environment for Japan, we cannot be overly optimistic about future demand for space in the center of Tokyo. However, if we take into account the demand for the rebuilding of existing structures, the supply of new offices

術者が作った鋼製の高速道路橋が重なる風景は、近代日本の履歴そのものとも言えるし、巨費を投じて首都高を消してみたところで、両側の風景はもはや日本橋が竣工したころの風景ではない(→ 246)。

　私たちの提案は、防災施設としての〈緑の網〉の連続性も確保しつつ、日本橋の記念碑性を考慮して、日本橋上空に限り首都高のための高架構造物をS字型の線形に変えて作り直すというものである。こうすることで、日本橋上空を開放して水面を美しく見せ、首都高に頸を挟まれた麒麟を救出し、同時にさまざまな時代の構造物が重層する現代的な河岸空間の風景を作ることができる。

245　かつての「未来の風景」
"Future landscape" from the past

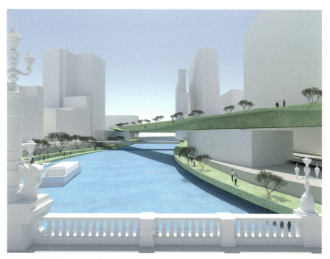

246　日本橋上空では〈緑の網〉を迂回させる
Green Web: Detour above Nihonbashi

and housing can be expected to continue to a certain extent. It is necessary in that case to induce a rational configuration of land uses.

If buildings along expressways were connected to Green Web, they would have access to an elevated transportation network-cum-disaster relief route, and be able to use energy in an economically and environmentally efficient way. To promote even more intensive use of Tokyo, FAR could be transferred to the areas in question, in coordination with the establishment of Green Partitions to be explained below. The aim would be to create a city equipped to prevent disasters and possessing an enhanced environmental character (→ 244).

New Place of Interest / Nihonbashi

The Metropolitan Expressway passes over a bridge called Nihonbashi built in 1911. Little respect is shown for the earlier structure—the girders of the elevated expressway hem in the bronze sculptures of mythical creatures called *kirin* that ornament the balustrades of the bridge (→ 245). Nihonbashi has been a public place in the "low city" (the area where commoners lived) since the Edo period, and traditionally the zero point from which distances to places in Japan have been measured. The Kilometer Zero marker is embedded today in the bridge. This is an important monument for Tokyo.[115] There have been calls to relocate the expressway and remove the elevated structure above Nihonbashi for those reasons.[116] However, the present townscape, with its juxtaposition of a Western-style bridge built by Meiji-period architects and a steel expressway built by Showa-period engineers can be said to perfectly encapsulate the history of modern Japan. Even if the expressway were eliminated, at great cost, the townscape that remained would be radically different from the townscape that existed at the time the bridge was originally built (→ 246).

Our proposal is to rebuild just the section of the elevated structure above the bridge and change it into an S-shape in plan, thus taking due account of the character of Nihonbashi as a monument while assuring the continuity of Green Web as a disaster-prevention facility. The bridge would have a clear view of the sky, the water surface would be visually more appealing, the *kirin* would be liberated from their present hemmed-in positions, and a contemporary riverside landscape in which structures from diverse periods are juxtaposed would be created.

数珠つなぎにする／緑の間仕切り

「数珠つなぎにする」は漸進的な介入を繰り返すことで線状要素（ファイバー）を作る操作である。

〈緑の間仕切り〉の主な目的は密集住宅市街地の防災性の改善である。延焼防止のために、小さい空き地を繋いで紐状の緑地帯を作り、地区を区切って延焼を抑え、被害を最小限にする（→247）。同時に緑地帯を避難経路にもする。そのうえ、緑地を増やせば居住環境の改善にもなる（→248）。

地震の被害は、地震による倒壊もあるが、事後の火災はそれ以上の大きな被害をもたらす。東京でいえば、山手線の外側に、都心を囲むように狭い道路のまま都市化されて密集した住宅地があり、このドーナツ状の地域での火災被害が懸念されている（→249）。

1923年に東京から横浜を襲ったマグニチュード7.9の大地震の被害は、低地で地盤が弱い下町地区に甚大な被害を与えた。震災後には、工場地帯からも遠く、比較的地盤が堅固な西側の台地が中産階級のための住宅地として好まれるようになった。この動きを捉えて、旧市街地の外縁の農村では道路も狭いまま、小規模な住宅地開発が広範囲に進められ、郊外住宅地が形成された。当初は敷地にそれなりの広さがあったが、その後の住民の世代交代時に細分化され、密集した市街地になってしまった。

公的な都市計画が考えるこの地域の改善方法は、都市計画道路で道路を広くし、宅地を共同化して耐火建築物で集合住宅にしようというものであるが、実現の見通しはない。〈緑の間仕切り〉はその代替案である。最大の特徴は、小さな空き地を繋げて作るので、どこからでも、また小さい部分からでも手をつけることができることである。完成すれば、緑地帯の平面形はうねうねとし、最終的な形は予見できないが、長い線状の防火

249 地震による火災被害の想定（東京都）
Estimated fire damage resulting from earthquakes

耐震化が進んだ日本では地震による倒壊被害より地震後に発生する火災被害が深刻である。

With Japan's highly advanced seismic technologies, the threat of fires occurring after an earthquake is more severe than that of buildings collapsing in the event of an earthquake.

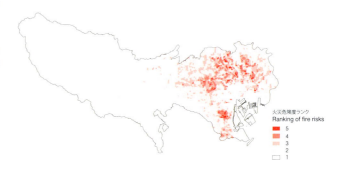

Stringing Together / Green Partitions

"Stringing together" is an operation that creates linear elements (fibers) through repeated gradual intervention.

The main objective of Green Partitions is to improve the disaster-prevention capability of densely built residential areas. To prevent fires from spreading, segmented green belts are created by connecting small empty lots; these green belts divide an area and minimize damage (→247). At the same time, these green belts provide evacuation routes. Furthermore, the increase in green areas also improves the residential environment (→248).

Earthquakes do damage by collapsing buildings, but the fires that occur after quakes are in fact more destructive. In the case of Tokyo, densely built residential areas with narrow streets lie outside the Yamanote line, effectively surrounding the center of the city. It is damage from fires in this ring of residential areas that is cause for concern (→249).

The Great Kantō Earthquake, which had a magnitude of 7.9, struck Tokyo and Yokohama in 1923. It caused enormous damage to shitamachi districts which were low-lying and had weak soil. After that disaster, hills in western Tokyo which had relatively solid ground and were at a distance from factory belts became the preferred places to build residential areas for the middle class. In response to this trend, small-scale residential developments took place over a wide expanse in farming villages on the periphery of existing urbanized areas—even though streets remained narrow—and formed suburban residential districts. Although such developments were not crowded at first, lots subsequently underwent subdivision as one generation of residents succeeded another, until eventually the areas became crowded with small houses.

The public, city-planning method for improving such areas is to make streets wider by means of so-called "city-planning roads," consolidate lots and construct fireproof apartment buildings. Prospects for realizing such projects are not good. Green Partitions represents an alternative approach. Most importantly, it involves connecting small empty lots. It can therefore be started anywhere, with small steps. When complete, a green belt will meander in plan; its final form is impossible to anticipate precisely, but it will provide a long, continuous fire barrier (→250).

The creation of green areas will result in a slight reduction in the total area of lots in the district, but the improvement of the area with respect to disaster prevention, environmental quality and comfort will raise the value of the area. In this way, balance can be maintained between the total value of assets in the area before the project and the total value of assets in the area after the project. The work of making adjustments through the exchange of land and undertaking projects to green empty lots is perhaps best performed by a community

帯として繋がる（→250）。

　緑地を作るために地区の宅地総面積は少し減るが、地域の防災性と環境性と快適性が高まるので、地域の価値が上がり資産総額としては従前従後の価値を平衡させることができる。土地交換の調整をしたり、空き地を緑地化する事業の展開は、地域の土地所有者全員が自分の宅地を現物出資して作るまちづくり会社が行なうのがよいだろう。土地の権利を証券化することができれば、土地の細分化を防ぎ、圏外に移転した人も土地の運用益を配当として受け取ることができる。事業費用の一部は、自治体の都市計画事業の代行として自治体から受託するか、出資者自身が都市計画税の減免を受けて、その分をまちづくり会社に運営費として支払うなどいくつかの方法が考えられる。地区内部の建物の建て替えについても、建築の形態のルールを自分たちで決め、地区計画制度117などを使ってまちづくり会社を核にして運用するのが望ましい。

緑の間仕切りの形態

　〈緑の間仕切り〉は密集市街地の防災性と居住性を向上させることが目的なので、幅員は建築防災学の研究から延焼遅延効果を期待できる4ｍを確保する。この値は、通風や採光、プライバシー確保のためにも必要であるし、建築基準法で求めている敷地が接するべき道の最低幅員の値でもある。また、防災上必要な位置に防火樹を配置すること（→250）が必要である。延焼抑止効果が高いのは照葉樹などである。緑地帯の線形は予想できないが、唯一の条件は緑地の一端を必ず避難場所かあるいは幹線道路に繋げることである。

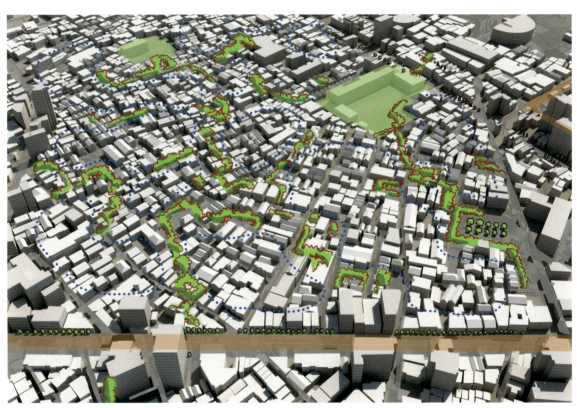

247　墨田区八広の〈緑の間仕切り〉
Green Partition in Yahiro, Sumida Ward

development company created through the investment of lots by all the landowners in the area. If rights to land can be securitized, the subdivision of land into small lots will be prevented and people who move out of the area in question can also receive benefits from the management of land in the form of dividends. Part of the cost of a project can be defrayed in a number of ways. For example, the community development company might receive funding from the local government inasmuch as it is acting as an agent of that government in undertaking a city planning project; the investors themselves might be granted a reduction in city planning taxes and pay that amount into the community development company to pay operating costs. With respect to the rebuilding of structures in the area, it would be desirable to have the people themselves establish the rules for building forms and undertake such work through the community development company, using measures such as the so-called "district planning system." [117]

Shape of Green Partitions

Since the objective of Green Partitions is to improve the disaster-prevention capability and livability of densely built

248 文京区根津の〈緑の間仕切り〉
Green Partition in Nezu, Bunkyo Ward

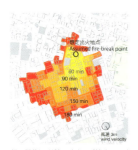

250 〈緑の間仕切り〉の延焼防止効果
Preventing the spread of fire with Green Partition

〈緑の間仕切り〉の延焼拡大防止効果。建物が全焼するまでの時間ををコンピュータ・シミュレーションした。

The effects of preventing the spread of fire with a Green Partition. The time elapsed until buildings burnt down were computer simulated.

緑の間仕切りが無いときの延焼時間予測
Estimated time for fire to spread without Green Partition

緑の間仕切りがあるときの延焼時間予測
Estimated time for fire to spread with Green Partition

residential areas, they ought to be four meters in width, a dimension that, according to research in the study of disaster prevention in buildings, can be expected to be effective in delaying the spread of fires. This dimension is also necessary to assure proper ventilation, lighting and privacy; moreover, it is the minimum width demanded by the Building Standard Law for the street on which a site has frontage (→ 250). In addition, "fireproof trees," that is, trees that are effective in preventing the spread of fires such as broad-leaved evergreen trees, ought to be planted where necessary from the point of view of disaster prevention. The linear form a green belt will ultimately take cannot be anticipated, but it must fulfill one condition—one end must always connect to either an evacuation area or an arterial road.

組み合わせる／暖かい網

「組み合わせる」は性格の異なる「流れ」を複数組み合わせることで新しい価値を生みだす操作である。

〈暖かい網〉の主な機能は、地域の鉄道網を持たない中小都市市民にとって使い勝手のよいバス交通サービスである（→ 251）。具体的には、都心部には路面電車と同様な効率性と快適性を備えたBRT(Bus Rapid Transit、バス高速輸送)方式[118]に、郊外部には利用者のきめ細かな要望に対応ができるオンデマンド方式バス[119]を組み合わせて公共交通網を作ることである（→ 252）。これは、中小都市で安心して老後の暮らしができるようにするうえで不可欠である。

半世紀前の日本の地方都市は、あらゆる面で今日より自立していた。多くの地方の中核都市には百貨店があり路面電車が走り、どんな都市にもバス網が細かく張り巡らされていた。しかし、その後それらは消えていった。多くの中小都市は経済面でも生活面でも就労面でも、地方の中枢拠点都市に依存するようになり、その町のなかだけでは生活が完結しなくなってしまった。さらに、自動車中心の空間構造に変わり、自家用車を複数台持つのが当たり前になり、経済的理由や身体的理由で自家用車を持てない世帯は移動の自由が奪われている。移動の自由が無ければ職を探すことも、学習することも、気晴らしをすることも、そして食べ物を手に入れることすらままならなくなる（→ 253）。移動の自由は、都市における市民的自由の基礎である。そのためには、都市は過半の市民の交通需要を賄えるような野心的な公共交通体系を目指さなければならない。それが都市の低炭素化に直結することは明らかである。

充実した公共交通があれば、多くの市民は、どこかに行くとき真っ先に公共交通の利用を思い浮かべるだろう。そのような公共交通を実現するために、車体の形式、乗り場の形、切符の買い方、路線と運行情報の提供の仕方などすべての施策を包括的に組み込んだ総合的な都市交通戦略が必要である。公共交通の方式では、路線設定や整備費用などを総合的に考えると、バスがもっとも現実的かつ優れた手段である。路面電車(最近はLRTと言い換えている)とバスを比較して、バスは運行の定時性に欠け、乗り心地も悪いなどと言われるが、これは不公平

251 交通計画には総合性が求められる
Synthesis is required in transportation plans

使い勝手のよい交通は、バス車体だけでなく、路線構成、路線案内地図、待合所など総合的に計画されることが必要である。

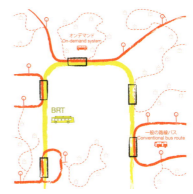

252 BRTとオンデマンドの組み合わせ
BRT combined with on-demand bus system

Transportation means that are easy to use must be planned comprehensively—in other words, they must include not only the bus vehicle itself but the bus network, route map and bus stops.

Combining / Orange Web

"Combining" is an operation that produces new value by merging multiple flows of different character.

The main function of an Orange Web is to provide convenient bus transportation service for residents of small and medium-size cities without a local rail network (→ 251). The idea is to create a public transportation network that combines a bus rapid transit (BRT) [118] system for the central area, which is comparable to streetcars with respect to efficiency and comfort, with an on-demand bus system[119] for suburban areas, which can respond to specific demands of users. Such a network is indispensable if seniors are to live in peace and comfort in small and medium-size cities (→ 252).

Local cities in Japan were more autonomous half a century ago than they are now. Many core cities in localities had department stores and streetcars, and bus routes formed a fine net covering every city. However, these have all since disappeared. Many small and medium-size cities have come to depend on central cities in their localities for economic support and places to live and work; they are no longer autonomous.

Moreover, their spatial organization has come to be centered on automobiles, and it is commonplace for people to have more than one car. Households that are without automobiles for economic or physical reasons have been robbed of freedom of movement. Without freedom of movement, looking for jobs, studying, relaxing, and even getting food become problematic (→ 253). Freedom of movement is a fundamental civic freedom. We must establish an ambitious public transportation system that meets the mobility demands of the majority of citizens. This will clearly lead to a lowering of carbon emission in cities.

If a fully developed public transportation system exists, the first thing most people will think of when they need to go somewhere is using public transportation. To realize

な比較に起因する。路面電車には専用軌道があり、プラットホームがあり、バスに比べて環境が充実している。バスも路面電車並みに、道路の真ん中にバス専用レーンを設け、付属設備を充実させれば乗り心地がよくなり、初期費用が安い分有利である。これがBRTである。BRTは海外に多くの事例があるが、日本では名古屋や東北の津波被災地にしかない。日本の多くの都市で実施されているバスレーンは歩道寄りに設置されているので沿道の商店の前に停まる車や自転車レーンと競合して効果を発揮しにくい。ただし、BRTは乗客が多い都心には向いているが、周辺部には向いていない。乗降客の最終目的地が拡散し乗降客数が少なくなるのでデマンドバスのほうが向いている。このような運行は、路線が決まっている路面電車にはできないことである。住宅地の奥まで入るために、バスの車体は小さい方が望ましい。

路面電車やバスや自転車など「小さい交通」の優遇は、脱自家用車依存の実現に向けて競い合っている都市の基本政策である。ところが、昨今、日本では路面電車だけが妙にもてはやされているのは奇妙である。新規のLRT建設は費用が嵩むだけ

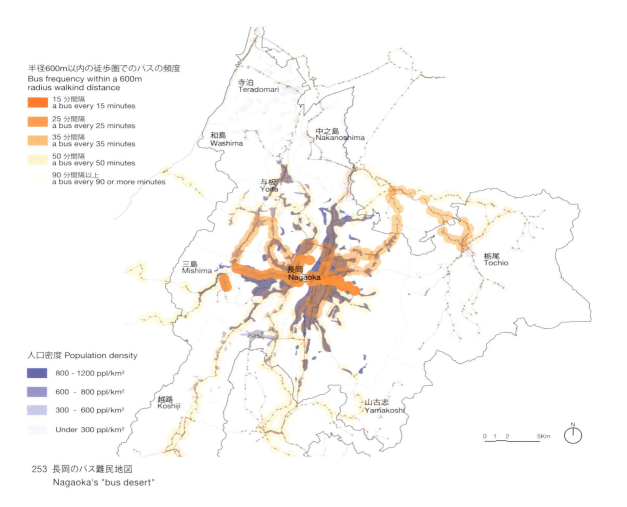

253 長岡のバス難民地図
Nagaoka's "bus desert"

such a system, a comprehensive urban transportation strategy that considers all aspects including the form of vehicle bodies, the arrangement of stops, the way tickets are purchased, routes and how information on service is provided needs to be devised. With respect to the means of public transportation, when factors such as the establishment of routes and the cost of development are considered comprehensively, buses are excellent and the most realistic means. Unlike streetcars (or light rail transit (LRT) as they are called now), buses do not operate punctually and are said to be less comfortable, but the comparison is unfair. Streetcars have a better equipped environment than buses; for example, they have their own tracks and platforms. If buses had a dedicated lane in the middle of the street and improved auxiliary facilities, they would be more comfortable and advantageous, given their low initial cost. That is the BRT system. There are many examples of BRT overseas, but in Japan systems exist only in Nagoya and in Tōhoku areas that were hit by the tsunami. In Japan, dedicated bus lanes created in many cities are located next to the sidewalk; full advantage cannot be taken of them because they must compete with vehicles that stop in front of stores and with bicycle lanes. BRT is suited to the center of a city where there are many passengers but not to peripheral areas. There, the final destinations of passengers are dispersed and fewer passengers get on or off; on-demand buses are more suited to those conditions. This sort of service cannot be

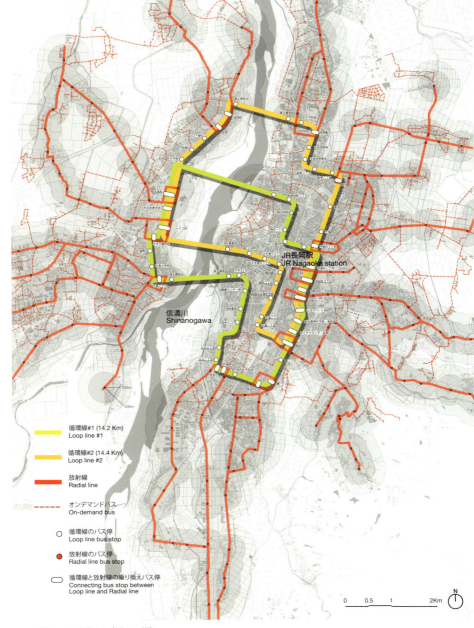

255 長岡市に適用した〈暖かい網〉
Orange Web as adapted to Nagaoka City

performed with streetcars with fixed tracks. Smaller buses that can go into every corner of a residential area are preferable.

Preferential treatment for Small Transportation such as streetcars, buses and bicycles is a basic policy for cities attempting to liberate themselves from dependence on automobiles. However, streetcars strangely get all the attention in Japan. New construction of an LRT system is not only more costly but lacks flexibility of routing. The shrinking of urban areas will often fail to proceed as expected. That being the case, buses which can serve any place that has streets has an overwhelming advantage over LRTs which are difficult to alter once a system is in place. The biggest problem in Japan in introducing new transportation systems such as these is that the agencies concerned with street traffic including buses still take the same old approach and prioritize automobile traffic and the use of automobiles by the public.

Superbuses

Superbuses, the next-generation system of public transportation we are advocating, have the following characteristics. This proposal takes the city of Nagaoka as an example but is applicable to many other cities (→ 254).

a. Combination of BRT system with on-demand bus system

A public transportation system ought to be like the circulatory system of an organism; it must have a thick

でなく路線に柔軟性がない。市街地の縮小は予想通り進まないことが多いから、一度敷設したら変更がしにくい鉄道より、道路さえあればどこでもサービスができるバスが圧倒的に優位である。こうした新しい交通システムを進めるうえで、日本の最大の問題は、バスを含めて道路交通に関する関係当局の姿勢は旧態依然とし、自動車交通そして自家用車優先から脱却できないでいることである。

スパー・バス

私たちが提案する次世代の公共交通システムであるスパー・バスは、以下の特徴を備えている。この提案は長岡市を事例としているが、多くの都市に適合するはずである（→ 254）。

a. BRT方式とデマンド方式のバスの組み合わせ

生物の循環組織がそうであるように、隅々までエネルギーを

254 〈暖かい網〉の幹線を担うBRT
　　BRT as the trunk line of Orange Web

BRTはBus Rapid Transit（バス高速都市交通）の頭文字を取った名称である。市内電車の軌道をバス専用レーンに置き換えたものと思えば良い。乗り心地、定時運行性、安全性など路面電車の良さに加えてバスの低コストと柔軟性が得られる。〈暖かい網〉では、日本の狭い道路の実情に適合できるように幅の狭い車体を提案している。

BRT is an abbreviation for Bus Rapid Transit. It can be thought of in terms of streetcar tracks replaced by dedicated bus lanes. In addition to the benefits of streetcars, such as comfort, convenience and safety, the BRT also combines the low cost and flexibility of a bus system. Narrow-bodied buses are proposed in Orange Web to fit to the narrow roads of Japan.

trunk and a fine network at the ends to deliver passengers to every corner of its service area. A network that is as rational as the circulatory system can be created by combining a BRT system in the center of a city and an on-demand bus system on the periphery.

b. Service route that combines rings and radials

Present bus routes in Nagaoka radiate from JR Nagaoka Station. This is often the case in local cities in Japan, but these bus routes are based on an activity pattern that existed half a century ago and do not reflect contemporary reality. An arrangement consisting of ring routes in the center of the city and multiple radial routes toward the suburbs better reflects the present-day pattern of activity (→ 255).

c. From buses that negotiate routes to buses that negotiate a network

A fare system for the use of a network, premised on transfers, similar to that generally used for subways, must be adopted. To increase buses' share of transportation, the fare needs to be based, not on the route taken, but on the total distance traveled (the method of calculation used by railways) or time. In addition, to enable passengers to transfer from a ring route to a radial route without having to walk from one stop to another, the routes will be organized so that ring route buses and radial route buses use the same bus stop.

d. Development of new bus vehicles (→ 256)

Narrow bodied buses: Circumstances on streets in Japan make it difficult to create dedicated lanes for buses. If such dedicated lanes could be made narrower,[120] more streets could have bus lanes installed. Theoretically, creating narrower bus vehicles would make it possible to create narrower bus lanes. The superbus is assumed to have a width of approximately 1.6 meters.

Doors on both sides: If vehicles were equipped with doors on both sides as railway cars are, then the platform for a bus stop could be made an island in the middle of the street or set on the side of the street as conditions dictate. Wide doors are desirable to facilitate getting on and off for wheelchair users and to shorten the time generally for

送り届けるためには、幹は太く、先端部のネットワークはきめ細かく張り巡らされていることが必要である。都心のBRT方式と縁辺部のデマンド方式を組み合わせることで、生物の循環組織のように合理的なネットワークを作ることができる。

b. 環状線と放射線を組み合わせた運行経路

長岡の、現在のバス路線はJR長岡駅を中心に放射状に組み立てられている。日本の地方都市ではよくあるが、これは半世紀前の人々の活動パターンであり現実を反映していない。現代の活動形態に合わせるために、中心部に環状路線を配し（→ 255）、郊外に向けて複数の放射路線を出すパターンが相応しい。

c. 路線に乗るバスからネットワークに乗るバス

地下鉄のように、乗り換えを前提に「ネットワークを使う」運賃システムにしなければならない。本腰を入れてバスの交通分担率を高めるためには運賃体系を路線制ではなく乗車した距離の合計（鉄道運賃の計算方法）か時間制を採用するべきである。また、乗客が環状路線と放射路線とのあいだで乗り換える場合、乗客がバス停間を移動しなくても済むように、放射路線のバスも環状路線のバスも同じバス停を使う路線構成にする。

d. 新しいバス車両の開発（→ 256）

狭幅ボディーバス：日本の道路事情からすればバス専用レーンの設置は容易ではない。もしバス専用レーンを狭くすることができれば[120]、バス専用レーンを設置できる道路が長くなる。バス専用レーンを狭くするためには、理論的にはバスの車体の幅を狭くすれば可能である。スパー・バスの車幅として約1.6 mを想定する。

両側ドア：鉄道と同じように車体の両側にドアを設ければ、バス停のプラットフォームの形式として、自由に島式と対向式を選べる。また車椅子利用者の乗降や乗降の時間短縮のために幅の広いドアが望ましい。

フラット低床：車内の安全な移動と身障者にとって使いやすい

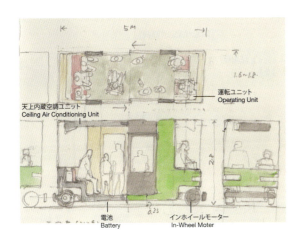

256 〈暖かい網〉を走るバス車輌
Bus vehicle on Orange Web

〈暖かい網〉を走るバスは、低床、段差のない床、両側幅広乗降口、小さい車輌の連結運転という条件を満たしたい。図はホイール・イン・モータ方式の電気バスを想定している。

Buses that run on Orange Web should be equipped with low, flat floors, doors on both sides, and small vehicles that are linked together. This figure is based on an electric bus with an in-wheel motor system.

getting on and off.

Low, flat floors: Absolutely flat floors are essential on buses to assure safety for passengers moving inside vehicles and to facilitate the use of buses by handicapped persons. Floors need to be made as low as possible because providing platforms for all bus stops is impossible.

Electric buses: Electric motor-driven vehicles meet the above conditions. Using electric motors contributes to a lowering of carbon emissions. In addition, if an in-wheel motor system (an arrangement in which a motor is directly connected to each wheel) can be used, low, completely flat floors are possible. The fact that the buses, operating on fixed routes, can be easily re-charged, is also an advantage. If wireless power transfer devices were installed under the pavement at bus stops, buses could be quickly re-charged surely and incrementally while at bus stops.

Utilization of intelligent transport systems (ITS) : ITS broadens the scope for buses immeasurably. First, accurate information on how buses are running can be provided to waiting people, dedicated lanes for buses can be made narrower because buses are precisely guided, and as a result, such bus lanes can be established over a wider area. Multiple vehicles can be arranged in a row and electronically controlled so that a fixed interval is maintained between any two vehicles; a driver to operate them is therefore needed only in the lead vehicle. Increasing or decreasing the number of linked vehicles according to demand facilitates the adjustment of passenger capacity and makes efficient management possible.

e. Efficient and low-priced fare collection system: Factors such as the effect on running time, the responsibility placed on the driver, and the cost of collecting fares ought to be considered comprehensively in determining the method of fare collection. Basically, collecting fares outside the vehicle eliminates one cause for delays and reduces the burden placed on the driver. An electronic card system might be used to reduce physical checks on passengers getting on and off. Another alternative is to take the plunge and provide the service free of charge if the objective is

バスのためには床が完全に平らであることが必須である。すべてのバス停にプラットホームを備えることは無理なので、低床を追求する必要がある。

電気バス: 以上の条件を満たすためには動力源は電気モーター駆動が適している。電動化すれば、低炭素化に貢献する。また、イン・ホイール・モーター・システム（各車軸に直結したモーターを配置する方式）が採用できれば、低床かつ完全に平らな床が実現できる。バスは決まった路線を運行するので充電がしやすいことも利点である。停留所の路面に非接触給電装置を設置すれば、停車中に少しずつ確実に急速充電できる。

ITS(Intelligent Transport Systems): ITSの活用は、バスの可能性を限りなく広げる。まず、バスを待つ乗客に的確な運行情報を提供することができる、正確な誘導をすることでバス専用車線の幅員を狭めることができ、その結果としてバス専用車線の設置可能範囲を延ばすことができる。複数車両が隊列を組み電子制御で車両間隔を一定に保つ運転をすることができるので、運転手は先頭車両にだけ搭乗して運行することができる。需要に応じて連結数を増減することで、乗客定員を適正化でき効率的な経営が可能になる。

e. 効率的で安価な料金徴収システム：料金徴収の方式は、運行

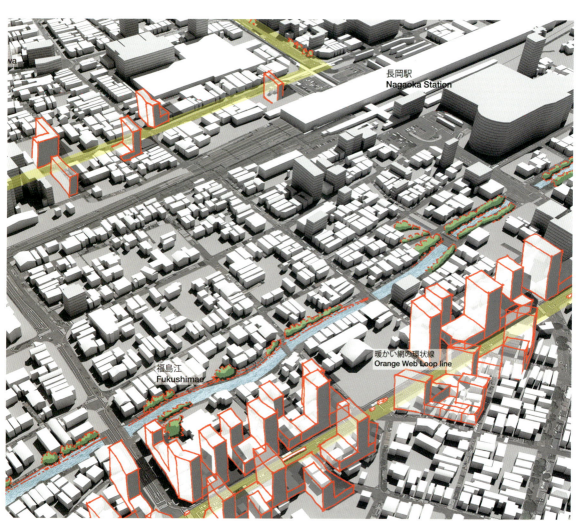

257 〈暖かい網〉と環状線沿道の高度土地利用
　　Intensive land use along Orange Web Loop line
公共交通の整備と都市の土地利用を関連づける都市政策をTODと呼ぶ。

Urban policies that link public transportation with urban land uses are called TOD (transit-oriented development).

時間への影響や、運転手への負担、料金徴収コストなど総合的に勘案すべきである。基本的には、車外で料金徴収をすれば遅延の原因と運転手の負担が減る。電子カードシステムを活用して物理的な乗降チェックを減らすか、脱自動車社会を目指して公共交通を社会インフラと位置づけ、思い切って無料とする政策選択もある。

f. 新しいバス停

バス停には屋根と風囲いを設ける。寒冷地ではバス停のベンチに暖房を組み込む。バス停をプラットホーム式とすれば、バスの床と水平になり、車椅子、自転車を容易に持ち込むことができる。バス停の待ち合い機能を、沿道の商業施設、たとえばコンビニエンスストアなどに組み込めば乗客に便利である。

g. 美しい路線地図

多くの人がスマートフォンの利用者である現代では、利用者のために、バスの旅程探索などの情報提供はウェブを通じて行なう。バス路線は美しく実用的にデザインされることが重要である。ロンドンの地下鉄路線図からわかるように、美しい路線図は、都市のイメージを向上させるだけではなく、それ自身が都市を象徴して人々の記憶に刻まれる

h. 都市の形態と交通を関係づけるTOD

利用者から見て便利なバス交通には高い路線密度と高い運行頻度が必要であり、経営者からみると高い利用率が必要である。そのためには、バス路線沿線にたくさんの人が住んでいると好都合である。公共交通の整備と都市の土地利用を連動させる都市計画が必要になる。このような考え方をTOD（Transit Oriented Development、公共交通指向型開発）と呼んでいる。これは都市のコンパクト化を側面支援する（→ 257）。

---

reduction of society's dependence on automobiles and public transportation is regarded as a part of society's infrastructure.

f. New bus stops

Bus stops will be equipped with roofs and windbreaks. Heating will be built into the benches of bus stops in colder regions. If bus stops are equipped with platforms, they will be level with bus floors, and wheelchairs and bicycles could be easily introduced into buses. Combining bus stops with commercial facilities such as convenience stores would be useful for passengers.

g. Beautifully designed route maps

Today, when smartphones are in wide use, information such as bus itineraries could be provided on the internet. It is important that bus routes be both beautifully designed and practical. As shown by the subway map of London, a beautiful route map not only improves the image of a city but makes an indelible impression on people as a symbol.

h. TOD relates urban form to transportation

From the point of view of users, convenient bus transportation requires high density of routes and high frequency of service; from the point of view of management, high rate of use is necessary. Having large numbers of people living along bus routes is convenient in that regard. City planning in which the development of public transportation is linked to urban land use becomes necessary. This approach is referred to as transit oriented development (TOD). This indirectly helps to make a city compact (→ 257).

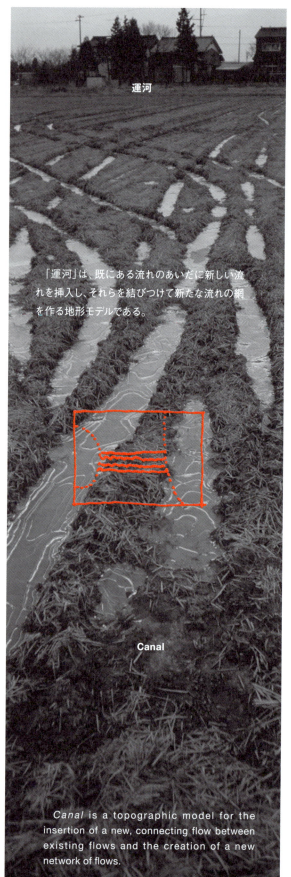

運河

「運河」は、既にある流れのあいだに新しい流れを挿入し、それらを結びつけて新たな流れの網を作る地形モデルである。

Canal

Canal is a topographic model for the insertion of a new, connecting flow between existing flows and the creation of a new network of flows.

結びつける／青い首飾り-品川

「結びつける」は独立した二つの流れを繋ぎあわせてセミラチス的な[121]ネットワークを作りだす操作である。

〈青い首飾り〉は東京湾の旅客舟運を充実させて、水際に近い主要な鉄道駅と東京湾を結びつけることで、都心の交通ネットワークの冗長性を高め、同時に水際都市としての東京の魅力を強化することを目指す。21世紀のメガシティに求められるのは、機能的であることを超えた感覚に訴える魅惑であり、その点で、東京湾とそれに注ぐ河川は大きな可能性を秘めている（→258）。

日本の主力産業であった重化学工業の工場が、70年代になって首都圏から地方都市や海外に転出し始め、広い土地が放出されるようになった。それが住宅地や大規模公園、大型ショッピングモールに引き継がれた。

1983年にはディズニーランドも開業し、赤錆色だった東京湾岸地帯が徐々に色彩豊かな商業空間に変わった。これにあわせて、一度は途絶えた水都としての伝統[122]の復活の機運が高

258 〈青い首飾り〉
Blue Necklace

Connecting / Blue Necklace—Shinagawa

"Connecting" is an operation that links two independent flows and creates a semi-lattice network.[121]

The aim of a Blue Necklace is to enhance the redundancy of the urban transportation network and reinforce the appeal of Tokyo as a city on the water by improving passenger boat transportation on Tokyo Bay and connecting major railway stations near the shore with Tokyo Bay. A mega-city of the twenty-first century must transcend the functional and appeal to the senses, and Tokyo Bay and the rivers flowing into it have great potential in that respect (→ 258).

Factories for the heavy chemical industry, a major industry in Japan, began to move from the Capital Region to local cities or overseas in the 1970s, freeing up large tracts of land. These were given over to residential use, large parks and large shopping malls.

Tokyo Disneyland opened in 1983, and the drab coastal region along Tokyo Bay was gradually transformed into a more colorful commercial space. An opportunity arose to restore Tokyo's character as a city of water.[122] In Tokyo,

まってきた。東京都では隅田川から東京湾周辺への観光船の航路を開設し、新造船も投入されて充実してきたが、現状は他の交通網と繋がっていないので孤立した観光ルートに過ぎない。

江戸時代に陸路や海路で西から江戸に入るなら品川からであった。だから、そこには関所がおかれていた。1872年には日本で最初の鉄道が横浜―品川間で開業し、やがて東京と大阪を結ぶ東海道線にまで延伸された。その後、1964年には並行して弾丸列車専用新線である東海道新幹線が開業し、2003年には東海道新幹線品川駅が新設され、東京駅とともにターミナル機能を分担している。さらに2027年には、中央新幹線が開業する予定であり、品川は首都圏側の始発駅になる。中央新幹線は、超電導磁気浮上式リニアモーターカーによって運行され、現在の新幹線は時速285キロで走り、100分かかる東京名古屋間を時速342キロに上げわずか40分で結ぶ超弾丸列車であり、東海道線のバックアップ機能も期待される。

提案は、品川駅の南420 mにある高浜運河から品川駅までの間の道路を開削し、品川駅港南口前まで水面を引き込み、駅を降りたらすぐに東京湾の船便に乗船ができるようにするものである（→ 259）。海外から来日する客を羽田空港でシャトル船で迎え、品川で新幹線に乗り換えて京都に行くという旅程を組むことができる（→ 260、261）。東京ディズニーランドや埋め

259 〈青い首飾り〉の品川
Shinagawa's Blue Necklace

品川駅には、山手線、首都圏の鉄道の主要幹線と、東海道新幹線、2027年開業予定の東海道リニア新幹線が集まる。

The Yamanote line and major railway lines in the metropolis, the Tōkaidō Shinkansen line and the Chūō Shinkansen line (schedued to open in 2027) all run through Shinagawa Station.

sightseeing boats began to ply the Sumidagawa River and Tokyo Bay, and service has since been improved by the introduction of new vessels. However, at present these are nothing more than isolated sightseeing routes since they are not connected to other transportation networks.

In the Edo period, Shinagawa was the point of entry into Edo for both land and water routes from western Japan, and a checkpoint was established there. In 1872, Japan's first railway opened between Yokohama and Shinagawa, and the line was eventually extended to form the Tokaidō line linking Tokyo and Osaka. In 1964, the Tōkaidō Shinkansen, a new line dedicated to bullet trains, was opened parallel to the Tōkaidō line. In 2003 Shinagawa Station on the Tōkaidō Shinkansen was newly created, becoming the second Shinkansen terminal after Tokyo Station. The Chūō Shinkansen is scheduled to open in 2027, and Shinagawa will be the station of origin in the Capital Region for that line. The Chūō Shinkansen will be a maglev super bullet train that travels at 342 kph and takes 40 minutes to go from Tokyo and Nagoya, a trip that now takes 100 minutes. It is expected to function as back-up for the Tōkaidō line.

Our proposal is to excavate the street between Shinagawa Station and Takahama Canal, which is 420

立て地にある三つの国際展示場、東京ビッグサイト、幕張メッセ、パシフィコ横浜に水路で行くこともできる。東京湾に近い主要駅で羽田と成田の二つの国際空港に鉄道便があるのは上野と東京と品川の三駅しかないが、上野駅も東京駅も海岸から遠すぎる。品川駅だけに可能性がある。

埋め立て地に建つ超高層住宅地にも船で直接行けると住宅地の魅力が増す。これらの集合住宅地は、いわゆるヤッピー好みなのだが、集合住宅地の設計に水際としての特性がまったく反映されてないので、郊外の新規開発地と変わらない風景になってしまっている。水上タクシー用のピアを持つ分譲住宅は幅広い居住者を満足させるだろう。また、水上交通の充実が防災性に貢献することも言うまでもない。帰宅困難者の移送、被災地の救援、物資の搬入、倒壊家屋の瓦礫の処理など、難問がかなり解決する。

260 ベニスのサンタルチア駅
Santa Lucia railway station in Venice

261 品川駅港南口に繋がる運河
Canal connected to Shinagawa Station Kōnan entrance
品川駅港南口の前に船が着くことを想像すると誰でもワクワクする
The idea of boats docking in front of Shinagawa Station's Kōnan entrance is exciting to any and everyone.

meters south of Shinagawa Station, and extend the canal to the Kōnan entrance to Shinagawa Station so that passengers getting off at the station can immediately board a boat plying Tokyo Bay (→ 259). The following itinerary is one possibility: A guest from overseas arriving in Japan could be met by a shuttle boat at Haneda Airport, transfer at Shinagawa to the Shinkansen and go to Kyoto (→ 260, 261). One could travel by water to Tokyo Disneyland or the three international exhibition centers on reclaimed land, Tokyo Big Sight, Makuhari Messe and Pacifico Yokohama. Of the main railway stations near Tokyo Bay, only three, Ueno, Tokyo and Shinagawa, offer service to the international airports at Haneda and Narita, and both Ueno and Tokyo are too distant from the coast. Shinagawa Station alone offers this possibility.

Direct access by boat would increase the appeal of high-rise residential areas on reclaimed land. These apartment buildings would appeal to young urban professionals but full advantage of being on the water's edge has not been taken; the landscape is not any different from that of new residential developments out in the suburbs. Condominiums with piers for water taxis would surely satisfy a wide range of residents. It also goes without saying that improved water transportation would contribute to an improvement of the area's disaster-prevention capability. Transportation of people who have difficulty getting home, relief of disaster-stricken areas, delivery of material, disposal of rubble from collapsed houses—the project could solve a number of difficult problems.

結びつける／青い首飾り-秋葉原

〈青い首飾り〉の鉄道駅との結節点として、秋葉原はもう一つの有力な候補地である（→ 262）。

江戸時代には、秋葉原は東北から江戸に入る炭、米、籾などの集積地であった。明治以降は鉄道交通の結節点となり、同時に神田川から引き込んだ船溜まりが作られ、舟運と鉄道便の結節点としての性格が加わり、秋葉原は市場の街として発展した。さらに、1935年には神田青果市場が設置された。第二次世界大戦後はラジオ部品の店舗が集まり、戦災復興も終わると、家電製品安売り店、その後は電子ゲーム店と続き、現在は、ホビーショップやアニメショップの集積地で、日本のサブカルチャーの聖地として外国人観光客にも人気がある。秋葉原は常に市場の街であった。

秋葉原の景観の最大の特徴は立体的に交差する鉄道であろう。東京の都心を環状に囲む山手線と首都圏を南北に縦断する京浜東北線と東西に横断する総武ー中央線、そして、地下鉄日比谷線とつくばエクスプレス線[123]の地下駅の計5線。そこに緑の網と〈青い首飾り〉が加わる。提案は、これらの複数の「流れ」を機能的にも視覚的にも繋ぐことである。この計画は、戦後になって埋められた船溜まりの記憶を甦らせることもできるだろう（→ 263）。

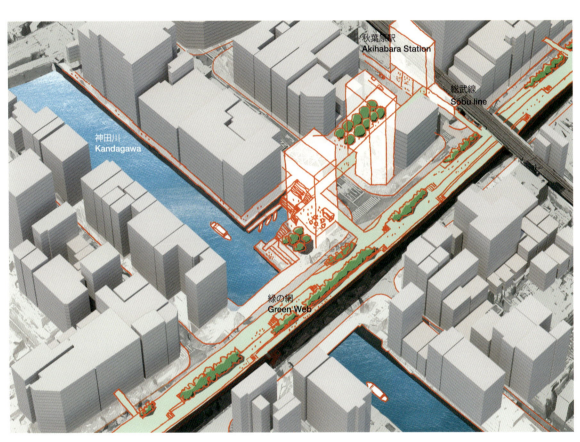

262 〈青い首飾り〉の秋葉原
Akihabara's Blue Necklace

Connecting / Blue Necklace—Akihabara

Akihabara is another possible site for a node linking a Blue Necklace and a railway station (→ 262). In the Edo period, Akihabara was an area where products entering Edo from the Tōhoku region such as coal and rice, both polished and unhulled, were collected. In the Meiji period, it became a railway node as well as a place where boats from Kanda River could tie up. The area's character as a node linking water transportation and rail service helped Akihabara develop as a market. In addition, the Kanda vegetable and fruit market was set up in 1935. After World War II, stores dealing in radio components gathered, and when postwar reconstruction was completed, discount stores selling household appliances appeared; these were followed by electronic game stores. Today, hobby stores and anime shops are concentrated in the area, which has become a center of Japanese subculture and a popular

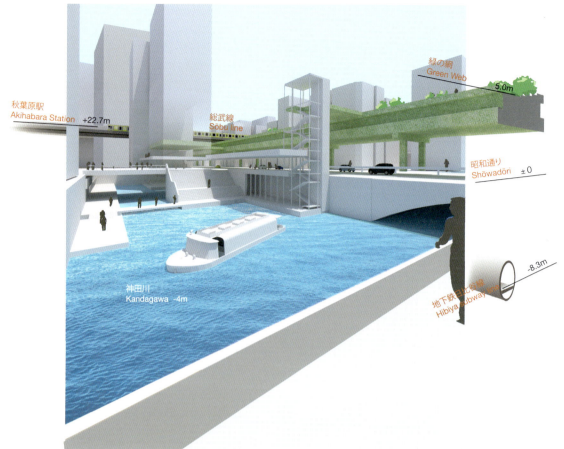

263 ビッグサイトと船で繋がる秋葉原
Akihabara connected to Big Sight by boat

交通網が積層して発展した秋葉原にもう一つの移動の層を加える。秋葉原もビッグサイトもオタクの聖地である。

Another transportation layer is added to Akihabara, where a transportation network has developed in multiple layers. Both Akihabara and Big Sight are holy grounds for the otaku.

attraction for foreign tourists as well. Akihabara has always been a market town.

The salient feature of Akihabara's townscape is the multilevel intersection of railways. There are five lines in all, the Yamanote line that loops around the center of Tokyo, the Keihin Tōhoku line cutting across the Capital Region north-south, the Sōbu-Chūō line cutting across the region east-west, the Hibiya subway line, and the underground station for the Tsukuba Express line.[123] To these would be added a Green Web and a Blue Necklace. The proposal calls for connecting these multiple flows functionally and visually. This project would also revive memories of the haven for boats that was filled in after the war (→ 263).

乱流

線形で機械的な「流れ」に支配されている都市に、螺旋的な「流れ」を招じ入れる地形モデルである。さまざまな外乱によって定常流が乱されると、そこに渦や瀬や澱みなど非定常的な「流れ」、乱流が発生し、それによって「場所」が活性化される。

現代都市は、無限に増加し成長する直線的な時間だけに関心を払い、自然界にある、生物の誕生と死、あるいは季節など、宇宙と生が本質的に備える螺旋的に進む時間の観念に無関心であった。

Turbulence

Turbulence is a topographic model for the introduction of a spiral flow into a city dominated by linear, mechanical flows. When a steady flow is disturbed by diverse, external forces, irregularities, that is, turbulences, such as whirlpools, rapids and stagnant pools are generated, activating place.

The contemporary city has been concerned only with linear time that endlessly increases and grows. It has ignored time unfolding in a spiral—found in the natural world in the birth and death of organisms and the cycle of the seasons—which is essential to the cosmos and to life.

縁飾りを付ける／緑の花輪

「縁飾りを付ける」は、曖昧な領域の縁を異化することで境界性を強調する操作である。強調された境界を挟んで、新たな流れを生み出すことを目指す。

日本の高齢化は、今後、団塊の世代によって加速され、やがて葬祭と墓地の需要を押し上げる。ところが近代都市は墓地を忌避してきたため、魂の安息場が不足している。〈緑の花輪〉は、東京都心部の西半分を載せる山の手台地の北縁である南北崖線の法面を公園墓地化して、東京に不足する公共墓地を増やすとともに、崖線下を走る鉄道の車窓から眺める「流れる庭園」を作る戦略である。この崖線は、現在は東京都内で一番長く、全長がコンクリート擁壁で覆われていて醜いので、この切り立った法面を一度撤去して法面の斜度を緩くし墓地として使えるようにすると同時に、斜面崩壊の危険から居住者と鉄道を守る（→264）。

現代社会は日常生活から死を遠ざけている。日本では祖父母と同居する子供は少ないうえに、約8割の人は病院で死ぬの

Edging / Green Garland

"Edging" is an operation that emphasizes an otherwise ambiguous border by differentiating it. The objective is to generate new flow along the emphasized boundary.

The aging of Japan will soon be accelerated by baby boomers, and eventually demand will rise for funerals and cemeteries. However, there is a shortage of final resting places because modern cities have avoided them. Green Garland is a strategy for turning the bluffs running north-south on the northern edge of the Yamanote Plateau on which the western half of the center of Tokyo sits into memorial parks, thereby increasing public cemeteries of which there is a shortage in Tokyo and creating a "flowing garden" observable from trains running along the foot of those bluffs. At present those bluffs are covered by the longest concrete embankment in Tokyo which is quite ugly. The idea is to remove that embankment, make the slopes gentler and usable as cemeteries, and at the same time protect residents and railways from the danger of landslides (→264).

In the course of everyday life contemporary society keeps death at arm's length. Many children in Japan grow

で、人の死を身近に体験しないまま成長する子供が多い。また、大都市では墓地が遠のく一方であり、墓参は一日仕事になる。大都市の住人には出身地の地方都市に墓がある人も多く、場所によっては墓参のために泊まりがけの旅行をしなければならない。

なぜ死を遠ざけるのか？　その理由の一つは、あきらかに宗教の影響力の低下で、死に行く人も死体も衛生的に忌まわしい生物学的現象になったからである。東京には都営墓地は8カ所で約417 haしかない。青山、雑司ヶ谷、谷中、染井は明治期、多摩は大正期に整備され、戦後の都営墓地は1948年の小平霊園と1971年の八王子霊園の二カ所に過ぎず、当然都民の要望に応えていない。

南北崖線は、荒川の河岸段丘で最大15 m程度の落差がある。その麓に沿って、山手線と東北線、東北新幹線などが走り、田端には広大な鉄道ヤードもあり、あたかもレールが織りなす様は琳派が大河を描けばかくもありなんという風景である。台地の上は山の手台地で学校や寺院があり、緑も多いが、住宅が立て込み擁壁の際まで押し寄せている。崖の下は下町低地と

264 北区上中里付近の〈緑の花輪〉
Green Garland near Kaminakazato, Kita Ward

266 鉄道のための公園〈緑の花輪〉
Park for a railroad in Green Garland

鮨詰めの通勤電車の乗客の目を楽しませることは、鉄道都市東京に求められる公共緑地の重要な役割ではないだろうか。

Contributing to the visual pleasure of commuting passengers packed like sardines in the train is another important role demanded of public green spaces in the railroad-based city of Tokyo.

up without any familiarity with death because few children live together with their grandparents and approximately 80 percent of people die in hospitals. In addition, cemeteries in metropolises are located in ever more distant places, and visiting a cemetery can take up an entire day. Many residents in metropolises have family graves in the local cities from which they originally came, and depending on the place, visiting a grave may mean an overnight stay in a local city. Why keep death at arm's length? One obvious reason is that the decline in the influence of religion has made dying and dead persons biological phenomena that are abhorrent from the perspective of sanitation. In Tokyo there are only eight cemeteries administered by the metropolitan government, totaling about 417 hectares. The cemeteries in Aoyama, Zōshigaya, Yanaka and Somei were created in the Meiji period, Tama Cemetery was developed in the Taishō period, and since the war, only two cemeteries have been created, Kodaira in 1948 and Hachiōji in 1971. The demand of the people of Tokyo is obviously not being met.

The north-south bluffs are tallest—about 15 meters in height—at the Arakawa river terrace. The Yamanote, Tōhoku and Tōhoku Shinkansen lines all run along the foot of the bluffs. At Tabata there is even a large rail yard, where the pattern of interweaving rails suggests a Rinpa-school painting of a river. On the Yamanote Plateau above the bluffs are schools and temples. There is a great deal of greenery, but houses have been built right up to the embankment and are pressing up against this verdure. In the area below the bluffs, referred to as "a low-lying working-class neighborhood," factories and stores stand out.

The people buried here will probably be mainly members of the baby boomer generation that made the period of intensive economic growth possible. Many of them came to Tokyo from cities and farming villages from northern Japan, and this enormous embankment was no doubt the first thing these youths, no doubt full of hope and anxiety, saw

呼ばれ工場や商店などが目立つ。
　埋葬されるのは日本の高度成長を支えた団塊の世代の人々が中心になるだろう。その中には東京の北の都市や農村から上京して来た人も多く、希望と不安をいだいて最初に目にした東京の風景は、この長大な擁壁の連なりだったはずである。彼らはこの場所に眠り、今度は、北に向かう列車の彼方に郷里と東京を隔てる遠い山並みを見ることであろう。まことに安息の地に相応しい場所ではないだろう（→ 265）。

この場所を毎日180万人の乗客[124]が通過し眺めるのだから、東京の巨大な公共空間野一つであると言ってよい。南北崖線は、ケビン・リンチに従えば「エッジ」としてイメージされる地形である。そのエッジが殺風景なコンクリートの擁壁では、それによって定義される近辺の地域イメージも一緒に悪くなる。〈緑の花輪〉は車窓から約10分間見える東京を再定義する緑地であり、都市の新しい景観としての流れる緑地の提案でもある（→ 266）。

265　高度成長を支えた戦士達の魂のための安息地
Resting grounds for the brave souls who supported the era of high economic growth

of Tokyo's townscape. They will now sleep here, this time looking north toward the homes they left behind and the distant mountains that stand between those homes and Tokyo. For them this is truly an appropriate place of repose (→ 265).

　The Yamanote and Keihin Tōhoku lines carry 1.8 million persons daily.[124] The landscape beyond their train windows, shared as it is by that many persons, can be said to constitute one of the biggest public spaces in Tokyo. The north-south bluffs are a topographical feature that serves as an "edge" in people's image of the city as defined by Kevin Lynch. If this edge remains a bleak concrete embankment, it will also spoil the image of the nearby areas defined by it. Green Garland is a proposal for a green area that will redefine Tokyo, which is visible from train windows for approximately ten minutes—a flowing green area that will become a new scenic feature of the city (→ 266).

Enclosing / Cloister of Life

　"Enclosing" is an operation that constructs a wall in the city and thus creates a clear boundary. Cloister of Life is a special part of Green Web and will be created through the change in use of the section of the Tokyo Expressway that bounds the Ginza district to the west. The upper portion of the expressway will be made into a Biome Greenhouse and the lower portion into a sports training facility. The Biome Greenhouse is a place for scientific experiments where climates and vegetation in different regions, from the Poles to the Equator will be recreated and simulations of the entire global ecology will be carried out.[125] The diverse sports existing today developed in different regions of the world and are closely tied to the climates in those regions. The training facilities for regional sports will recreate the suitable climate for each sport. Heat will be exchanged with the Biome Greenhouse on the upper level so that environmental adjustments are made efficiently (→ 267).

囲う／生命の回廊

「囲う」は、都市のなかに壁をつくり、明確な境界を作る操作である。

〈生命の回廊〉は〈緑の網〉の特別な部分であり、銀座地区の西側を限る東京高速道路の用途変更によって作る。高架道路の上部に「バイオーム温室」を新たに建設し、下部を地域的なスポーツのトレーニング施設に用途を変えるのである。「バイオーム温室」とは極地から赤道直下までの気候と植生を再現して地球の生態全体のシミュレーションを行う科学実験の場である[125]。一方、多様な現代スポーツの種目は地球のさまざまな地域で生まれたものであり、気候と分ち難く結びついている。地域的なスポーツのトレーニング施設にはそれぞれの種目に適した気候が再現され、上階のバイオーム温室とのあいだで熱のやり取りをして効率的に環境調整をする（→267）。

この二つの施設が合体した紐状の全く新しい施設は、銀座に高い象徴性を付加し、同時に銀座地区の西端を魅力的でなおかつ明確な境界を築き、国際的ブランドの旗艦店がならぶ銀座をグローバルシティに相応しい「島」に変える都市ブランディング戦略である（→268、269、270、271）。対象の高架道路は、首都高速道路の一部のように見えるが、実は当初から民間企業が建設、管理する特別な道路である[126]。全長約2kmで高架下は商業施設として利用され、土木構造物というより「非常に長い建築」である（→272）。

国際的なファッションブランドのブティックが並ぶ地区の存在はグローバルシティにとって欠かせない。現代東京を代表する

267 生命の回廊
Cloister of Life

This entirely new resource, linear in form, in which these two facilities will be combined, is an urban branding strategy to change the Ginza district, an area already filled with the flagship stores of international brands, into an "island" befitting a global city by giving it a highly symbolic character and creating an attractive and clear boundary at the western end of the district (→ 268, 269, 270, 271). The expressway in question may seem a part of the Metropolitan Expressway but in fact has been from the start a special road constructed and administered by private enterprise.[126] It is approximately two kilometers in total length, and the space underneath it is used by commercial facilities. It is more an extremely long building than a civil engineering structure (→ 272).

Every global city needs a district lined with the fashionable boutiques of international brands. At present, the two fashion-oriented streets in Tokyo are Ginza and Omotesandō. These are centers of commercial culture where tourists as well as local residents gather. That, however, is not enough. The representative main street of a city requires a mythical quality validating that city, for example, some anecdote related to the history of a king, the history of the seizure of political power or the creation of the city. In recent years, sports and greenery have also become compositional elements of the myth of a city. Greenery is a spice than enlivens any myth, and sports and greenery are the essence of life. Since the Meiji period, Ginza has been a center of culture, one closely associated with Western culture, and until the 1970s it was the most famous shopping area in Tokyo. Omotesandō, an avenue located between the Outer Gardens of Meiji Shrine and the Yoyogi district, the two principal venues of the 1964 Olympics, subsequently emerged and displaced the Ginza district in the top spot. Omotesandō has the advantage of proximity not only to the biggest green network in central Tokyo but to southwestern suburban residential areas where many affluent people live. However, the gentrification of areas along the coast of Tokyo Bay in recent years has contributed to a revival of the Ginza district. Cloister of Life is a scheme to take this one step further and restore the Ginza district to its former glory.

ファッションストリートは銀座と表参道である。そこは、住民だけでなく旅行客も集め、商業的な文化の発信地となる。ただ、それだけでは不十分であろう。都市を代表する目抜き通りはその都市を正統づける神話性が欠かせない。王の歴史や政権奪取の歴史、あるいは都市の創成に関わる逸話などである。都市の神話の構成要素に近年はスポーツと緑が加わってきた。緑は、どんな神話をも豊かにする香辛料であり、スポーツと緑は生命のエキスである。銀座は明治時代以来欧米文化と結びついた東京の文化発信の地であり、1970年代までは東京の繁華街を代表していた。その後、1964年の東京オリンピックの主会場であった神宮外苑と代々木に挟まれた位置にあった表参道が銀座の挑戦者として登場し銀座の栄光の地位を奪った。表参道は東京都心で最大の緑地ネットワークの中にあるだけではなく、何より豊かな人が住む西南の郊外住宅地が後背地として控えている強みがあったからである。ところが、近年湾岸地域のジェントリフィケーションを背景に再び銀座が表参道を追い上げてきている。生命の回廊は銀座の挽回策である。

268 新橋
Shinbashi

269 数寄屋橋
Sukiyabashi

270 銀座三丁目
Ginza 3 chōme

271 京橋
Kyōbashi

強化された銀座の領域性
Strengthening the boundary of the Ginza district

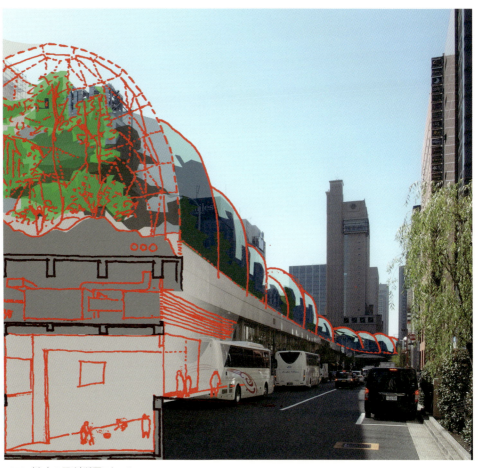

272 〈生命の回廊〉断面スケッチ
Sectional diagram of Cloister of Life

高速道路の路面の上にバイオーム温室を作る。下の商業施設は世界の地域スポーツの練習場に変える。上下で熱のやりとりをする。
A biome is created under the floor of the Expressway. The commercial facilities below it are transformed into practicing grounds for regional sports from around the world. Heat is exchanged between the upper and lower levels.

「庭」は、自然物と人工物の対話によって形成される人間の居住地の全体像を示す地形モデルである。庭は、気候と地質と地形に支配され、構成する植物の相互関係に依存し時に群をなし、多様性を保持しつつ複雑な秩序をもって継続する。それは自然の恵みでありながら同時に人間の知性の所産である。人間の居住地という庭は地球にとっていまや脅威となり、地球を危機に陥れている。庭師がきちんと管理しなければやがて衰退する庭である。

Garden is a topographical model indicating the overall image of the human habitat that is being created through dialogue between nature and artifact. Garden is influenced by climate, soil quality and topography; depending on the interrelationships between constituent vegetation, it will at times form a group, and maintaining diversity, it will continue to exist, possessing a complex order. Even though it represents the blessings of nature, it is also a product of the human intellect. Gardens that are human habitats have become threats to the earth, and have brought earth to a state of crisis. These are gardens that will eventually decline unless properly managed by gardeners.

灌漑／暖かい食卓

どんなに上手に庭をデザインしても水遣りを欠かせば樹々は枯れてしまう。「灌漑」は居住地を維持するうえで基本的な関わり方である。〈暖かい食卓〉は、住まいから歩ける距離に値打ちな価格で食事ができる場所を、かつて日本の都市のどこにもあったように銭湯のように遍在させることを目指す。

〈暖かい食卓〉で提供される料理は、栄養のバランスが良く、年金生活者を含めて多くの人が利用できるように、値段は学生食堂並みの価格に設定する。独居している人もたまには近隣の人と食卓を挟んで顔を合わせて世間話をすることもできるし、話をしなくても、お互いの存在を感じとることで満足を得ることもできるだろう。

実現に向けて制度の整備も必要であるが、同時に地域の食文化の問題としての考慮も必要である。大都市の都心であればコンビニがその役割を担ってもよいが、地方都市では、人口が少なく民間企業に頼ることができない。税制上の支援、場所の借料の無償化（公的補助、寄付）などの経済的な支援策を前提

Irrigation / Orange Tables

No matter how skillfully a garden has been designed, trees will wither and die unless they are properly watered. "Irrigation" is a basic method of involvement in the maintenance of a habitat. The goal of Orange Tables is to make places that are within walking distance of homes and where people can partake of good-value meals as ubiquitous in Japanese cities as public bathhouses once were.

Meals served at Orange Tables will be offered at prices at the level of student cafeterias to enable many people including pensioners to eat nutritious, well-balanced meals. People, including those living alone, will be able to engage in occasional small talk with neighbors across a table, and even if there is no conversation, being in each other's company will surely be satisfying.

Achieving this will require not only institutional measures but consideration of problems of local food culture. In the center of a metropolis, convenience stores may be able to serve such a role, but in a local city, we cannot depend on private enterprises because of the small size of the

に非営利での運営が基本となるだろう。地元の食材を使い、地元の人々で運営すれば、小さいながらも地域経済に貢献する。

文化的課題は外食に対する抵抗感である。食事は極めて文化的な行為である。外食の意味は地域によって大きく異なる。中国南部から東南アジアなどでは、外食は日常的であり一般に家庭の厨房は小さい。日本では内食が基本であり外食は非日常的である。特に女性が無闇に外食することは、好ましくないという考えが根強い。しかし、一方で、祭礼時などに共食することは日本でも盛んに行なわれ、地域社会の紐帯の強化に貢献してきた。新しい共食の文化は「寂しい社会」を「暖かい社会」に変える原動力の一つになるだろう。つまり、経済的支援と同時に、外食し共食することは楽しく、ファッショナブルだという文化的キャンペーンを展開して〈暖かい食卓〉を後押しすることが必要である。

目指すは経済的に効率的な社会構造の追求のみにあるのではなく、住まうに値する社会と環境を育てることである。

小さい庭／緑の指＋パッチワーク

小さい庭は、都市を庭に見立てたときに、都市を一つの大き

273 〈暖かい食卓〉
Orange Tables

40%の人口が高齢者となる社会では、生活のインフラとして、廉価にお美味しい食事ができる場所がどこにも必要である。空き家を活用しNPOによって、公的な支援を得て運営する。

In a society where 40% of the population will be elderly, there will be a need everywhere for places that provide inexpensive, good food. Such places can be provided through the use of vacant homes and can be managed by an NPO through public funds.

population. In all likelihood it will basically be a non-profit operation premised on economic support in such forms as tax exemptions and subsidies (public assistance or contributions) for rent. If local produce is used for preparing meals and local people are in charge of the operation, the endeavor will contribute in a small way to the local economy.

Meanwhile, one cultural problem is resistance to dining out. Dining is a supremely cultural act. The meaning of dining out differs considerably depending on the region. In southern China and Southeast Asia, dining out is an everyday experience, and household kitchens tend to be small. In Japan, dining in is the norm and dining out is the exception. There is a deep-rooted aversion, particularly among women, to dining out excessively. However, sharing meals at festive times is common, even in Japan and has contributed to the reinforcement of ties in local society. A new culture of shared meals will become a driving force in the transformation of a "lonely society" into a warm and friendly society. That is, Orange Tables will require, not only economic support, but the organization of a cultural campaign to suggest to people that dining out and sharing meals can be enjoyable and fashionable.

The goal is not only the pursuit of an economic and efficient social structure but the building of a society and environment worth living in.

い庭として考えるのではなく、住民の手に負える小さい庭の集合として考える地形モデルである。

〈緑の指〉は、2006年のファイバーシティv.1.0で提案した郊外の再組織化の空間像であり、居住地選択は住民の利便性で決まるという功利的な前提を押し出している。具体的には、鉄道郊外の原理をより徹底して、鉄道駅の歩行圏に高い人口を集め、その外の圏域の低密度化あるいは緑化を目指す。これを全体的に俯瞰すると、駅の周りに集まった住宅地が都心から放射状に延び、一方緑地は都市圏周辺から都心に向かって郊外線と郊外線の間に延びる。住宅地と緑地の二つの逆方向に延びる線状の連なりが、二つの手の指どうしを組みあわせたように見えるので〈緑の指〉と呼んだ（→ 274）。鉄道が通勤通学の足として普及している日本の大都市では、歩行圏外に住む人は都市インフラに負担を掛けているので、高い税金を課すなどすれば、多くの人が駅の歩行圏内に住み、その外は緑地となるコンパクトな居住地が駅を核としてできる。これによって、現在ある充実した鉄道網を人口減少下でも維持でき、車依存を減らしてCO₂削減に貢献し、高齢社会においても高いモビリティを失わずに済む（→ 275）。

主要な通勤通学手段が鉄道である大都市郊外では〈緑の指〉は有効であろう。しかし、鉄道などをもたない地方都市では、高齢社会を迎えるにあたり新たに公共交通を整備する必要があるが、それができたとしても、現在の分散的な構造は残るだろう。また、成熟し高齢化する社会では、今までほど住むことに効率を重視するわけでもないだろう。だから、大都市でさえ、〈緑の指〉はすべての郊外居住者に等しく魅力的だとは限らない。しかし、このことは、個々の居住者の自由な選択にまかせればよいということを意味するわけではない。都市全体が無秩序で非効率なパターンに拡散して、全体としては郊外居住の社会的コストは高くつくだけではなく、居住者自身にとっても魅力的な居住地とはならない。

この難しい問題を解決する一つの方法は、住民が自分たちの居住地の経営を自ら行ない、将来像を決めることではないだろうか。そのためには、自治体が行なっている公共サービス提供業務の一部を近隣住民組織に相当の地方税とともに委譲する。たとえば、負担を重くしても高い公共サービスを受けたいと

274 〈緑の指〉
Green Fingers
田園と都市が交互に噛み合うように接する都市形態が好ましい。
A preferable urban form is that in which rural and urban districts intertwine.

Small Gardens / Green Fingers + Patchwork

Small Gardens is a topographical model that likens the city, not to a single large garden, but a collection of small gardens that can be managed by residents.

"Green Fingers" is a spatial image of the reorganization of the suburbs proposed in Fibercity v.1.0 in 2006 and takes as its premise the utilitarian idea that convenience for residents is the deciding factor in the selection of places of residence. Specifically, the objective of Green Fingers is to apply the principle behind suburban development by railways but in a thoroughgoing way; that is, to concentrate the population within walking distance of railway stations and to lower the density or to green areas beyond that distance. The overall result would be residential areas concentrated around stations extending radially from the center of the city; meanwhile, green areas would extend from the outskirts of the city toward the city center between suburban lines. Linear residential areas and green areas will alternate and extend in opposite directions much like the fingers of two hands clasped together, hence the name Green Fingers (→ 274). In Japan's metropolises where railways serve as the main means of commuting to work and to school, if, for example, high taxes were levied on people living beyond walking distance of stations since they put a large burden on the urban infrastructure, then many people will choose to live near stations and areas beyond walking distance can be made green; compact habitats centered on stations can be created. In this way, the existing railway system can be maintained even as the population declines; dependence on cars can be reduced, contributing to lower carbon emission, and high mobility can be retained despite the aging of society (→ 275).

Green Fingers is likely to be effective in metropolitan suburbs where railways are the main means of commuting. However, new public transportation systems will need to be developed in local cities without railways as society ages. Even if such systems are created, the present dispersed structure will probably remain. In a mature, aging society, not as much importance is likely to be placed on efficiency in choice of dwelling. Therefore, even in a metropolis, Green Fingers will not necessarily appeal to the same degree to all people living in the suburbs. However, that does not mean individual residents should be left to choose

いう選択肢もあれば、逆もある。人口減少を逆手にとって、戸当たりの敷地面積を拡大してゆったりした生活を実現する選択肢もあるだろう。独自の土地利用規制を決めたり、エネルギー供給会社の選択なども自治的に行なうこともできるだろう。

こうした新たな仕組みが浸透すれば、茫漠と広がる郊外住宅地は、あたかも色彩豊かな〈パッチワーク〉の様相を呈することになろう。

21世紀の成熟した都市では、その将来像を前もって市民に示し、それに沿って現実の都市を再組織化することが必要である。特に縮小期には、最善の政策でさえ現状の維持に過ぎな

いかもしれない。そのような状況を皆がよく理解し、それより悪くならないようにするために必要な措置についての理解を共有することが必要である。努力をすれば損害を最小限にすること、放置すれば惨めな状態になることを市民が認識しなければならない。ところが対象地域が広がれば広がるほど、利益構造が複雑になり合意に達することが難しくなり何もできないか、さもなくば、政治家が人気取りの政策で後世に借金を残すことになる。それを避けるためには、再組織化の事業に取り組む地域を狭くとり、成員がお互いに認識できる程度に関係者の数を絞り込んで、地域の実情を肌身で感じている人たちが集まり、独自の運営をする方が地域の利益になり、しかも全体でも利益

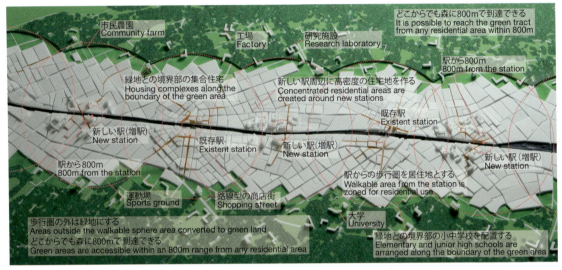

275 緑の指＋モザイク
Green Finger＋Patchwork

便利さをもとめて駅の周辺に住みたい人、緑豊かな場所にひろびろと住みたい人、それぞれ応分の都市費用を負担をすることを前提に住みかたの選択ができる郊外のゾーニングパターン。

A suburban zoning pattern in which those who seek the convenience of areas near the station and those who seek green spaces can select how they live by shouldering the appropriate costs.

for themselves. The entire city will spread out in a disorderly, inefficient pattern, resulting in not only a higher social cost for suburban habitats but less attractive habitats for the residents themselves.

One way to solve this difficult problem may be to have the residents themselves manage their own habitats and decide the future image for those habitats. To enable them to do this, some of the public service operations local governments now perform, as well as appropriate local taxes, should be transferred to neighborhood organizations. For example, receiving a higher level of public services at a higher cost is a choice, and so is its opposite. Advantage might be taken of the decline in population to enlarge individual lots and make dwellings more spacious—that too could be a choice. Each area might also decide on its own land-use regulations or choose its own energy supply company.

If such mechanisms were adopted, suburban residential areas that are now a boundless expanse will surely take on the appearance of a richly colored patchwork.

In a mature city of the twenty-first century, it is necessary to suggest to people in advance the future image of the city and to reorganize the actual city along those lines. Particularly in a period of shrinkage, the best policy may be nothing more than maintaining present conditions. Everyone must clearly understand what those conditions are and have a shared understanding of what measures need to be taken if conditions are not to worsen. People must perceive that with effort, damage can be minimized, and that if the situation is ignored, things may become unpleasant. However, the bigger the target area, the more complex the profit structure and more difficult the achievement of an agreement. Either nothing gets done or politicians adopt a measure that is popular but leaves future generations with debt. To avoid this, the target area for a reorganizational project needs to be kept small, the persons involved ought to be limited to a number such that members recognize each other. If people who are well acquainted with actual conditions in the area gather together and manage things on their own, the area is likely to benefit more and the possibility of maximizing overall profit will be greater. The objective of "Patchwork" is to have the residents of each area form their own autonomous

を最大化できる可能性が高い。

　パッチワークは地域住民による自治的な運営によって地区ごとに独自な近隣社会形成を目指す。パッチワークの最終的な形態は緑の指と違って、前もって明確に描くことはできない。その代わり、明確な地域の運営の原則は示されなければならないが、われわれが思い描いている地区ごとの運営組織は、分譲集合住宅（マンション）の管理組合を強化したものである。

　郊外の住宅地再組織化は、功利性と効率性に重きを置いた〈緑の指〉と自治性と共同性に重きを置いた〈パッチワーク〉の二つの戦略の組み合わせで行なうのが合理的であろう。〈緑の指〉が成立する地域では、〈パッチワーク〉の上位に緑の指の政策が被さるであろう。〈緑の指〉の政策のもとでは、駅から離れた緑地地域に住もうとすると、高いインフラ負担金が課せられることになるが、それでも緑地地域にゆったり住みたいと願う人がいるだろう。それを許容するためにも〈緑の指〉と〈パッチワーク〉の組み合わせが必要である（→276）（→277）。

neighborhood society. It is different from Green Fingers in that it is impossible to anticipate what the ultimate form of Patchwork will be like. To make up for that, local management principles must be clearly indicated. The organization managing each area, as we envision it, will be a more empowered version of the association of owners in a condominium.

In reorganizing suburban residential areas, the rational approach in all likelihood will be to combine the two strategies, namely Green Fingers, which emphasizes practicality and efficiency, and Patchwork, which emphasizes autonomy and communality (22622). In those areas where Green Fingers are formed, the Green Fingers policy will probably be layered over that of Patchwork. Under the Green Fingers policy, high infrastructure charges would be levied for living in green areas at a distance from stations, but there are likely to be people who want to live spaciously in green areas nevertheless. Green Fingers and Patchwork will need to be combined to permit that to happen (→276)(→277).

277（次頁）　ファイバーシティ長岡2050

次頁に示す図は2050年の長岡市のための計画である。
計画範囲である旧長岡市の人口は現在の19万5000人が2050年には30%減り約13万4000人になると想定している。また、人口減少は地理的にランダムに発生すると仮定している。
空間の再編成は以下の3つの方法を主軸に据えている。
1）脆弱な公共交通を補強することを目的として、長岡市全体に〈暖かい網〉を適用する。環状路線を走るBRT沿いに集合住宅開発を誘導し、沿道に多くの人が住むことになるだろう。
2）住民が、学校区毎に自らの居住地の経営に積極的に関わり、居住地の行方を自ら決める仕組みを想定している。このようなシステムは、都市の将来像を描くことと根本的に矛盾することでもある。
3）都市長岡が秘めている可能性を引き出すことを目的とした小さい線状の都市デザインプロジェクトをいくつか提案した。これらのプロジェクトを進める主体は、自治体政府、民間企業、住民組織、そしてそれらの組み合わせが考えられる。

278　都心を流れる農業用水福島江沿いに水面近くに設ける遊歩道

279　もう一つの都心の小河川柿川沿いに設ける市場

280　撤退した百貨店のニッチな需要に応える住宅への用途変更

281　コルテン鋼の車止めのある駐車場。鉄分を含んだ融雪水により長岡の道路は赤く染まっている。赤は長岡の色である。

282　〈暖かい食卓〉

283　廃寺の改修による〈暖かい巡回〉の拠点

284　TODの考え方により都市活動を〈暖かい網〉の環状幹線路線沿いに集中させる

285　〈暖かい網〉のためのバス路線網案内図

286　雁木の再生

287-291　信濃川河川敷の水位の変化によって表情を変える公園、水位は上から順に－4m、－2m、平均水位、2m

276　〈緑の指＋モザイク〉
Green Finger + Patchwork

便利さをもとめて駅の周辺に住みたい人、緑豊かな場所にひろびろと住みたい人、それぞれ応分の都市費用を負担をすることを前提に住みかたの選択ができる郊外のゾーニングパターン。

A suburban zoning pattern in which those who seek the convenience of areas around the station and those who seek spacious green areas can select how they live by shouldering the appropriate costs.

277　(next page) Fibercity Nagaoka 2050

The figures on the following page show a proposal for Nagaoka City in the year 2050.
The population of 195 thousand today in the planned area (Nagaoka before the merger) is estimated to decrease by 30% in 2050, or to 134 thousand. Shrinkage is assumed to occur randomly.
The following were the three principal methods used in the spatial reorganization process:
1) Orange Web is applied throughout Nagaoka City to supplement its weak public transportation network. Construction of mass housing complexes will be induced along the BRT route on the Loop line and many people will live along this line.
2) A mechanism in which residents are actively involved in the management of their residential districts based on school area is assumed, although this kind of system fundamentally conflicts with depicting a future urban vision.
3) A number of small linear urban design projects are proposed to draw out the hidden potential of Nagaoka City. These projects may be led by the local government, private corporations, residential groups and/or combinations of such organizations.

278　Promenade along Fukushimae, an agricultural canal running through the cental area of Nagaoka

279　Market along Kakigawa River, another warterway in the cental area of Nagaoka

280　Change in use of a declining department store into housing that responds to niche demands.

281　Parking lot with car stops in corten steel. The roads of Nagaoka are stained red from snowmelt runoff containing iron. Red is the color of Nagaoka.

282　Orange Tables.

283　An Orange rounds base converted from an abandoned Buddhist temple.

284　Intensification of urban activities along the Orange Web following TOD principles.

285　Bus route map for Orange Web.

286　Renewal of gangi (covered alleys)

287-291　A park designed on the riverbed of the Shinano River that changes in appearance in accordance to changes in water height

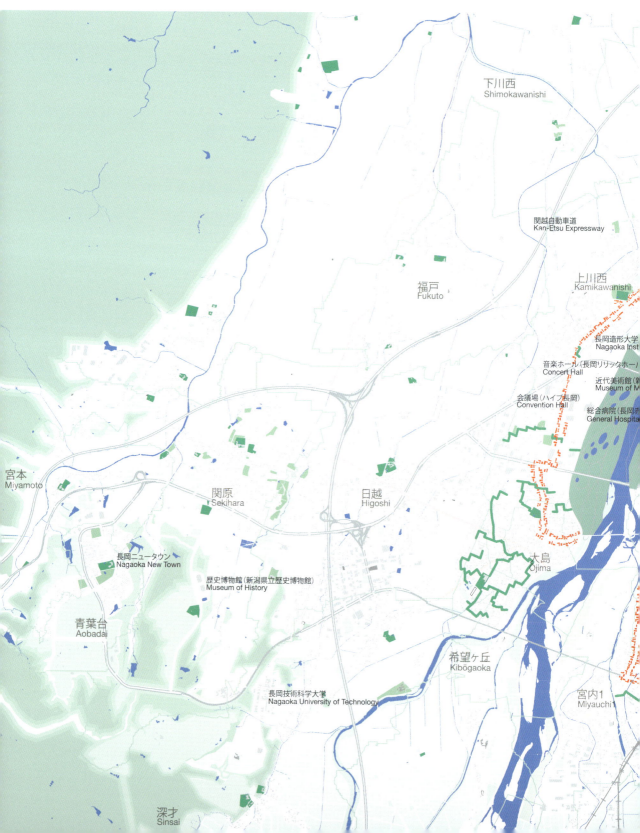

	河川 River
	既存緑地 Existing green tracts of land
	暖かい網 環状線沿道の高度土地利用 Intensive land use along the Loop line of Orange Web
	緑の間仕切り Green Web
	既存の近隣組織 （想定する中学校区を基礎とする住宅地管理単位） Existing neghiborhoods or assumed residential communities based on the junior highschool district system to be managed by their habitants.

277 ファイバーシティ長岡2050
Fibercity Nagaoka 2050

結語
「重建設主義」と「大きい流れ」に打ち勝つために

本書で提案した「小さい線的介入による都市の再組織化」は、効率が低いものや時代遅れのもの、あるいは長く使ってガタがきたものを壊して、新しい開発で置き換える方法ではない。そうではなく、既に目の前にあるものの可能性を引き出すことで、その環境に慣れ親しんだ人々の記憶に敬意を払い、人々の活動する環境の総合力を高める発想である。これが実現する条件は、日本の都市計画が基礎をおいている「重建設主義」を克服し、「大きい流れ」の支配に対抗して「小さい流れ」を支援することである。

1つの時代に輝く都市は、その時代の社会構造と思想に相応しい物的環境(土木構造物や建築、さまざまな輸送施設そして公共空間などを総称)が整えられることで実現する。したがって、時代が変われば、物的環境も変わらなければならないことになるが、現実には、それは容易ではない。しかし、近代日本はこれまで、それを実行してきた。すなわち、不都合な部分を新品に交換するという外科的処置の発想で対応してきた。それはまさに近代主義と成長の時代を体現していたと言えよう。近代主

Conclusion: Overcoming "Constructionalism" and "Big Flow"

The reorganization of cities through small linear interventions that is proposed in this book is a way of paying respect to the memories of people who are familiar with an environment and strengthening the overall capacity of the environment in which people are active by tapping the potential of what already exists instead of simply destroying what has become inefficient, out-of-date, or dilapidated through long use. To achieve this we must overcome the "constructionalism" in which Japanese city planning is complicit and resist the dominance of Big Flow by supporting Small Flow.

When a city's physical environment (by which I mean its public works and buildings, means of transportation and public spaces) is appropriate for the social structure and philosophy of that age, it is radiant. When times change, the physical environment too must change, but that is not easily achieved. Nonetheless, modern Japan has managed to carry that out up to now. That is to say, it has responded by replacing unsuitable parts with new parts, adopting as it were a surgical approach to the problem. That approach can be said to be the essence of modernism and of the

義を構成する思想的要素には、工学的発想、言い換えれば、歴史の否定がある。この点でも、日本は率先して実践し、常に都市計画を白紙の上に描いてきた。これは、全面的刷新を好む性向と言える。つまり、建設することにありとあらゆる期待をする傾向である。建設は、景気刺激と雇用の創出、福祉政策の推進、文化の発展など一振りで何でも生み出す魔法の杖のようにみなされてきた。全面的刷新ということになると、改革派も保守派も、官僚も企業人も、そして地方都市の住人も大都市の住人の誰もが歓迎し、魔法の杖の威力を疑うことがない。その結果、何時の時代も、国も企業も個人も、まるで追い立てられるようにそこらじゅうで建設活動を展開し、この国の自然と歴史的環境を壊してきた。これを私は、西欧近代初期における「重商主義」にならって「重建設主義」と呼ぼうと思う。

「重建設主義」を克服しなければならない理由は、まずもって、縮小の時代に対応するためであるが、それだけではない。「重建設主義」の克服は、物的環境に文化的連続性を復活するためでもある。今更いうことでもないが、日本やアジアにおける近代化は、ほぼ欧化であり、これは自動的に西欧崇敬と自らの伝統文化の否定によって成立している。その結果、現にある物的環境を否定するという心性を育ててきた。この心性は、日本が欧化によって明治維新という政治文化革命を成功させたことと第二次世界大戦の敗戦で決定的になった。西欧崇敬と伝統文化の否定は、今日でも世代を超えて受け継がれ、アジア全体に蔓延している。一方、21世紀には、経済活動の中心は西欧からアジアに移行することが確実である。問題は、富がもっとも集中する場所で、自己否定と全面的刷新が繰り返されることである。

人間の思考も活動も具体的な物的環境を離れては成立しない。社会科学系の学者やその教えを受けた実務家には社会も経済も物的環境から独立して展開できると信じている人が多く、理解されづらいかもしれないが、物的環境は人間の思考と活動の舞台であり、それらの鋳型であるということは記憶されるべきである。

日本を含めアジア地域では、今、この自己否定をどう克服するかが問われていると私は考えている。そのためにそして最初に、われわれがすべきことは、現に目の前にあるものをまず受け入

period of growth. The philosophical element of modernism is an engineering-oriented approach, or to put it another way, a rejection of history. Japan has taken the lead in implementing that and always practiced city planning on a blank slate. It has a predilection for complete change—an expectation that construction will solve any conceivable problem. Construction has been regarded as a magic wand, the waving of which can produce virtually anything from stimulating the economy and creating employment to promoting welfare policies and enhancing culture. On that one point—the power of the magic wand of complete change—everyone is a believer, both reformers and conservatives, bureaucrats and corporate executives, the residents of local cities and residents of metropolises. As a result, in every age, the government, corporations and individuals have been in a rush to expand construction activities everywhere and destroyed the natural and historical environment of this country. Following the example of "mercantilism," I have chosen to call this belief in the benefits of construction, *constructionalism*.

This must be overcome, first of all, to deal with the age of shrinkage, though that is not the only reason. Constructionalism must also be overcome to revive cultural continuity in the physical environment. Modernization in Japan and Asia in general has been virtually synonymous with Westernization and brought about by the automatic worship of the West and rejection of traditional culture. As a result, it has fostered a mentality that rejects the existing physical environment. Japan's success in carrying out the Meiji Restoration through Westernization and the country's defeat in World War II were decisive in the evolution of that mentality. Worship of the West and rejection of traditional culture are still very much in evidence in all generations, not only in Japan but all across Asia. Meanwhile, the center of economic activities is certain to shift from the West to Asia in the twenty-first century. The problem is repeated self-rejection and complete change in the places where wealth is most concentrated.

Human thoughts and activities cannot develop in isolation from actual physical environments. There are many scholars in the social sciences and those outside academia that have been influenced by the teachings of those scholars who believe that society and the economy are somehow independent of the physical environment. *It ought to be remembered that the physical environment is the stage for, and the shaper of, human thoughts and activities*. I believe the question now for the Asian region including Japan is how to overcome that self-rejection. What we must first do is accept what exists in front of us.

What has been proposed in this book is to practice that in city planning. None of the proposals preserves, or looks back nostalgically on, the past. What we have tried is, first to accept everything that the generations preceding ours have constructed (including the results of the worship of the West and the rejection of traditional culture in the age of modernization), then to select what can be used, and finally to focus our energy on utilizing those things that have been selected. Only then, that is, only when *constructionalism* has been overcome, will a soil in which continuity of culture can grow be created in Asia.

Since the 1970s, the sell-by date for urban policies founded on modernism, the forces of capitalism have launched an aggressive counterattack, asserting that the market ought to decide urban forms and that city planning regulations are unnecessary and ought to be relaxed. That assertion has been translated into policy to some extent in every country. The relaxation of regulations is intended to increase the speed and volume of *flow* by playing down the need for, and minimizing, control on flow through public policies. However, as this book has explained, the world

れることである。本書で提案したことは、いずれも過去を凍結して保存することでも過去を懐古することでもない。私たちが試みたのは、近代化の時代も含めて、われわれに先行する世代が築いたものすべて（そのなかには西欧崇敬と伝統文化の否定の結果も含まれる）を、まずは受け入れ、次に利用できるものを選び出し、それを活用することに力を注ぐことである。そのとき、はじめて、アジアにも、「文化の連続性」ともいうべき何かが生まれるだろう。

　近代主義による都市政策が賞味期限切れになった1970年代以降、資本主義側からの反撃の勢いが増した。つまり、都市形態は市場が決めるべきであり、都市計画の規制は無用で緩和すべきであるという主張である。それは各国で多かれ少なかれ政策化されている。都市計画の規制緩和は、公共政策による「流れ」の制御を無用視し、それを最小限にすることで流速と流量を増やそうというものである。しかし、本書で述べたように世界は既に限界に達しており、「流れ」の総量を増やして豊かにするという策の有効性は懐疑的である。都市計画の規制緩和による成長は、いずれ閉塞することが自明な策であり先がない。

　本書で提案した「線的介入による都市の再組織化」の構想は、「小さい流れ」に市民権を与え、それの占有率を高めることで、人々が住み、働き、移動する環境の改善を当事者の手に取り戻そうというものである。それゆえ、出来上がる物的環境の形態だけを議論しても不十分であり、環境の管理運営にまでに視野を広げる必要がある。そして、その一つの可能性は、行政と企業との連携を保ちつつ、住民が主導する地域の自然環境および物的環境の管理・運営である。本書では、「数珠つなぎにする／緑の間仕切り」と「小さい庭／緑の指+パッチワーク」で、その可能性を示唆したが、まだ十分とは言えない。今後の日本の都市にとって最大の課題の一つである。

has already reached its limits, and the effectiveness of any plan to increase the total volume of flow is limited. It is clear that growth achieved through the relaxation of city planning regulations will eventually subside.

The concept of the reorganization of cities through linear interventions proposed in this book is intended to restore to people the capacity to improve the environment in which they live, work and move by giving due recognition to, and increasing the share of, *small flows*. That is why discussing merely the form of the resulting physical environment is not enough and why we must broaden our perspective to include the management of the environment. One possibility is that, while cooperation is maintained between government and corporations, residents will take the lead in the management of the physical environment of a community including the natural environment. That is suggested in this book by "Stringing Together / Green Partitions" and "Small Gardens / Green Fingers + Patchwork."

Notes

- The websites listed, unless otherwise noted, were accessed between August and September 2015.
- For Japanese references, only the author's name and date of publication are noted in the English citation.
- Japanese translations were referenced for sources written in languages other than Japanese, unless otherwise noted. The English citation, however, notes page numbers that correspond to the English versions of the source cited.

Acknowledgments

1. Architectural Institute of Japan Special Committee for a Low Carbon Society, established with the Ministry of the Environment's General Fund for the Promotion of Research on the Global Environment (Hc-08).

2. Architectural Institute of Japan Special Committee for Constructing a Re-organization Model for Cites in the Depopulation Age.

Introduction

3. Tange Kenzō Laboratory, the University of Tokyo, "Tokyo Plan 1960", *Shinkenchiku*, March 1961, pp.79-120.

4. Along with the Osaka Expo, the government also held a competition for designing a future vision for Japan and its citizens in the 21st century. For this, Tange Kenzō formed the "21st Century Japan Research Group" and submitted a proposal that was selected together with that developed by Yoshizaka Takamasa's team. In his proposal, Tange Kenzō furthered his ideas from "Tokyo Plan" 1960, introducing information as the first element that will define Japan in the 21st century, together with free time, transportation and energy. This could precisely be called a plan that dealt with flows in the city, but Tange's main interest lay in megastructures and Big Flows.
Japan of the 21st Century Committee (1971). Reference is written in Japanese only. (Tange Kenzō was the representative director of the committee.)

Part I

5. Herman Daly says that, because the benefits gained through economic growth will fall less than the requisite costs in a world that surpasses the earth's "carrying capacity," we should think of development as opposed to growth. He argues that the world should aim towards a stable-state economy in the future, "with constant stocks of people and artifacts, maintained at some desired, sufficient levels by low rates of maintenance."
Herman Daly and Edahiro Junko (1996). Reference is

6. 国際連合の推計による("World Population Prospects The 2015 Revision", Department of Economic and Social Affairs, Population Division, United Nations, http://esa.un.org/unpd/wpp/Publications/Files/Key_Findings_WPP_2015.pdf)。	written in Japanese. 6. Based on estimates by the United Nations. United Nations, Department of Economic and Social Affairs, Population Division, "World Population Prospects: The 2015 Revision, Key Findings and Advance Tables," *Working Paper No. ESA/P/WP, 241*, 2015, http://esa.un.org/unpd/wpp/Publications/Files/Key_Findings_WPP_2015.pdf
7. 世界自然保護基金による("World Planet Report 2014", http://wwf.panda.org/about_our_earth/all_publications/living_planet_report/ 日本語要約版は、https://www.wwf.or.jp/activities/lib/lpr/WWF_LPRsm_2014j.pdf)。	7. World Wide Fund for Nature, *Living Planet Report 2014*, 2014, http://wwf.panda.org/about_our_earth/all_publications/living_planet_report/.
8. 長岡の家庭での電気器具の普及率は1970年で洗濯機95%、冷蔵庫92%、テレビ(白黒)86%、掃除機66%であった。長岡市『長岡市史 通史編 下巻』長岡市、1996、p. 817	8. The percentage of electrical appliance ownership by households in Nagaoka was 95% for washing machines, 92% for refrigerators, 86% for TVs (black & white) and 66% for vacuums in 1970. City of Nagaoka (1996). Reference is written in Japanese.
9. ナオミ・クラインは、現代のグローバル企業がどのようにして消費者の欲望を操作しているかを詳細な調査によって明らかにしている。ナオミ・クライン『ブランドなんか、いらない』松島聖子訳、2001、はまの出版	9. Naomi Klein clarifies the details of how modern global corporations manipulate consumer desires. Klein, *No Logo*. (New York: Picador, 2000).
10. ジェニーグスタフソンによる(Jenny Gustavsson et al., "GLOBAL FOOD LOSSES AND FOOD WASTE, EXTENT, CAUSES AND PREVENTION", 2011, Food and Agriculture Organization of the United Nations, PP.4-9, http://www.fao.org/docrep/014/mb060e/mb060e.pdf)。	10. Food and Agriculture Organization of the United Nations, *Global Food Losses and Food Waste: Extent, Causes and Prevention* (Rome: 2011), 4-9, http://www.fao.org/docrep/014/mb060e/mb060e.pdf.
11. 水野和夫『資本主義の終焉と歴史の危機』集英社、2014、pp.12-25	11. Mizuno Kazuo (2014). Reference is written in Japanese.
12. 国立社会保障・人口問題研究所による、平成24年1月推計の参考推計のなかから、死亡中位出生中位としたときの封鎖人口推計(参考値)である(http://www.ipss.go.jp/syoushika/tohkei/newest04/sh24sanko.html)。	12. This number comes from the projected closed population (reference value) at medium-fertility and medium-mortality variants based on estimates from the National Institute of Population and Social Security Research as of January, 2012. Population Projection for Japan, http://www.ipss.go.jp/syoushika/tohkei/newest04/sh24sanko.html. Reference is written in Japanese.
13. 日本の教育に対する公的負担の低さは以前から議論されており、OECD諸国の平均を下回っている。文部科学省『平成21年度文部科学白書』pp.19-23(http://www.mext.go.jp/b_menu/hakusho/html/hpab200901/1295628_005.pdf)。	13. The low level of public expenditures on education in Japan has been discussed for some time, as it has fallen below the average of the OECD countries. Ministry of Education, Culture, Sports, Science and Technology, 2009, http://www.mext.go.jp/b_menu/hakusho/html/hpab200901/1295628_005.pdf. Reference is written in Japanese.
14. 国立社会保障・人口問題研究所の人口統計資料集「特定年齢の平均余命:1921〜2060年」、2014による(http://www.ipss.go.jp/syoushika/tohkei/Popular/Popular2014.asp?chap=5&title1=%87X%81D%8E%80%96S%81E%8E%F5%96%BD)。	14. Based on the National Institute of Population and Social Security Research's population statistics (2014) regarding average remaining life expectancies for various age groups: 1921~2060. National Institute of Population and Social Security Research, 2014, http://www.ipss.go.jp/syoushika/tohkei/Popular/Popular2014.asp?chap=5&title1=%87X%81D%8

E%80%96S%81E%8E%F5%96%BD. Reference is written in Japanese.

15. Ariyoshi Sawako's *Kōkotsu no Hito* (Shinchōsha, 1972, trans., *The Twilight Years*) forged the opportunity for social debate on the problems of aging. In 1973, the following year, the Tanaka Kakuei Cabinet attempted to enrich social welfare policies, calling the year *fukushi gannen* (the first year of the Welfare era).

16. Donella Meadows, Dennis Meadows, Jørgen Randers, and William W. Behrens III, *The Limits to Growth* (Universe Books, 1972).

17. Daly and Edahiro (1996). Reference is written in Japanese only.

18. Serge Latouche, *Survivre au dévelopment* (2004).

19. Both construction and management of urban facilities are necessary for developing and sustaining a city — and these are the tasks of urban planning. The term "urban planning" was often used with an emphasis only on the aspect of construction. This book therefore uses the term "(urban) reorganization" in place of "urban planning." At the same time, the term "reorganization" is also an intent to distance the idea from "(urban) regeneration," a term used extensively over the past ten years in Japan. This is because the term "regeneration" embodies a sense of nostalgia for the era of heavy construction during the period of high economic growth wrought with excessive excitement.

20. There were some differences in the architectural styles adopted by Japan in its colonized lands prior to and after the founding of Manchukuo (1932). Prior to its founding, Western-style architecture was the main style adoped, as in Taiwan and the Korean Peninsula. After Manchukuo was founded, however, Teikan-style architecture, with an Asian-style roof laid over a Western-style classical building, started to appear. Koshizawa Akira suggests that this is related to the founding concept behind Manchukuo, which celebrates the harmony of the five races (Japanese, Han Chinese, Manchus, Mongols and Koreans) (Koshizawa 1988). There seem to be, however, no such definitive guidelines in colonization policies, and we should see this as the shift in architectural ideologies on the mainland simply being reflected onto colonial architecture.

21. Koshizawa (2005). Reference is written in Japanese.

22. In both Tokyo and Nagaoka, large-scale changes to the implementation of postwar reconstruction plans were met with resistance, making some backtracking necessary for parts of the plans.
    Koshizawa (1991). City of Nagaoka (1996). Both references are written in Japanese.

---

15. 有吉佐和子の『恍惚の人』(新潮社、1972)は、高齢者問題を社会的な争点とするきっかけを作った。翌年の1973年を田中角栄内閣は、福祉元年と呼び社会福祉政策の拡充をはかった。

16. ドネラ・H・メドウズ『成長の限界—ローマ・クラブ人類の危機レポート』大来佐武郎監訳、ダイヤモンド社、1972

17. デイリー、前掲書、p102

18. セルジュ・ラトゥーシュ『経済成長なき社会発展は可能か？—<脱成長>と<ポスト開発>の経済学』中野佳裕訳、作品社、2010

19. 都市が発展し持続するためには、都市施設の建設と管理の両方が必要であり、それが都市計画の任務である。「都市計画」という用語は、日本においては、建設的側面だけが強調して使われることが多かった。そこで、本論では「都市計画」の代わりに「(都市空間の)再組織化」を使っている。また「再組織化」は、この10年間日本で盛んに使われている「(都市)再生」から距離をとる意図もある。「再生」という用語には、高度経済成長下に異常に高まった都市の大建設時代への郷愁が感じられるからである。

20. 日本が植民地で建てる建築の様式は、満州国(1932年)以前で満鉄(南満州鉄道株式会社)が主導した時代と建国以後で幾分違いがある。建国以前では、台湾や朝鮮半島と同じく、基本的に西欧の様式建築であった。満州国建国以降にはアジア建築の屋根を西欧古典建築に載せる帝冠様式の建築が現れる。越澤明は、五族(日、漢、満、蒙、鮮)の協和を謳う満州国の建国のコンセプトとの関係を示唆している(越澤明『満州国の首都計画』日本経済評論社、1988)。ただし、植民地政策にそれほど明示的な方針があったわけでもないようで、本土における建築思想の転換が植民地建築にそのまま反映したと見るべきであろう。

21. 越澤明『復興計画』中公新書、2005、p262

22. 東京でも長岡でも、戦災復興計画の実施は大規模な変化は抵抗を受け、一部は計画の後退を余儀なくされた。越澤明『東京都市計画物語』日本経済評論社、1991、pp.154-170、長岡市『長岡市史 通史編 下巻』長岡市、1996、pp.723-730

23. 第二次世界大戦後の急激な人口増と産業の復興を受けて都市部の発展と農村部の発展の不均衡が目立ってきた。そのなかで、政府は1962年に、全国総合開発計画(全総)を打ち出し、都市部の住宅供給の必要性を強調している。経済企画庁編『全国総合開発計画』大蔵省印刷局、1962

24. 総理府住宅統計によれば、1968年に戦後はじめて住宅戸数が世帯数を上回った。全国住環境整備推進協議会編集『住環境整備三十年のあゆみ』社団法人全国市街地再開発協会、1991

25. 多摩ニュータウン(事業認可年と計画人口(以下同じ);1966年、34万人)、港北ニュータウン(1974年、22万人)、千葉ニュータウン(1966年、34万人)、そして、筑波研究学園都市(1963年、30万人)のいずれも当初の計画人口に達することができず、その後、計画人口の下方修正を行っている。この事情は長岡ニュータウンでも同じである。

26. 長岡の西部にある丘陵地の総合的な開発で、工場団地、流通センター、住宅団地、レクリエーション施設)などを含み、計画人口4万人で1975年に事業認可を受け、事業予算870億円で着工した。

27. 政府によって立案された国土の将来像である。1962年の第一次全総のあと、1969年(新全総)、1977年(三全総)、1987年(四全総)、1998年(五全総)と続く。その後の政府による国土計画も含めて、表現は変わっても、基本的な目的は首都圏一極集中の是正であった。

28. 1925年に都市計画法が成立し、3年後に市街地建築物法が制定されて長岡市は用途地域を指定した。その時の配分は、工業地域が14.9%、商業地域が11.8%、住居地域が71.7%、未指定地域が1.6%であった。第二次世界大戦前から、長岡が工業都市であったことがわかる。

29. 藤森照信『日本の近代建築(下)』岩波書店、1993, pp.133-135

30. 正確を期せば、超高層建築が実現されるためには、都市計画規制の変更が伴わなければならない。建設省は、都市の建物の形態を絶対高さで規制する仕組みから延べ床面積の敷地面積に対する比率(容積率と言う)の制限で形態を規制する仕組みを導入した。

23. The sudden rise in population after World War II and the revival of industries brought about a significant imbalance in the development of urban areas as opposed to villages. Under such conditions, the government issued the Comprehensive National Land Development Plan in 1962, emphasizing the necessity for supplying housing in urban areas.

24. According to the Bureau of Statistics in the Office of the Prime Minister, the number of housing units in the postwar period first outnumbered the number of households in 1968.

25. Tama New Town (project authorization year: 1966; planned population: 340 thousand), Kōhoku New Town (1974; 220 thousand), Chiba New Town (1966; 340 thousand) and Tsukuba Science City (1963; 300 thousand) were all unable to reach their initial planned populations and revised, lowered planned populations was set thereafter. The same occurred for Nagaoka New Town.

26. Nagaoka New Town was a comprehensive development project located along the hilly areas of western Nagaoka. The development plan included an industrial complex, distribution center, housing complex and recreation facilities and was planned for a population of 40,000. The project was approved in 1975 and construction began with a project budget of 87 billion yen.

27. This was a future vision established by the government. After the first Comprehensive National Land Development Plan in 1962, the New Comprehensive National Land Development Plan was proposed in 1969, followed by the Third Comprehensive National Land Development Plan (1977), the Fourth Comprehensive National Land Development Plan (1987) and the Fifth Comprehensive National Land Development Plan (1998). While the designs changed throughout these plans as well as other national plans issued by the government thereafter, the basic purpose remained to rectify the extreme concentration in the metropolis.

28. Nagaoka City designated its land-use zones through the City Planning Act concluded in 1925 and the Urban Building Law established three years later. The distribution of zones at the time was: 14.9% industrial, 11.8% commercial, 71.7% residential and 1.6% undefined. Nagaoka was thus an industrial city since pre-WWII.

29. Fujimori Terunobu (2005). Reference is written in Japanese.

30. In more precise terms, skyscrapers could only be realized with accompanying changes to urban planning regulations. The Ministry of Construction therefore shifted its policy from a mechanism for regulating the form of buildings in the city through absolute height to

that for regulating form by limiting the floor area ratio (FAR).

31. アメリカ合衆国の大陸横断鉄道の開通は1869年である。

31. The transcontinental railroad in the United States was completed in 1869.

32. 社会における経済的不平等さを測る指標としてジニ係数があるが、それで比較すると、日本の社会はスウェーデンより不平等であり、アメリカやイギリスより平等な社会であり、ドイツやフランスと同じグループに属し、ドイツとともに社会的格差が開きつつある。一方学歴でみると、日本では高等教育を受けた人口が44%を占め、OECDの平均30%を大きく上回り、アメリカより高い。

32. According to Gini's coefficient, an index for measuring economic inequalities in a society, Japanese society has more inequalities than the Swedish, while it is more equal than American or English societies. It belongs to the same group as German and French societies and its social disparities are widening together with those of Germany. On the other hand, regarding the generalization of higher education, 44% of the Japanese population has attended institutions of higher education, higher than the percentage for the United States and largely surpassing the OECD average of 30%.

33. 吉田和男は、日本の所得税制は「非常に厳しい累進課税になっており、……平等を重視した税構造になっています。……(税率と控除額を低くした1987、8年の税制改正後でも)世界的な水準と比較すれば、課税最低限度も高く、累進度も高い……」括弧内引用者。吉田『入門現代日本財政論―公共部門の現実と理論』有斐閣、1991、p.113

33. Yoshida Kazuo says that Japan's income tax assumes a steeply progressive system with a structure that stresses equality (even after the 1987-88 tax reform that lowered deductions and the tax rate). Compared to global standards, its minimum taxable ceiling as well as its progressivity rank high.
Yoshida (1991). Reference is written in Japanese.

34. 新雅史『商店街はなぜ滅びるのか―社会・政治・経済史から探る再生の道』光文社、2013、p80

34. Arata Masafumi (2013). Reference is written in Japanese.

35. フレデリック・ロー・オルムステッドとカルバート・ボーによって、1869年にイリノイ州に計画された住宅地。オルムステッドは、マンハッタンの中心にあるセントラルパークの設計者として知られる造園家にして都市計画家である。

35. This was a housing district planned in Illinois in 1869 by Fredrick Law Olmstead and Calvart Voux. Olmstead is an urban planner and landscape architect known as the designer of Central Park, located in the center of Manhattan.

36. フランク・ロイド・ライトが構想する都市ビジョン。私塾タリアセンで1932年に模型を製作した。農本主義的な思想を背景にした一戸当たり1エーカー(4000 m²)という超低密都市である。移動は自動車を想定している。

36. Broad Acre City is an urban vision proposed by Frank Lloyd Wright, a model of which was built in 1932 in his home and private studio of Taliesin. It is an ultra low-density city based on agrarian principles, with 1 acre (4,000 m^2) assigned to each housing unit. The mode of transportation assumed in this plan was the automobile.

37. 宇都宮深志編『サッチャー改革の理念と実践』三嶺書房、1990、p.10 および内田勝敏編『イギリス経済 サッチャー革命の軌跡』世界思想社、1989、p.2

37. Utsunomiya Fukashi et al. (1990) and Uchida Katsutoshi et al. (1989). Both references are written in Japanese.

38. デヴィッド・ハーヴェイ『新自由主義―その歴史的展開と現在』渡辺治監訳、作品社、2007、pp.220-222

38. David Harvey, *A Brief History of Neoliberalism* (New York: Oxford University Press, 2005).

39. JR京葉線は主に千葉県内の東京湾岸を走る鉄道路線。1975年に開設された時は曽我駅と千葉港を結ぶ貨物線であったが、その後順次延伸し1986年には旅客営業を始め1990年には東京駅まで乗り入れた。

39. The JR Keiyō Line is a railway line that runs along Tokyo Bay, mainly through Chiba Prefecture. While it was initially opened in 1975 as a freight line connecting Soga Station and the Port of Chiba, it was gradually extended to carry passengers in 1986 and reach Tokyo Station in 1990.

40. 1975年に伝建地区として選定されたのは、妻籠(長野県)、角館(秋田県)、白川村荻町(岐阜県)、産寧坂(京都市)、祇園新橋(京都市)、堀内地区(萩市)、平安古地区(萩市)であった。第2回歴

40. The preservation districts selected in 1975 were: Tsumago (Nagano Prefecture), Kakunodate (Akita Prefecture), Shirakawa-go Ogimachi (Gifu Prefecture),

史的記念物の建築家・技術者国際会議において、建築保存の精神と具体的な取り扱いの方向付けを規定したベネチア憲章が採択されたのは1964年である。イタリアの建築家カルロスカルパによる歴史的建造物の改修設計作品は、ベネチア憲章の精神を具体的な建築デザインとしてみせ、以後の改修の規範となった。アメリカでも同時期に、ジラデリースクエアなど20世紀前半の産業遺構のリノベーションが実現するようになっていた。

Sanneizaka (Kyoto City), Gion Shinbashi (Kyoto City), Horiuchi district (Hagi City) and Hiyakomachi (Hagi City). The Venice Charter, adopted in 1964 at the Second International Congress of Architects and Technicians of Historic Monuments, offered a guide to the spirit of building preservation as well as a means for their concrete management at the Second International Congress of Architects and Technicians of Historic Monument. The Italian architect Carlo Scarpa endowed the spirit of the Venice Charter with concrete of the architectural expression through his renovations of historical buildings. They have since become models for renovations thereafter. Renovations simultaneously begun to appear in the United States, such as Ghirardelli Square, a renovated industrial remain from the first half of the 20th century.

41. 藻谷浩介は、新しい取り組みとして、岡山県真庭市の木質バイオマス発電（銘建工業の中島浩一郎氏）、オーストリアのギュッシングなどを取材している。藻谷、NHK広島取材班『里山資本主義』角川書店、2013、p102

41. Motani Kōsuke has investigated some trials in Güssing, Austria as well as a wood biomass generator in Maniwa City, Okayama Prefecture (Nakashima Kōichirō from Meiken Lamwood Corporation) as examples of such new approaches.
Motani (2013). Reference is available in Japanese.

42. 広井良典『人口減少社会という希望―コミュニティ経済の生成と地球倫理』朝日新聞社、2013、p.36

42. Hiroi Yoshinori (2013). Reference is available in Japanese.

43. 地方自治体の公共施設投資は1960年代から増え始め80年代から更に加速して増えている。一方、総務省は公共施設の耐用年数を60年としているので、2020年以降になると建て替える必要が出てくる。耐用年数とは別に、1981年の建築基準法施行令改正で耐震基準が厳しくなったので、それ以前に作られた建物の多くは法が求める強度を満たさず建て替えが求められている。

43. Investments in public facilities by local governments began to increase from the 1960s and further accelerated in the '80s. On the other hand, the Ministry of Internal Affairs and Communications has determined the lifespan of public facilities to be 60 years – therefore, after 2020, there will be a need to rebuild these facilities. Apart from building lifespans, the 1981 amendments to the Building Standards Law strengthened seismic standards and made buildings constructed prior to it insufficient and in need of reconstruction.

44. 長岡市『長岡市史 通史編 下巻』長岡市、1996、pp.61-75

44. City of Nagaoka (1996). Reference is written in Japanese.

45. 桐敷真次郎「天正・慶長・寛永期江戸市街地建設における景観設計」、『東京都立大学都市研究報告』24、1971

45. Kirishiki Shinjirō (1971). Reference is written in Japanese.

46. 辻惟雄『日本美術の歴史』東京大学出版会、2005、p.77

46. Tsuji Nobuo (2005). Reference is written in Japanese.

47. 法政大学エコ地域デザイン研究所編『外濠　江戸東京の水回廊』鹿島出版会、2012、p.21

47. Laboratory of Regional Design with Ecology, Hosei University (2012). Reference is written in Japanese.

48. 環境庁水質保全局『かけがえのない東京湾を次世代に引き継ぐために』大蔵省印刷局、1990、p.36

48. Environment Agency, Water Quality Bureau (1990). Reference is written in Japanese.

49. オルムステッドが言ったとされている。幸田露伴も「公園は都府の肺臓なり」と述べている。幸田露伴『一国の首都』岩波書店、1993、p.102

49. Olmstead is believed to have coined this term. Kōda Rohan has also written about parks are the lungs of a city.
Kōda (1993). Reference is written in Japanese.

50. 新井博・榊原浩晃編著『スポーツの歴史と文化―スポーツ史を学ぶ』道和書院、2012

50. Arai Hiroshi et al. (2012). Reference is written in Japanese.

51. 山本武利、西沢保編『百貨店の文化史：日本の消費革命』世界思想社、1999、p.178

52. ロバート・フィッシュマン『ブルジョワ・ユートピア—郊外住宅地の盛衰』小池和子訳、勁草書房、1990、p.69

53. 和田菜穂子『近代ニッポンの水回り—台所・風呂・洗濯のデザイン半世紀』学芸出版社、2008、p.25

54. 現在の代々木公園。吉見俊哉は戦後の米軍の施設と、その後の若者の地区として発展する歴史のあいだの関係を分析している。吉見『親米と反米—戦後日本の政治的無意識』岩波書店、2007、pp.125-149

55. カール・ポランニー（著）『経済の文明史』平野健一郎、玉野井芳郎編訳、日本経済新聞社、1975

56. 経済企画庁『平成元年度・経済白書／平成経済の門出と日本経済の新しい潮流』1989では、「高度化」「グローバル化」と並んで「ストック化」が挙げられている。

57. 松原隆一郎は景観について積極的に発言している。首都高は、松原が難じる日本の景観の一つである。松原『失われた景観　戦後日本が築いたもの』PHP新書、2002、pp.207-208

58. NHKが「新製品の研究開発、今も記憶に残る社会的事件、日本人の底力を知らしめた巨大プロジェクトなどに焦点を当て、その成功の陰の知られざるドラマを伝えるドキュメンタリー番組」と称する『プロジェクトX　挑戦者たち』第168回/177回　2005年4月5日放送で「首都高速　東京五輪への空中作戦」と題して首都高建設が取り上げられ称えられている。

59. 幸田露伴は明治32（1899）年に、「江戸の美風は地を払って、東京の俗悪にわかに発達せるもの」と書いている。幸田『一国の首都』岩波書店、1993

60. モダニズム美学で再解釈された和風建築。建築家吉田五十八、村野藤吾、堀口捨巳らによって開拓、発展され、現代の「和風」デザインの源泉となっている。

61. 新幹線は、おそらく英語の表現にならったのだと思われるが、当初「弾丸列車」と呼ばれていた。

62. 日本の路面電車の最盛期は1932年で、全国の67都市を走っていた。事業者数83、路線延長は1479 kmあった（以上「世界の路面電車」東京新聞、2011年2月13日）。東京の路面電車である都電について言えば、1日の乗客数は、1919年に108万人に達した。都電の最盛期は1962年で、路線は213 kmに延び、1日の乗客数

51. Yamamoto Taketoshi (1999). Reference is written in Japanese.

52. Robert Fishman, *Bourgeois Utopias: The Rise and Fall of Suburbia* (New York: Basic Books, Inc., 1987).

53. Wada Nahoko (2008). Reference is written in Japanese.

54. This is the present-day Yoyogi Park. Yoshimi Shunya analyzes the relationship between postwar U.S. military facilities and their history later as districts for the young. Yoshimi Shunya (2007). Reference is written in Japanese.

55. Karl Polanyi, "The Self-regulating Market and the Fictitious Commodities: Labor, Land, and Money," in *The Great Transformation* (Boston: Beacon Press, 1944). Polanyi, "Our Obsolete Market Mentality," *Commentary*, Feb. 1, 1947.

56. The Economic Planning Agency (1989) notes that "stock" is necessary along with "advancement" and "globalization" in the economic white papers of 1989.

57. Matsubara Ryūichirō speaks extensively on scenic value. The Metropolitan Expressway is one example that he criticizes.
Matsubara (2002). Reference is written in Japanese.

58. NHK's *Project X: Challengers*, a documentary series that features huge projects that reveal the latent power of the Japanese people and aims to tell of the unknown dramas behind these successes, introduces and praises the Metropolitan Expressway in its 168th/177th episode "The Metropolitan Expressway: Tokyo Olympics Air Operations" broadcast April 5, 2005.

59. Kōda Rohan writes in Meiji 32 (1899) that the fine customs of Edo have vanished and Tokyo has advanced rapidly into vulgarity.
Kōda (1993). Reference is written in Japanese.

60. The modern Japanese style was formulated through a reinterpretation of traditional Japanese architecture with modernist aesthetics by architects Yoshida Isoya, Murano Togo, and Horiguchi Sutemi. It has since provided inspiration to contemporary "Japanese" designs.

61. The Shinkansen (bullet train) was likely initially called the *dangan ressha* (literal translation of "bullet train"), following the English expression.

62. The golden age of streetcars in Japan was around 1932, when the streetcar ran through 67 of the nation's cities. There were 83 operators and the tracks ran a total of 1,479km (according to a Feb. 2011 article in *Tokyo Shimbun*). The total number of passengers per day on

150万人であった。地下鉄は1950年代までは二路線しかなかったが60年代以降に急速に路線数を増やし、それに合わせるように、都電の路線が漸次廃止され1972年に二系統だけになり今日に及んでいる。

toden, the streetcar service in Tokyo, reached 1.08 million in 1919. The *toden* reached its peak in 1962, when its tracks stretched a total of 213km and its total passengers per day numbered 1.5 million. In contrast, the subway system consisted of only two lines until the 1950s, but this number grew rapidly after the 60s. And, as if to respond to this, *toden* was gradually discontinued until only two lines remained in 1972, which continue to be operated today.

63. 日本の交通事故死者数で史上もっとも多い1970年には、総数1万7765人、1万台あたり約1人を数えたが2013年には、それぞれ、4,373人、0.5人まで減っている。これには長年に亘る道路施設の改善、運転者マナーの向上、自動車の交通事故回避技術の革新などが貢献しているが、救急医療体制の整備などが死者数を減らしているという側面もある。交通事故による負傷者数も重度後遺障害者数は増えているのが現実である。

63. The road traffic death rate in 1970 was 17,765, or 1 person per every 10,000 cars. It was the year that marked the greatest number of traffic-related deaths in the history of Japan. By 2013, however, these numbers decreased to 4,373 and 0.5 respectively. Long years of work on improving road facilities and driving manners as well as innovations in accident prevention technologies have likely contributed to these numbers. Additionally, another aspect of this equation is the improved emergency medical system that has also contributed to reducing the number of traffic-related deaths. In reality, however, the total number of injured passengers and those who suffer severe, long-term effects have increased.

64. 宇沢弘文は、ある活動が第三者や社会に対して直接間接に与える被害(外部不経済という)の内発生者が負担しない分を「社会的費用」という概念で捉えようとしている。宇沢が計上してる費用は、道路建設費用、交通事故の被害、公害による被害、交通混雑による被害などである。宇沢『自動車の社会的費用』岩波書店、1974。同様な主張は上岡直見も行っている。上岡『クルマの不経済学』北斗出版、1996

64. Uzawa Hirofumi uses the concept of "social costs" in an attempt to understand the damages wrought by a certain activity directly or indirectly to any third person or society (external diseconomies) that are not shouldered by the instigator. The costs that Uzawa describes here include road construction costs, damage caused by traffic accidents, the effects of pollution and the damages that ensue from traffic congestion.
Uzawa Hirofumi (1974). Reference is written in Japanese. The same claim is made by Kamioka Naomi: Kamioka (1996). Reference is available in Japanese only.

65. 国土交通省「国土交通省道路統計年報2014　道路の現況」による(http://www.mlit.go.jp/road/ir/ir-data/tokei-nen/2014/nenpo02.html)。

65. Based on report: Ministry of Land, Infrastructure, Transport and Tourism, http://www.mlit.go.jp/road/ir/ir-data/tokei-nen/2014/nenpo02.html. Reference is written in Japanese.

66. 環境省「「自動車NOx・PM法の車種規制について」パンフレット｜自動車NOx・PM法」(http://www.env.go.jp/air/car/pamph/)。

66. Based on pamphlet: Ministry of the Environment, http://www.env.go.jp/air/car/pamph/. Reference is written in Japanese.

67. 首都圏では関東鉄道の常総線と竜ヶ崎線、神奈川中央交通の本厚木と辻堂駅発のバス路線の一部で自転車の持ち込みが可能である。長岡には自転車を持ち込めるバス路線はない。

67. In the Capital Region, bicycles are allowed on Kantō Railway's Jōso Line and Ryūgasaki Line as well as some of Kanachū's bus routes departing from Hon'atsugi and Tsujidō stations. There are no bus routes in Nagaoka that allow passengers to carry on bicycles.

68. 国土交通省関東地方整備局「東京湾水環境再生計画(案)—美しく豊かな東京湾のために」2006の「2. 東京湾及びその流域の概要」による(http://www.ktr.mlit.go.jp/ktr_content/content/000010108.pdf)。

68. Based on document: Ministry of Land, Infrastructure, Transport and Tourism. Kanto Regional Development Bureau, http://www.ktr.mlit.go.jp/ktr_content/content/000010108.pdf. Reference is written in Japanese.

69. 都市の水辺の重要性、魅力についてもっとも持続的かつ精力

69. The architectural historian Jinnai Hidenobu has been

的な研究と活動をつづけるのが建築史学者陣内秀信である。法政大学エコ地域デザイン研究所所長をつとめる陣内秀信は、イタリア、日本、中国などの都市を対象に、都市史と文化人類学的関心をもって、都市空間と水に関する共同研究を進め、膨大な研究蓄積を築いてきた。陣内の研究は、この分野の研究に明確な輪郭を描き、現実の都市政策にも大きな影響を与えている。

70. 政府系シンクタンクである日本生産性本部のなかの政策発信組織「日本創成会議」の人口減少問題検討分科会（座長・増田寛也元総務相）は、2014年に、若年女性の流出により2040年に全国の896市区町村が「消滅」の危機に直面するという予測を発表した。内容には異論もあるが「縮小」問題の認識を一般化した功績は大きい。増田寛也編著、『地方消滅─東京一極集中が招く人口急減』中央公論新社、2014

71. 冨山和彦『なぜローカル経済から日本は甦るのか』PHP研究所、2014

72. 首都圏だけで見ると、集合住宅居住世帯は1968年に36%で、そのうち3/4が木造で狭小なアパートであった。2010年には集合住宅居住世帯は55%に増えた。その52%が分譲住宅である。分譲住宅はほぼ例外なく非木造である。

73. 1921年から分譲の大船田園都市（現鎌倉市、東海道線）で平均約495 m²、1921年から分譲の藤沢片瀬西浜（現藤沢市、東海道線）で平均約590 m²、1925年から分譲のひばりが丘（現西東京市、西部池袋線）で平均約830 m²など。片木篤他『近代日本の郊外住宅地』鹿島出版会、2000

74. 国土交通省土地・水資源局の2003年の調査によれば、東京23区に住む年間収入が500～700万円未満の世帯（日本の世帯の平均所得は528.9万円）の平均所有敷地面積は75 m²である。

75. 長岡では、地区計画制度を使って最低敷地面積を200 m²と決めている分譲住宅地が多い。

76. もう一つの労働力の不足を補う方法は、多くの先進諸国のように移民を受け入れることである。日本政府は、制度的には非常に慎重に対応している。しかし、実態は既に多数の外国人労働者が日本国内で働き、首都圏でも地方都市でも、特定の民族居住地区が徐々に形成されはじめている。1990年代までの日本における不法移民については、サスキア・サッセンも触れている。サッセ

the most consistent and vigorous researcher and advocate regarding the importance of watersides in the city. Jinnai, the Director of the Laboratory of Regional Design with Ecology, has cultivated an enormous accumulation of research by advancing joint studies on urban spaces and water in Italy, Japan and China through a cultural anthropology and urban history outlook. Jinnai's research gives clear definition to this field and has had great influence on current urban policies.

70. In 2014, the Japan Productivity Center's Japan Policy Council, a government-affiliated think tank that announces government policies (headed by former Minister of Internal Affairs and Communications Masuda Hiroya), published a projection that 896 of the nation's municipalities will face the danger of "death" by 2040 due to the outflow of young women. While there have been some objections to what it purports, it has achieved the huge feat of popularizing awareness towards the issue of "shrinkage."
Masuda (2014). Reference is written in Japanese.

71. Tomiyama Kazuhiko (2008). Reference is written in Japanese.

72. Looking only at the Capital Region, 36% of households lived in apartment buildings in 1968. Of these, 3/4 were small, wooden apartment buildings. This percentage of apartment dwellers increased to 55% in 2010, of which 52% were purchased. These privately-owned apartment units were almost, without exception, not wooden buildings.

73. The apartment units (for sale) in Ōfuna Garden City from 1921 were approx. 495 m² in area: in Fujisawa Katase West Beach (currently Katase City, Tokaidō Line) from 1921, they were approx. 590 m²; and in Hibarigaoka (currently Nishitokyo City, Seibu Ikebukuro Line) from 1925, they were approx. 830 m², among other examples.
Katagi et al. (2000). Reference is written in Japanese.

74. According to a 2003 survey by the Ministry of Land, Infrastructure, Transportation and Tourism's Water Resources Department, the average owned lot size for households in the 23 wards of Tokyo with an average annual income of under 5 - 7 million yen was 75m² (the average income of households in Japan is 5.289 million yen).

75. In Nagaoka, many housing lots for sale are fixed with a minimum site area of 200m², in accordance to the district planning system.

76. Another means of supplementing the shortage of labor would be to accept immigrants, like many other areas of the developed world. The Japanese government has approached this in a very cautious manner. However, in reality, many foreign workers have already begun to work in Japan and communities of foreign residents have

gradually formed both in the metropolis as well as local cities. Saskia Sassen also mentions illegal immigrants in Japan up to the 1990s in her book.
Sassen, *Globalization and Its Discontents* (1998).

77. There were, at one point in time, five movie theaters in Nagaoka, however, the last one was closed in 1990. Later, in 2007, a large cinema complex equipped with 10 screens was constructed in the newly developed urban area on the left bank of the Shinano River.

78. In 1990, the European Commission published the *Green Paper on the Urban Environment*, within which the compact city was recommended as a sustainable urban form. The 1998 opinion paper on "Towards an Urban Agenda in the European Union" clarified the EU's position in promoting the compact city.
Kaidō (2001). Reference is written in Japanese.

79. Kaidō *Ibid*.

80. The compact city concept first appeared in government policies under the name "intensive urban structure" in 2006 at the Ministry of Land, Infrastructure, Transport and Tourism's Panel on Infrastructure Development. By 2009, a spatial vision called the "eco-compact city" was mentioned in a report by a subcommittee established within the panel for investigating basic issues and directionalities in urban planning, and this became the vision for the future city. It clearly stated that, based on the theories of selection and concentration, new investments would not be made in areas other than core areas. This vision was later toned down with investigations related to reconstruction efforts after the Great East Japan Earthquake. By 2014, a system for suitable land use planning was established based on the Act on Special Measures concerning Urban Reconstruction. According to this system, it is important to advance the "compact city plus network" and rethink the entire structure of the city including welfare and transportation – this includes medical, welfare, commercial and residential districts located in close quarters as well as accessibility to necessary facilities through public transportation for all residents including the elderly.
Ministry of Land, Infrastructure, Transport and Tourism, http://www.mlit.go.jp/common/001050341.pdf.
Reference is written in Japanese.

Part 2

81. Fiber comes from the English word meaning filaments or threadlike elements. Fibercity thus denotes a fiber-like city, named so from the richly versatile quality of fibers that become cloth when woven together.

82. Matsubara Ryūichirō is an economist who harshly criticizes the cityscapes of Japan. Matsubara believes

れは「清潔な廃墟を思わせる奇妙な景観」と糾弾されている。松原『失われた景観―戦後日本が築いたもの』PHP研究所、2002、p.26

83. クリスチャン・ノルベルグ・シュルツやエドワード・レルフなどのハイデガーに多かれ少なかれ影響を受けた場所論は日本の建築家にはよく読まれている。これらの著者は、現代都市には場所が喪失したことを強調し、レルフはその原因として情報通信と交通、つまり「流れ」の高速化と大衆化であるとしている。レルフ『場所の現象学―没場所性を越えて』高野岳彦他訳、筑摩書房、1999、p.165

84. レルフ上掲書、p.165

85. 首都圏を都心から70 km圏とすると、この範囲内に3600万人余が住む。このうち、10〜50 km圏を郊外とするなら、そこには76.5％の人が住む。さらに、10〜20 km圏は鉄道郊外と言える地域であり、その外は自動車郊外であり、そこに54.5％の人が住む。ただし、首都圏の郊外人口を同心円状の土地利用モデルに基づいて算出することは限界がある。その理由は、第一に首都圏は多芯的だからであり、第二に、都心から遠くても鉄道沿線は鉄道を利用し、都心に近くても駅から遠ければ自動車依存の生活になるからである。（各圏域の人口は、国勢調査2010年に基づいて高津伸司が作成した資料による。「第25回ハイライフセミナー郊外に明日はあるか」公益財団法人ハイライフ研究所、2014 http://www.hilife.or.jp/wordpress/?cat=137）

86. ジョルジュ＝ウジェーヌ・オスマンは1853年から1870年までセーヌ県知事を務め、ナポレオン3世の構想に沿った大規模なパリの改造事業を率いた。フランソワーズ・ショエは「まさにサーキュレーションと呼吸のシステムと言う二重のコンセプト」と呼んでいる。ショエ『近代都市―19世紀のプランニング』彦坂裕訳、井上書院、1983、p34

87. パリの大改造とパリコンミューンを結びつけるのはベンヤミンである（ヴァルター・ベンヤミン『パサージュ論Ⅰ―パリの原風景』今村仁司他訳、岩波書店、1993）。ショエは、このような軍事的、政治的説明を退ける。彼はオースマンのパリの近代性を強調し、交通と衛生上の合理性を強調する（ショエ上掲書）。

88. 森鴎外は、文芸誌『スバル』（1909-1913発行）に、ドイツ在住の「無名氏」を装って、毎号「椋鳥通信」という文章を寄せていた。これは、世界の出来事を1、2カ月の時間差で伝えるという速達性が目指された。マリネッティの未来派宣言はフランスの新聞「フィガロ」の1909年2月20日の第一面を飾ったが、同年5月1日発行の

suburbs to be the greatest abomination and denounces them as bizarre landscapes reminiscent of an immaculate ruin.
Matsubara (2002). Reference is written in Japanese.

83. Christian Norberg-Schulz and Edward Relph's theories of placeness, influenced more or less by Heidegger, have been read widely by architects in Japan. These authors emphasize the fact that places have been lost in the modern city. Relph further cites the causes of this as telecommunications and traffic – in other words, the acceleration and popularization of "flows."
Edward Relph, *Place and Placeness* (London: Pion, 1976).

84. *Ibid.*, 65-66.

85. If we consider the Capital Region to be the range within 70km from the city center, over 36 million people live in this area. And if the 10-50km range within this is considered the suburbs, 76.5% of the people in the Capital Region are suburbanites. Furthermore, the 10-20km range can be called railway suburbs, while areas outside this can be considered automobile suburbs, where 54.5% of the population live. However, there is a limit to computing populations of suburbanites in the Capital Region simply by using a concentric land use model. The reasons for this are: first because the Capital Region is polycentric; second, because people who live along the railroad can use trains even if they live far from the city center and those who live far from the station must depend on cars even if they live close to the city center. The populations of each area have been calculated from on documents created by Takatsu Shinji based on the 2010 National Census.
Research Institute for High-Life, http://www.hilife.or.jp/wordpress/?cat=137. Reference is written in Japanese.

86. Georges-Eugène Haussmann was the Prefect of the Seine Department in France from 1853 to 1870 and led the large-scale renovation of Paris in accordance to Napoleon III's plans. Françoise Choay calls this plan "the dual concept of a circulatory and respiratory system."
Françoise Choay, *The Modern City: Planning in the 19th Century*, trans. Marguerite Hugo and George R. Collins (New York: G. Braziller, 1969), p. 19.

87. Walter Benjamin ties the great renovations of Paris to the Paris Commune (Walter Benjamin, *Das Passagen-Werk*, ed. Rolf Tiedemann (Frankfurt am Main: Surkamp Verlagg, 1982).). Choay avoids this kind of military/political explanation, emphasizing instead the modernity of Haussmann's Paris and its practicality in terms of transportation and hygiene (Choay 1969).

88. Marinetti's *Futurist Manifesto* decked the front page of the French newspaper *Le Figaro* on February 20, 1909.
F.T. Marinetti, *Critical Writings*, ed. Gunter Berghaus, trans. Doug Thompson (New York: Farrar, Straus and Giroux, 2006), p. 13.

It was translated into Japanese by Mori Ōgai, one of the representative writers of the Meiji era, and was published in the literary magazine *Subaru* on May 1, three months after its initial publication in France. Ōgai published a series entitled "Mukudori Tsūshin (Correspondence from the Grey Starling*)" in every monthly issue of *Subaru* (published 1909-1913) under the guise of an anonymous writer living in Germany. "Correspondence from the Grey Starling" aimed to provide the speedy delivery of world news 1 to 2 months after their occurrence.
Tōru Yamaguchi (2005). Reference is written in Japanese.
*The title is a transliteration of the original Japanese title.

89. CIAM is an acronym for the French term Congrès International d'Architecture Moderne, meaning the International Congress of Modern Architecture. It was an international conference of architects and critics who led the modern architecture movement and was held 11 times between 1928 and 1959.

90. Metabolism was a Japanese urban movement that argued that architectural and urban changes should accompany social changes and population growth. Kurokawa Kishō, Kikutake Kiyonori and Kawazoe Noboru led the publishing of what may be called its manifesto, *Metabolism/1960 – Proposals for a New Urbanism*, at the World Design Conference in 1960.

91. Archigram was a British avant-garde architecture group formed in 1961. It was also the name of a magazine that the group organized and published in London. While their architectural theories were similar to those of the Metabolists, their unique graphical expressions were outstanding and were a vanguard of their times.

92. Jane Jacobs, *The Life and Death of Great American Cities* (New York: Random House, 1961).

93. Christopher Alexander, "A City is Not a Tree," *Architectural Forum* 122, no.1 (1965), 58-62. (Part I) Alexander, "A City is Not a Tree," *Architectural Forum* 122, no.2 (1965), 58-62. (Part II).

94. Tim Ingold is an anthropologist who has studied the various cultural meanings of lines. In his book *Lines: A Brief History*, he differentiates between peripatetic walking, in which the route itself has meaning, and transport, in which only the destination matters. He then takes this to differentiate networks that join places of activity with lines from meshworks composed of knots that appear as a result of the different trails taken. Textures are a concept very similar to that of meshworks. Tim Ingold, *Lines: A Brief History* (New York: Routledge, 2007).

95. The poet Ishikawa Takuboku from Iwate Prefecture sang of Ueno, the final stop on the Tohoku line: "I slip into the crowd/ Just to hear the accent/ Of my faraway hometown."

Roger Pulvers, *The Illusions of Self: 200 Tanka by Ishikawa Takuboku* (Tokyo: Kawade Shobo Shinsha, 2015).

96. ミルチャ・エリアーデは「居住地や都市の防御施設は、おそらく初めは魔術的目的のたであったに違いない」と述べている。エリアーデ『聖と俗―宗教的なるものの本質について』風間敏夫訳、法政大学出版局、1969、p.41

96. Mircea Eliade states that "it is highly probable that the fortifications of inhabited places and cities began by being magical defenses."
Mircea Eliade, *The Sacred and the Profane* (New York: Harcourt Brace & World, 1959), p.49.

97. 日本の伝統的空間を設計の方法論として分析した研究書でもっとも優れたものである。そこでは「形成の原理」として、方位、重畳、布石、天地人、真行草、さおび、ま(間)、かいわい、を上げている。その一つである「重畳」は「中心核のまわりに一定の性質をもった空間を付加またはたたみこんで出来た空間である」としている。都市デザイン研究体『日本の都市空間』彰国社、1968、p.32

97. The leading study on traditional Japanese spatial design methodologies notes the fllowing elements: *hōi* (orientation), *chōjō* (heirarchical accessibility), *fuseki* (placement due to circumstances), *tenchijin* (esthetical triangle), *shin-gyō-sō* (formal-informal), *saobi* (process designing), *ma* (imaginary space) and *kaiwai* (activity space). Of these, *chōjō* is considered a space that is formed by adding or folding spaces with a certain quality around a core.
Toshi Dezain Kenkyūkai (1968). Reference is written in Japanese.

98. 槇文彦の集合体の研究は滞米時代になされた。日本語訳は「集合体に関するノート」の第二章、第三章が相当する。『JA16号 槇文彦』新建築社、1994

98. Maki Fumihiko's study on collective groups was conducted during his time in the United States.
Fumihiko Maki, *Notes in Collective Form* (Washington University, 1964).

99. 藻谷も「国全体としての方向転換は一朝一夕には行かないだろう。だからこそ、市町村単位、県単位、地方単位での取り組みを先行させることが、事態の改善につながっていく」と述べ、改革は身近なスケールから進めるべきだと主張している。藻谷上掲書、p.137

99. Motani states that the whole nation cannot simply change its gears overnight; this is why a situation can be improved through approaches on the municipal, prefectural and regional scales. He emphasizes that reforms should be advanced from a more local scale.

100.「日本法人であるアマゾンジャパン株式会社はシステム運営・顧客サービスを担当する」にすぎず、商品の売主は、「米国ワシントン州法人であるAmazon.com Int'l Sales, Inc.であり……日本で稼得した利益に対して日本の法人税を支払っていなかった」。国税庁は「2003年から2005年について140億円の追徴課税を行った」がその後協議により大幅に減額されたという。国際企業の租税回避問題がここでも起こっている。括弧内の引用は、ウィキペディア「Amazon.co.jp」による (http://ja.wikipedia.org/wiki/Amazon.co.jp)。

100. Amazon Japan is a Japanese subsidiary of Amazon that is only in charge of systems management and customer service. The vendor of the products is Amazon.com Int'l Sales head quatered in Washington (US), and no corporate taxes are paid to Japan for the profits earned in Japan. The National Tax Agency required the payment of 14 billion yen in additional taxes for the period between 2003 and 2005, which was later reduced greatly after discussions. This demonstrates another instance of the problem of tax avoidance by international corporations.
Wikipedia, http://ja.wikipedia.org/wiki/Amazon.co.jp. Reference is available in Japanese only.

101. ムハマド・ユヌス『貧困のない世界を創る―ソーシャル・ビジネスと新しい資本主義』猪熊弘子訳、早川書房、2008

101. Muhammad Yunus, *Creating a World Without Poverty* (New York: Public Affairs, 2007).

102. 現代日本の建築家の間では、小さい介入に対する関心が高まっている。そのなかでも、「小さな矢印」という概念で場から微気候まで含めて連続的に捉えようとする小嶋一浩がいる。小嶋『小さな矢印の群れ―「ミース・モデル」を超えて』TOTO出版、2013
隈研吾は長らく小さい要素を組み上げることで建築を作ることに挑戦して、建築表現に新境地を切り開いた。隈『小さな建築』岩波書店、2013

102. There is an increasing interest in small flows among modern Japanese architects. Kojima Kazuhiro attempts to understand everything from places to microclimates continually through the concept of "small arrows" (Kojima, 2013). Kuma Kengo has long attempted to make architecture through the combination of small elements and has opened up a new frontier in architectural expressions (Kuma, 2013). Both references written in Japanese.

103. アメリカの政治学者フランシス・フクヤマは経済活動における信頼の役割の重要性を論じている。フランシス・フクヤマ『「信」無くば立たず―「歴史の終わり」後、何が繁栄の鍵を握るのか』加藤寛訳、三笠書房、1996

104. 広井によれば、神社は約8万1000、お寺は8万6000あり、コンビニ5万よりはるかに多いという。広井はそれを利用して「鎮守の森・自然エネルギーコミュニティ構想」を提唱している。広井、上掲書、p.74

105. 宗教建築や境内の森などは宗教法人の、いわば私的な資産なので、宗教法人が収入を増やす目的から、境内の貴重な建物や境内を売却する恐れがある。それらは、ときには、文化資産であり、境内は都市の公共空地としての価値が高いことがあるだろう。その喪失は都市空間にとって取り返しがつかないことになる。また、大都市では、墓地などを未利用容積と捉えて、容積の一部を買い取ろうという開発業者も現れるであろう。そうなると都市の過密化が進行する。いずれにしろ一宗教法人の問題を超えて、都市の公共的な価値の毀損である。それゆえ、公的な経済支援でこれらを守ることが必要な場合があるだろう。

106. 首都高速道路の長さは2011年、通行車輌数(日平均)は2010年、走行台キロは1995年の調査による。「首都高速道路構造物の大規模更新のあり方に関する調査研究委員会(第1回委員会資料)」首都高速道路株式会社、2011 (http://www.shutoko.co.jp/company/interview/~/media/pdf/corporate/share2011/company/info/interview/doc/01_04.pdf)。

107. 政府の中央防災会議の予測(2005)による (http://www.bousai.go.jp/kaigirep/chuobou/senmon/shutohinan/1/pdf/shiryou_2.pdf)。

108. 内閣府の調査(2012)による (http://www.bousai.go.jp/jishin/syuto/kitaku/2/pdf/4.pdf)。

109. 都道府県の公安委員会は、首都高などの自動車専用道路を緊急自動車専用路に指定し、地上の幹線道路を緊急交通路に指定している。路上にナマズのキャラクターをあしらった看板が表示されている。

110. 東京都は、2011年4月に施行した「東京における緊急輸送道路沿道建築物の耐震化を推進する条例」を制定して、沿道の建物が倒れて道を塞がないように耐震診断や耐震改修に関する実施状況報告書の提出を求め、改修の補助金なども用意している

103. American political scientist Francis Fukuyama discusses the importance of trust in economic activities. Francis Fukuyama, *Trust: The Social Virtues and The Creation of Prosperity* (New York: Free Press Paperbacks, 1996).

104. According to Hiroi, there are 81,000 shrines and 86,000 temples, far outnumbering the 50,000 convenience stores in Japan. Hiroi further advocates a ChinjunoMori (Grove of the Village Shrine) and natural energy community plan using these shrines and temples. Hiroi Yoshinori (2013). Reference is written in Japanese.

105. The forests and precincts of religious buildings belong to religious corporations and are private property, so to speak. Thus there is a danger that these religious corporations may sell the precious buildings on their precincts or the precinct itself for profit. These are, at times, cultural properties, and temple/shrine precincts are highly valuable as public spaces in the city. The loss of such spaces will cause irreparable damage to urban spaces. At the same time, in the metropolis, developers may begin to buy up unused FAR (floor area ratios) from graveyards. This will advance urban congestion. In any case, this goes beyond the simple issue of any single religious corporation and becomes damaging to the public value of the city itself. Therefore, in some cases, such spaces must be protected with public funds.

106. The length of the Metropolitan Expressway is based on a 2011 survey, the number of vehicles per day is based on a 2010 survey, and traffic flow in vehicle-kilometer is based on a 1995 survey. Metropolitan Expressway Company, 2011, http://www.shutoko.co.jp/company/interview/~/media/pdf/corporate/share2011/company/info/interview/doc/01_04.pdf. Reference is written in Japanese.

107. Based on predictions by the government: Central Disaster Prevention Council, 2005, http://www.bousai.go.jp/kaigirep/chuobou/senmon/shutohinan/1/pdf/shiryou_2.pdf. Reference is written in Japanese.

108. Based on the Cabinet office's survey (2012):Cabinet Office, 2012, http://www.bousai.go.jp/jishin/syuto/kitaku/2/pdf/4.pdf. Reference is written in Japanese.

109. The prefectural public safety commission has specified expressways like the Metropolitan Expressway as emergency expressways and arterial roads at ground level as emergency traffic routes. These roads are equipped with signs that show a catfish character on them.

110. Tokyo established an ordinance in April 2011 promoting seismic retrofitting of roadside buildings to secure emergency transportation roads during times of crisis. The ordinance required the submittal of a report

（http://www.shutoko.co.jp/efforts/prevention/earthquake/）。

道路に関する防災計画を調べてみると、信じられないような縦割りを発見する。上述の「緊急輸送道路」は各自治体の道路管理（建設関係）部局が指定する道路であるが、公安委員会（交通管理部局）は同じ道路に別の名称「緊急交通路」を与え、別の防災体制を組んでいる。後者に比べて前者が多くの道路を指定している。市民にとっては混乱のもとでしかない。

111. 「首都高の取り組み」首都高速道路株式会社による（http://www.shutoko.co.jp/efforts/prevention/earthquake/）。

112. 都内には84カ所の地域冷暖房システムがあり、そのうち64カ所の地域冷暖房システムが〈緑の網〉の沿道にある。下水処理場と清掃工場もエネルギー供給の可能性をもった施設である。下水処理場は対象地域内には16カ所あり、清掃工場は10カ所ある。

113. エネルギー源として化石燃料だけではなく、再生可能エネルギーとの融合を目指すことが必要である。具体的には、熱効率、環境性を高くできるようコージェネレーション、燃料電池、太陽光・風力発電、バイオマスなどを複合的に利用するホロニックエネルギーシステムの考え方を取り入れる。同システムは東京大学浅野研究室と東京ガスで共同研究を進めている。このようなプラントの能力を試算すると、プラント規模はコージェネレーション、熱供給、バイオマスの各プラントにそれぞれ約2,000 m²の規模を想定すれば、供給能力は、電力量20,000 kW、熱量18,000 kWになる。このシステムを導入した場合、従来の個別空調と比較して約37％の省エネルギー、約36％のCO_2削減効果が得られる。また、配管敷設やプラント建屋建設について、約50％のコストダウンが期待できる（東京ガス株式会社の技術的検討による試算、2006年時、fc 25JE）。

114. 政府は、2001年に「大深度地下の公共的使用に関する特別措置法」を定めて、地上の土地所有権が地中に深く及ばないようにして、将来の大深度地下空間の公共利用に向けた準備を進めている。

regarding seismic capacity evaluation and retrofitting so that roadside buildings do not fall on the road and become roadblocks. They also provide grants related to retrofitting.
Metropolitan Expressway Company, http://www.shutoko.co.jp/efforts/prevention/earthquake/. Reference is written in Japanese.
A survey of road-related disaster prevention plans reveals an unbelievably overcompartmentalized bureaucracy. While the abovementioned emergency transportation roads are designated by municipal road management (construction-related) departments, the public safety commission (traffic department) gives a different name to the same roads (emergency traffic routes) and forms a different disaster prevention system for them. As opposed to the latter, the former designates more roads. For the people, they only cause more confusion.

111. Metropolitan Expressway Company, Efforts of Shutoko, http://www.shutoko.co.jp/efforts/prevention/earthquake/. Reference is written in Japanese.

112. There are 84 district heating and cooling facilities in Tokyo, 64 of which are located along Green Web. Sewage and waste incineration plants are also potential energy supply facilities. There are 16 sewage treatment plants and 10 waste incineration plants in the target area.

113. In an attempt to combine fossil fuels and renewable energy for energy sources, a Holonic Energy System that employs a combination of the concepts of cogeneration, fuel cells, photovoltaic power generation, wind power generation, and biomass is adopted to raise thermal efficiency and apply ecological principles. This system is currently under development through joint research between Professor Asano's Laboratory at the Department of Mechanical Engineering at The University of Tokyo and Tokyo Gas Co., Ltd.
If the size of each plant assumes a scale of about 2,000m² for cogeneration, heat supply and biomass, it can generate about 20,000kW of electrical power and 18,000kW of thermal energy. When this system is introduced, compared to conventional individual air conditioning systems, there will be a 37% reduction in energy and a 36% reduction of CO_2 emissions. Moreover, a cost reduction of approximately 50% for piping installation and plant construction can be expected. (Based on technological studies made by Tokyo Gas Co., Ltd. in 2006, fc24JE)

114. The government established a special measure on the public use of the deep underground in 2001. This limits land ownership above ground so that it does not extend to the deep underground. With this, preparations have been underway for the public use of deep underground spaces in the future.

115. 首都高の評価は時代によって変わってきた。神村崇宏によると、建設当初は、高度経済成長期でもあり、観光ガイドに「美しい曲線」とか「都会的」として賞賛されていたのだが、70年代になると、公害問題が頻出し、否定的な評価に変わった。神村他「首都高速道路のイメージ変遷に関する研究」環境システム研究、土木学会、1996
また、篠原修は、首都高が「日本橋と弁慶橋の景観を破壊した」と難じている。篠原他「首都高速道路の景観評価」第4回土木史研究発表論文集、1984

116. 地元の企業を中心に「日本橋地域ルネッサンス100年計画委員会」が1999年に組織され、日本橋上空の首都高撤去を訴え、それを梃に地域の再開発を提案している。

117. 地区計画は1980年の都市計画法改正で定められた制度である。一定の地区の住民の合意に基づいて、地区内の建築物の形態の制限を一般的な形態規制に上乗せの規定を課すことができる。

118. BRTは簡単に言えば、路面電車の軌道をバス専用レーンに置き換えたものである。道路中央部に専用レーンを設けることで乗り心地が改善され、他の交通に邪魔されず定時運行性や高い安全性など、路面電車の良さをもったバス路線が実現できる。

119. 乗客の要求に対応して運行する乗合形態の公共交通システム。時間もルートも希望に応じる方式からルートと停留所が決まっていて選択する方式まで幅広くある。また車輌でも、バスを使うものが主流であるがタクシーを使うものもある。

120. 道路構造令で車線(レーン)幅は、2.75 mから3.5 mと規定されている。ただし、状況により0.25 mの加減が認められている。

121. セミラチスという系の特性を表す用語は、アメリカの建築家クリストファー・アレグザンダーによって導入された。アレグザンダーは、セミラチスをツリーと対比的に用い、現実の都市はセミラチスであるのに、近代都市計画はツリー的な組織で思考すると批判し、1960年代の近代建築批判を代表する著作となった(原著"A CITY IS NOT A TREE"は1965年に発行された)。アレグザンダー『形の合成に関するノート／都市はツリーではない』稲葉武司、押野見邦英訳、鹿島出版会、2013

122. 1920(大正9)年に東京市には12の渡船場があり年間227万人が利用していた。東京都『都史紀要三十五　近代東京の渡船と一銭蒸汽』東京都公文書館、1991、p.1

123．つくばエクスプレス線は、つくば学園都市と秋葉原(将来計画では東京駅まで)を結ぶために、2005年に開業した鉄道路線で

115. Evaluations of the Metropolitan Expressway have changed over time. According to Kamimura Takahiro, it was initially praised for its beautiful curves and urban nature in tourist guidebooks when it was first opened during the period of high economic growth. However, in the 1970s, there came to be a heightened awareness towards the problem of environmental pollution, and opinions took a negative turn (Kamimura et al., 1996).
At the same time, Shinohara Osamu criticizes the Metropolitan Expressway as having destroyed the landscape of Nihonbashi and Benkeibashi. Shinohara et al. (1984). Reference is written in Japanese.

116. The Committee for the 100-Year Renaissance Plan for Nihonbashi, led by local companies, was established in 1999. They call for the removal of the Metropolitan Expressway above the Nihonbashi and propose a redevelopment of the region based on this removal.

117. The "district planning system" was a system established with the revised City Planning Act in 1980. It allows the additional regulation of building form in the district on top of general restrictions on building form.

118. BRT is a system that replaces streetcar tracks with dedicated bus lanes. Setting aside a lane dedicated to buses in the middle of the road would make them more comfortable, heighten safety and improve punctuality, as there would be no other form of transportation in the way. Such a system would realize a bus route that incorporates the benefits of streetcars.

119. On-demand buses are a public transportation system that operates in response to passenger needs. A wide range of such buses exist, including systems that respond to both time and route demands as well as those in which bus stops are predetermined. Most of these systems use buses, but some also use taxis.

120. The Road Structure Ordinance sets the width of car lanes between 2.75 and 3.5m. However, adjustments in a 0.25m range is allowed in certain cases.

121. The term semi-lattice to express this quality was first employed by the American architect Christopher Alexander. Alexander contrasts the semi-lattice structure with the tree structure and criticizes modern city planning for being based on a tree-like organization. He has become a representative author of modern architectural criticism in the 1960s.
Alexander, *A City is Not a Tree* (1965).

122. There were 12 ferry landings in Tokyo in 1920 (Taishō 9), used by 2.27 million people per year.
Tokyo Prefecture (1991). Reference is written in Japanese.

123. The Tsukuba Express is a railway established in 2005 to connect Tsukuba Science City with Akihabara (to

ある。首都圏ではもっとも新しい郊外鉄道路線である。沿線開発と鉄道を一体化した開発は、現在までのところ概ね成功し、沿線居住者が着実に増え、鉄道会社の経営状況も良好である。

124. JR東日本「路線別利用状況(2012)による(http://www.jreast.co.jp/rosen_avr/pdf/1987-2012.pdf)。

125. バイオーム(Biome)は生物群系と訳され、ある気候条件(特に気温と降水量)下で安定した動植物の集りのことである。熱帯雨林、サバナ、ツンドラなど、それぞれの気候帯ごとに、異なるバイオームがある。また、そのようなバイオームを再現する人工施設のこともバイオームと呼ぶことがある。バイオームの先行事例としては、アメリカ合衆国のバイオスフィア2(1991)、イギリスのエデンプロジェクト(2001)がある。

126. 首都高速道路は、政府組織である首都高速道路公団によって作られ運営されていたが、2005年に民営化され、道路と付属施設を日本高速道路保有・債務返済機構が所有し、首都高速道路株式会社が同機構から借り受けて運営している。これに対して〈生命の回廊〉の対象としている東京高速道路は当初から民間会社が保有・運用してきた。

extend to Tokyo Station in the future). It is the newest suburban railway line in the Capital Region. Unifying railway development with development of the neighborhoods along it, it has for the most part succeeded at present; the number of residents along the railway has steadily increased and the railway company enjoys good economic success.

124. Based on railway use conditions surveyed by the East Japan Railway Company in 2012:
JR East, 2012, http://www.jreast.co.jp/rosen_avr/pdf/1987-2012.pdf. Reference is written in Japanese.

125. Biomes are communities of plants and animals in stable states under certain climate conditions (especially temperature and rainfall). Different biomes exist for each of the climatic zones, including tropical rainforests, savannas and tundras. At the same time, the term biome is also used to describe manmade facilities that reproduce these kinds of biomes. Examples include Biosphere2 (1991) in the United States and the Eden Project (2001) in the United Kingdom.

126. The Metropolitan Expressway was built and operated by the Metropolitan Expressway Public Corporation (a government organization), which was privatized in 2005. The road and accessory facilities belong to the Japan Expressway Holding and Debt Repayment Agency, and the Metropolitan Expressway Company rents and operates them. As opposed to this, the Tokyo Expressway under consideration in the Cloister of Life has been owned and operated by a private corporation from the outset.

図版出典

序

001 東京大学丹下健三研究室、1961 模型撮影：川澄明男

002 内閣府「平成25年版 高齢社会白書」(http://www8.cao.go.jp/kourei/whitepaper/w-2013/zenbun/index.html) 掲載のグラフを編集し描き直した。

Illustration Credits

Preface

001 Tange Kenzō Laboratory, The University of Tokyo (1961), Photo: Kawazumi Sumio.

002 Graph redrawn with data adapted from:
Government of Japan Cabinet Office, http://www8.cao.go.jp/kourei/whitepaper/w-2013/zenbun/index.html. (Reference is written in Japanese.)

第一部

101 United Nations Department of Economic and Social Affairs, Revision of the World Population Estimates and Projections,1998

102 下記資料掲載のグラフを描き直した。
United Nations Department of Economic and Social Affairs, *World Economic and Social Survey 2009: Promoting Development, Saving the Planet.* (2009), http://www.un.org/en/development/desa/policy/wess/wess_archive/2009wess.pdf

103 下記資料掲載のグラフを描き直した。
United Nations Environmental Programme, *The Emissions Gap Report 2014*, http://www.unep.org/publications/ebooks/emissionsgapreport2014.

104 内閣府『平成25年版 少子化社会対策白書』掲載のグラフを編集し、新たにグラフを作成した(http://www8.cao.go.jp/shoushi/shoushika/whitepaper/measures/w-2013/25pdfhonpen/25honpen.html)。

105 下記資料掲載の資料からグラフを作成した。
United Nations Department of Economic and Social Affairs, *2015 Revision of World Population Prospects* http://esa.un.org/unpd/wpp/

106 下記資料掲載の資料からグラフを作成した。
OECD.Stat, "General government gross financial liabilities, % of nominal GDP," *OECD Economic Outlook* No.94 (November 2013).

107 厚生労働省「平成26年簡易生命表」2015 http://www.mhlw.go.jp/toukei/saikin/hw/life/life14/index.html

108 総務省統計局「人口推計」(http://www.stat.go.jp/data/jinsui/)、東京都「東京統計年鑑」(http://www.toukei.metro.tokyo.jp/tnenkan/tn-index.htm)、長岡市「長岡市統計年鑑」(http://www.city.nagaoka.niigata.jp/shisei/cate12/h27/)掲載の資料からグラフを作成した。

110 都立中央図書館特別文庫室所蔵
111 南満洲鉄道株式会社編集『大連』1941
113 独立行政法人都市再生機構提供
114 各時代の鉄道路線地図は、正井泰夫編『アトラス東京 地図で読む江戸〜東京』平凡社、1986を基に著者が作成した。各時代の海岸線の形状は、谷謙二「今昔マップ on the web:時系列地形図閲覧サイト」(http://ktgis.net/kjmapw/kjmapw.html?lat=35.55534674744165&lng=139.90931483178713&zoom=11&dataset=tokyo50&age=1&map1type=roadmap&map2type=roadmap&dual=true&mapOpacity=10&altitudeOpacity=4) を基に著者が作成した。
下記以外の各時代の人口および長岡の人口は総務省統計局「人口推計」(http://www.stat.go.jp/data/jinsui/)による。
1632年の首都圏の人口については、斎藤誠治が1650年の江戸と小田原と川越の合計の人口を推計しているので、この数字を用いた(斎藤『江戸時代の都市人口』、地域開発(240)1984)。
1888年の首都圏の人口は、内務省編『国勢調査以前 日本人口統計集成2 (明治19-21年)』(東洋書林、1992)による

Part 1

101 United Nations Department of Economic and Social Affairs, *Revision of the World Population Estimates and Projections* (1998).

102 Graph redrawn with data adapted from:
United Nations Department of Economic and Social Affairs, *World Economic and Social Survey 2009: Promoting Development, Saving the Planet* (2009), http://www.un.org/en/development/desa/policy/wess/wess_archive/2009wess.pdf.

103 Graph redrawn with data adapted from:
United Nations Environmental Programme, *The Emissions Gap Report 2014*, http://www.unep.org/publications/ebooks/emissionsgapreport2014 .

104 Graph redrawn with data adapted from:
Government of Japan Cabinet Office, http://www8.cao.go.jp/shoushi/shoushika/whitepaper/measures/w-2013/25pdfhonpen/25honpen.html. (Reference is written in Japanese.)

105 Graph drawn with data adapted from:
United Nations Department of Economic and Social Affairs, *2015 Revision of World Population Prospects*, available from: http://esa.un.org/unpd/wpp/.

106 Graph drawn with data adapted from:
OECD.Stat, "General government gross financial liabilities, % of nominal GDP," *OECD Economic Outlook* No.94 (November 2013).

107 Japan Ministry of Health, Labour and Welfare, 2015, http://www.mhlw.go.jp/toukei/saikin/hw/life/life14/index.html. (Reference is written in Japanese.)

108 Graph drawn with data adapted from:
Statistics Bureau of Japan, http://www.stat.go.jp/data/jinsui/. (Reference is written in Japanese.)
Tokyo Metropolitan Government, http://www.toukei.metro.tokyo.jp/tnenkan/tn-index.htm. (Reference is written in Japanese.)
Nagaoka City Government, http://www.city.nagaoka.niigata.jp/shisei/cate12/h27/index.html. (Reference is written in Japanese.)

110 Tokyo Metropolitan Library Archives.
111 South Manchuria Railway Co.(ed.), Dairen (1941).
113 Provided by Urban Renaissance Agency.
114 Railway network maps for each year drawn based on:
Masai Yasuo, *Atlas Tokyo: Edo/Tokyo through Maps* (Heibonsha, 1986)
Coast line for each year drawn based on data from:
Tani Kenji , http://ktgis.net/kjmapw/kjmapw.html?lat=35.55534674744165&lng=139.90931483178713&zoom=11&dataset=tokyo50&age=1&map1type=roadmap&map2type=roadmap&dual=true&mapOpacity=10&altitudeOpacity=4. (Reference is written in Japanese.)
All population data, except that mentioned below, adapted from:
Statistics Bureau of Japan, http://www.stat.go.jp/data/jinsui/. (Reference is written in Japanese.)
Population total for 1632 is from Saito Seiji's calculation for the population in Edo, Odawara, and Kawagoe in

1914年の首都圏の人口は、内閣統計局編『国勢調査以前　日本人口統計集成16（大正2年-4年）』（東洋書林、1993）による。

115　東京大学施設部蔵
116　野田正穂他編『日本の鉄道 成立と展開』日本経済評論社、1986を基に地図を作成。
117　ファイバーシティに関する著者による公表物一覧 25JE
118　運輸政策研究機構『平成23年版　都市交通年報』(2013)を基に地図を作成。
119　Peter Newman、Jeffrey Kenworthy, *Sustainability and Cities*, Island Press (1999).
120　ファイバーシティに関する著者による公表物一覧 105JE、25JE
121　Ludwig Hilberseimer, *Groszstadt Architektur (Die Baubücher Band 3)* Stuttgart: Hoffmann, 1927).
122　国の経済成長統計の計算法はしばしば変わってきた。1998年までの成長率は1990年基準により、2014年までの成長率は2005年基準によって集計、発表されている。
国民経済計算確報1998年度国民経済計算（http://www.esri.cao.go.jp/jp/sna/data/data_list/kakuhou/files/h10/12annual_report_j.html）。
国民経済計算確報2014年度国民経済計算（http://www.esri.cao.go.jp/jp/sna/data/data_list/kakuhou/files/h26/h26_kaku_top.html）。

123　ル・コルビュジエ、ピエール・ジャンヌレ著、W. ボジガー、O. ストノロフ編『ル・コルビュジエ全作品集第一巻』吉阪隆正訳、A.D.A. EDITA (1979).
© FLC / ADAGP, Paris & JASPAR, Tokyo, 2016 G0305

124　撮影：大野秀敏、2004
126　経済産業省「商業動態統計」(http://www.meti.go.jp/statistics/tyo/syoudou/result-2/index.html) 掲載の資料からグラフを作成した。
127　東京については下記データを基に地図を作成。東京都コンビニエンスストアの緯度経度データ、高橋三雄「フリーソフトによるデータ実践GIS」第12回、http://www.sinfonica.or.jp/kanko/estrela/refer/s29/index.html
長岡については、『i-townpage』を基に地図を作成。

128　撮影：大野秀敏、2013
129　撮影：大野秀敏、2016
131　曽根悟『新幹線50年の技術史：高速鉄道の歩みと未来 』(講談社、2014)の他JR各社のホームページ を基に地図を作製。

132　撮影：大野秀敏、2016
133　撮影：大野秀敏、2010

1650. Saito Seiji (1984). (Reference is written in Japanese)
Population total used for 1888 is from Japan Ministry of Home Affairs, ed. (1992). (Reference is written in Japanese.)
Population total used for 1914 is from Japan Ministry of Home Affairs, ed. (1993). (Reference is written in Japanese.)

115　Tokyo University Facility Department Archives.
116　Map drawn with data adapted from:
Noda Masaho (1986). (Reference is written in Japanese)
117　See list of Author's Publications on Fibercity. 25JE.
118　Map redrawn with data adapted from:
Institution for Transport Policy Studies (2013). (Reference is written in Japanese.)
119　Peter Newman and Jeffrey Kenworthy, *Sustainability and Cities* (Island Press, 1999).
120　See Author's Publications on Fibercity. 105JE, 25JE.
121　Ludwig Hilberseimer, *Groszstadt Architektur (Die Baubücher Band 3)*（Stuttgart: Hoffmann, 1927).
122　Methods for calculating economic growth statistics for the nation have repeatedly changed over the years. The growth rate up to 1998 was tabulated and published based on standards from 1990, while the growth rate up to 2014 was based on standards from 2005.
Government of Japan Cabinet Office, 1998, http://www.esri.cao.go.jp/jp/sna/data/data_list/kakuhou/files/h10/12annual_report_j.html. (Reference is written in Japanese)
Government of Japan Cabinet Office, 2014, http://www.esri.cao.go.jp/jp/sna/data/data_list/kakuhou/files/h26/h26_kaku_top.html. (Reference is written in Japanese)
123　Le Corbusier et Pierre Jeanneret, W. Boesiger, O. Stonorov, *Le Corbusier et Pierre Jeanneret : Oeuvre Complète 1910-1929*, （Les Editions D'Architecture Erlenbach, 1948).
© FLC / ADAGP, Paris & JASPAR, Tokyo, 2016 G0305
124　Photo: Ohno Hidetoshi, 2004.
126　Graph drawn with data adapted from:
Japan Ministry of Economy, Trade and Industry, http://www.meti.go.jp/statistics/tyo/syoudou/result-2/index.html. (Reference is written in Japanese)
127　Maps for Tokyo drawn based on the following data:
Tokyo convenience store data adapted from:
Takahashi Mitsuo, http://www.sinfonica.or.jp/kanko/estrela/refer/s29/index.html. (Reference is written in Japanese.)
Nagaoka convenience store data adapted from:
NTT Directory Services Co., http://itp.ne.jp/?rf=1. (Reference is written in Japanese.)
128　Photo: Ohno Hidetoshi, 2013.
129　Photo: Ohno Hidetoshi, 2016.
131　Map drawn with data adapted from:
Sone Satoru (2014). (Reference is written in Japanese.)
Also used: the websites of each JR company.
132　Photo: Ohno Hidetoshi, 2016.
133　Photo: Ohno Hidetoshi, 2010.

134	ファイバーシティに関する著者による公表物 3J
135	撮影：大野秀敏、2013
136	撮影：澁谷 達典、2014
137	ファイバーシティに関する著者による公表物 25JE.
138	歌川広重「名所江戸百景 する賀てふ」(太田記念美術館蔵)
139	撮影：守屋佳代、2015
140	ボストン美術館蔵
141	明治大学工学部建築学科神代研究室『日本のコミュニティ』(SD別冊No.7)、鹿島出版会、1975
142	撮影：大野秀敏、2014
143	撮影：大島耕平、
144	撮影：向山裕二、2016
145	撮影：久保秀朗、2016
146	撮影：久保秀朗、2015
147	亀倉雄策(アートディレクション)、村越 襄(フォトディレクション)、早崎 治(写真)「第2号東京オリンピックポスター：スタートダッシュ」
148	日本百貨店協会『日本百貨店協会統計年報2013』2014
149	撮影：大野秀敏、2016
150	建設経済研究所「建設経済レポートNo.62」、2014 (http://www.rice.or.jp/regular_report/pdf/construction_economic_report/レポート全文/No.62/No.62.pdf) 掲載の資料からグラフを作成した。
151	建設経済研究所「建設経済レポートNo.59」、2012 (http://www.rice.or.jp/regular_report/pdf/construction_economic_report/レポート全文/No.59_201210.pdf) 掲載の資料からグラフを作成した。
152	ル・コルビュジエ、ピエール・ジャンヌレ著、W. ボジガー、O. ストノロフ編『ル・コルビュジエ全作品集第一巻』吉坂隆正訳、A.D.A. EDITA (1979)。 © FLC / ADAGP, Paris & JASPAR, Tokyo, 2016 G0305
153	「写真の中の明治・大正　東京編」国立国会図書館／近代デジタルライブラリー
154	撮影：大野秀敏、2016
155	国土交通省社会資本整備審議会道路分科会「第7回基本政策部会資料：公共交通に対する考え方」、2002 (http://www.mlit.go.jp/road/ir/kihon/sir74.pdf) 掲載のグラフを編集し、新たにグラフを作成した。
156	国土交通省「国土数値情報鉄道データ」(http://nlftp.mlit.go.jp/ksj/gml/datalist/KsjTmplt-N02-v2_2.html) を基に地図を作成した。
157	地図データはZmapTownII東京都データセット2008年版(株式会社ゼンリン)を使った。 「スーパーマーケット」の位置はiタウンページ(2015年、NTTタウンページ株式会社)を使った(http://itp.ne.jp/)。
158	地図データはZmapTownII新潟県データセット2008年版(株式会社ゼンリン)を使った。 「スーパーマーケット」の位置はiタウンページ(2015年、NTTタウンページ株式会社)を使った(http://itp.ne.jp/)。

134	See Author's Publications on Fibercity. 3J.
135	Photo: Ohno Hidetoshi, 2013.
136	Photo: Shibuya Tatsunori, 2014.
137	See Author's Publications on Fibercity. 25JE.
138	Ōta Memorial Museum of Art.
139	Photo: Moriya Kayo, 2015.
140	Fenollosa-Weld Collection © Museum of Fine Arts, Boston.
141	Kojiro Laboratory, Department of Architecture, School of Science and Technology, Meiji University (1975). (Reference is written in Japanese.)
142	Photo: Ohno Hidetoshi, 2014.
143	Photo: Oshima Kohei.
144	Photo: Mukaiyama Yuji, 2016.
145	Photo: Kubo Hideaki, 2016.
146	Photo: Kubo Hideaki, 2015.
147	Kamekura Yūsaku (art direction), Murakoshi Jō (photo direction) and Hayasaki Osamu (photography), *The Start of the Sprinters' Dash*.
148	Japan Department Stores Association, (2014). (Reference is written in Japanese.)
149	Photo: Ohno Hidetoshi, 2016.
150	Graph drawn with data adapted from: Research Institute of Construction and Economy (2014), http://www.rice.or.jp/regular_report/pdf/construction_economic_report/レポート全文/No.62/No.62.pdf. (Reference is written in Japanese.)
151	Graph drawn with data adapted from: Research Institute of Construction and Economy (2012), http://www.rice.or.jp/regular_report/pdf/construction_economic_report/レポート全文/No.59_201210.pdf. (Reference is written in Japanese.)
152	Le Corbusier et Pierre Jeanneret, W. Boesiger, O. Stonorov, *Le Corbusier et Pierre Jeanneret : Oeuvre Complète 1910-1929*, (Les Editions D'Architecture Erlenbach, 1948). © FLC / ADAGP, Paris & JASPAR, Tokyo, 2016 G0305
153	*The Meiji and Taisho Eras in Photographs*, Tokyo Volume, The National Diet Library.
154	Photo: Ohno Hidetoshi, 2016.
155	Subcommittee on Roads, Panel on Infrastructure Development, Japan Ministry of Land, Infrastructure, Transport and Tourism (2002), http://www.mlit.go.jp/road/ir/kihon/sir74.pdf. (Reference is written in Japanese.)
156	Map drawn with data adapted from: Japan Ministry of Land, Infrastructure, Transport and Tourism, http://nlftp.mlit.go.jp/ksj/gml/datalist/KsjTmplt-N02-v2_2.html. (Reference is written in Japanese.)
157	Maps based on data adapted from: ZmapTownII Tokyo Dataset (2008), distributed by Zenrin. Supermarket locations based on: iTownpage (2015), distributed by NTT TownPage Corporation, http://itp.ne.jp/. (Reference is written in Japanese.)
158	Maps based on data adapted from: ZmapTownII Niigata Dataset (2008), distributed by Zenrin. Supermarket locations based on: iTownpage (2015), distributed by NTT TownPage

159　厚生労働省『平成25年度介護保険事業状況報告（年報）』（www.mhlw.go.jp/topics/kaigo/osirase/jigyo/13/index.html）掲載の資料からグラフを作成した。

160　撮影：大野秀敏、2016

161　国土交通省都市局「都市における人の動き—平成22年全国都市交通特性調査集計結果から」、2012（http://www.mlit.go.jp/common/001032141.pdf）掲載の資料からグラフを作成した。

162　地方史研究協議会編『日本産業史大系4　関東地方篇』（東京大学出版会、1959）

163　総務省「平成20年住生活総合調査　表143　住まいに関する意向1」（https://www.e-stat.go.jp/SG1/estat/GL08020103.do?_toGL08020103_&tclassID=000001028375&cycleCode=0&requestSender=search）掲載の資料からグラフを作成した。

164　グーグルマップから事例を採取し、それらを実測した。

165　撮影：大野秀敏、2016

166　国立社会保障・人口問題研究所「日本の世帯数の将来推計（全国推計）」（2013年1月推計）（http://www.ipss.go.jp/pp-ajsetai/j/HPRJ2013/t-page.asp）掲載の資料からグラフを作成した。

167　ファイバーシティに関する著者による公表物 35JE、表紙

168　撮影：大野秀敏、2015

169　撮影：大野秀敏、2008

170　撮影：大野秀敏、2016

171　国土交通省「コンパクトシティの形成に向けて（未定稿）」2015（http://www.mlit.go.jp/common/001086645.pdf）

172　ファイバーシティに関する著者による公表物 4J

173　平成16年度都市計画基礎調査（長岡市）を基に2009年に大野研究室で実地調査をして地図を作製。

174　撮影：大野秀敏、2015

第二部

201　久隅守景画「夕顔棚納涼図屏風」東京国立博物館蔵
202　撮影：大野秀敏、2014
203　撮影：大野秀敏、1997
204　撮影：大野秀敏、2008
205　撮影：大野秀敏、2001
206　Fredrick Law Olmstead. "Riverside." http://www.riverside.il.usより転載

207　撮影：大野秀敏、2007
208　ル・コルビュジエ、ピエール・ジャンヌレ著、W. ボジガー、O. ストノロフ編『ル・コルビュジエ全作品集第二巻』吉坂隆正訳、A.D.A. EDITA (1979)。
　　© FLC / ADAGP, Paris & JASPAR, Tokyo, 2016 G0305

Corporation, http://itp.ne.jp/. (Reference is written in Japanese.)

159　Graph drawn with 2013 statistics data adapted from: Japan Ministry of Health, Labour and Welfare, www.mhlw.go.jp/topics/kaigo/osirase/jigyo/13/index.html. (Reference is written in Japanese.)

160　Photo: Ohno Hidetoshi, 2016.

161　Graph drawn with data adapted from: City Bureau, Japan Ministry of Land, Infrastructure, Transport and Tourism (2012), http://www.mlit.go.jp/common/001032141.pdf. (Reference is written in Japanese.)

162　Society of Local History, ed. (1959). (Reference is written in Japanese.)

163　Graph drawn with 2008 statistics adapted from: Japan Ministry of Internal Affairs and Communications, https://www.e-stat.go.jp/SG1/estat/GL08020103.do?_toGL08020103_&tclassID=000001028375&cycleCode=0&requestSender=search. (Reference is written in Japanese.)

164　This survey consisted of sampling and measuring examples selected from Google Maps.

165　Photo: Ohno Hidetoshi, 2016.

166　Graph drawn with 2013 statistics data adapted from: National Institute of Population and Social Security Research, http://www.ipss.go.jp/pp-ajsetai/j/HPRJ2013/t-page.asp. (Reference is written in Japanese.)

167　See Author's Publications on Fibercity. 35JE: cover.

168　Photo: Ohno Hidetoshi, 2015.

169　Photo: Ohno Hidetoshi, 2008.

170　Photo: Ohno Hidetoshi, 2016.

171　Japan Ministry of Land, Infrastructure, Transport and Tourism (2015), http://www.mlit.go.jp/common/001086645.pdf. (Reference is written in Japanese.)

172　See Author's Publications on Fibercity.4J.

173　Map drawn by Ohno Laboratory in 2009 based on field surveys and data adapted from:
　　Nagaoka City Government. (Reference is written in Japanese.)

174　Photo: Ohno Hidetoshi, 2015.

Part 2

201　Tokyo National Museum Archives.
202　Photo: Ohno Hidetoshi, 2014.
203　Photo: Ohno Hidetoshi, 1997.
204　Photo: Ohno Hidetoshi, 2008.
205　Photo: Ohno Hidetoshi, 2001.
206　Fredrick Law Olmstead. "Riverside, Illinois." http://www.fredericklawolmsted.com/riverside.html.
　　"Riverside." http://www.riverside.il.us
207　Photo: Ohno Hidetoshi, 2007.
208　Le Corbusier et Pierre Jeanneret, W. Boesiger, O. Stonorov, *Le Corbusier et Pierre Jeanneret : Oeuvre Complète 1929-1934*, (Les Editions D'Architecture Erlenbach, 1948).
　　© FLC / ADAGP, Paris & JASPAR, Tokyo, 2016 G0305

209	Photo: Kitajima Toshiharu, 2010.
210	Photo: Ohno Hidetoshi, 2015.
211	See Author's Publications on Fibercity. 24JE, 25J.
212	Photo: Ohno Hidetoshi, 2014.
213	Photo: Ohno Hidetoshi, 1985.
214	National Diet Library Digital Collection Archives.
215	Photo: Ohno Hidetoshi, 2016.
216	Kröller Müller Museum Archives.
217	Provided by Maki Fumihiko. Original drawing shown in: Fumihiko Maki, *Notes in Collective Form* (Washington University, 1964).
218	Photo: Ohno Hidetoshi, 2011.
219	Photo: Ohno Hidetoshi, 2011.
220	Photo: Ohno Hidetoshi, 2014.
221	See Author's Publications on Fibercity. 24JE, 25J.
223	Toshi Dezain Kenkyūkai ed. (1968). (Reference is written in Japanese.)
224	Lynch Kevin, *The Image of the City*, Figure 38, p. 147, ©1960 Massachusetts Institute of Technology, by permission of The MIT Press.
230	See Author's Publications on Fibercity. 100JE.
231	See Author's Publications on Fibercity. 100JE.
233	See Author's Publications on Fibercity. 101JE.
234	See Author's Publications on Fibercity. 100JE.
237	See Author's Publications on Fibercity. 25JE, 26J.
238	Japan Ministry of Economy, Trade and Industry, Agency for Natural Resources and Energy, Electricity and Gas Industry Department, Policy Planning Division, supervised and Japan Heat Supply Business Association, ed. (2010). (Reference is written in Japanese) Clean Authority of TOKYO, http://www.union.tokyo23-seisou.lg.jp/index.html (Reference is written in Japanese.)
242	See Author's Publications on Fibercity. 44J.
244	See Author's Publications on Fibercity. 25JE.
245	Photo: Ohno Hidetoshi, 2014.
249	Data adapted by Ohno Laboratory from: The Tokyo Disaster Prevention Council ed. *Tokyo Local Disaster Prevention Plan: Earthquakes* (revised 2003).
250	Simulation based on Disaster Mitigation Community Development Planning Support System, available from: http://www.bousai-pss.jp (Reference is written in Japanese.) See Author's Publications on Fibercity. 25JE, 26J.
251	See Author's Publications on Fibercity. 99JE.
252	See Author's Publications on Fibercity. 99JE.
253	See Author's Publications on Fibercity. 99JE.
254	See Author's Publications on Fibercity. 99JE.
255	See Author's Publications on Fibercity. 99JE.
256	See Author's Publications on Fibercity. 99JE.
260	Photo: Ohno Hidetoshi, 2014.
273	See Author's Publications on Fibercity. 104JE.
274	See Author's Publications on Fibercity. 49JE.
277	See Author's Publications on Fibercity. 47-104JE.

ファイバーシティに関する著者による公表物
Author's Publications on Fibercity

・冒頭の文献番号に付すアルファベットは、言語を示す。邦文、英文、仏文、独文、中文に対応して、J、E、F、G、Cを付す。二つ以上の言語で書かれた文献は対応する記号を重ねる。
・MPF PRESSは、大野研究室の出版組織である。MPFはMetropolis Forum の略称である。

・The letters next to the numbers stand for languages. J for Japanese, E for English, F for French, G for German and C for Chinese. Items written in more than two languages have been listed with multiple letters.
・MPF Press is the Ohno Laboratory's self-publishing organization. MPF is an acronym for Metropolis Forum.

論文/Articles

fc1E. Ohno Hidetoshi, "Fibercity – An Urban Reorganization Theory for an Age of Shrinkage." In *Cities in Transition: Power, Environment, Society* edited by Wowo Ding, Arie Graafland, Andong Lu. Rotterdam: nai010 publishers, 2015.

fc 2E. Ohno Hidetoshi, Interview with Leon Spita. "Hidetoshi Ohno TOKYO2050 FIBERCITY." In *A Japanese Anthology — Cutting-edge architecture* (Italian title: *Antologia Giapponese Un'architettura d'avanguardia*). Estero, Italy: Giagemi Editore, 2015.

fc 3J. 大野秀敏、佐藤和貴子、斉藤せつな『〈小さい交通〉が都市を変える—マルチ・モビリティ・シティをめざして』NTT出版、2015。

fc 4J. 和田夏子、大野秀敏「都市のコンパクト化のCO_2排出量評価—長岡市を事例とした都市のコンパクト化の評価に関する研究その1」日本建築学会環境系論文集、第76巻、第668号、2011、pp.935-941。和田夏子、大野秀敏「都市のコンパクト化の費用評価—長岡市を事例とした都市のコンパクト化の評価に関する研究その2」日本建築学会環境系論文集、第78巻、第687号、2013、pp.419-425。

fc 5E. Ohno Hidetoshi, "Fibercity as a Paradigm Shift of Urban Design." In lecture proceedings *The New Urban Question: Urbanism Beyond Neo-Liberalism* edited by Juregen Roseman, Lei Qu and Diego Sepulveda, 49-52. Delft: 2010.

fc 6E. Ohno Hidetoshi, "Fibercity – Designing for Shrinkage." In *Future Asian Space* edited by Limin Hee, Davisi Boontharm and Erwin Viray, 171-189. Singapore: NUS Press, 2010.

fc 7E. Ohno Hidetoshi, "Lecture by Ohno, Hidetoshi: Fibercity 2050 – Designing for Shrinkage." In Refabricating City: A Reflection Hong Kong-Shenzhen Bi-City Biennale of *Urbanism/Architecture* edited by Weijen Wang and Thomas Chung, 288-290. Hong Kong: 2010.

fc 8E. Ohno Laboratory, "Fibercity 2050: Designing for Shrinkage." In *Refabricating City: A Reflection, Hong Kong-Shenzhen Bi-City Biennale of Urbanism/Architecture* edited by Weijen Wang and Thomas Chung, 191-192. Hong Kong: 2010.

fc 9J. 大野秀敏「地方都市の魅力を実感できるまちづくり」、『ユニバーサルデザイン 健康都市 輝く都市から健康都市へ』株式会社ユーディ・シー、2010、pp. 76-79

fc 10J. 大野秀敏、福川裕一、藻谷浩介、伊藤俊介、饗庭伸「郊外の仕切り方—捨てずに縮小するための方法論を考える」、『建築雑誌 125(1603)』日本建築学会、2010、pp.26-33

fc 11J. 大野秀敏、伊藤友隆、天野裕「21世紀の地方都市の空間像の研究」、『財団法人住宅総合研究財団研究論文集No.36』財団法人住宅総合研究、丸善、2010、pp. 45-58

fc 12J. 大野秀敏(取材記事)「人口減少時代の都市ビジョンを考える」、『国土交通 特集：エコでコンパクトなまちへ』国土交通省、2010、p. 10

fc 13E. Ohno Hidetoshi, "Fibercity/ Designing for Shrinkage." In *Eco-Urbanity: Towards Well-Mannered Built Environment* edited by Darko Radovic, pp.79-91. London: Routledge, 2009.

fc 14J. 大野秀敏「21世紀の都市ビジョン」第29回住総研シンポジウム「縮小都市における居住」財団法人住宅総合研究財団、2009、pp. 23-28

fc 15J. 大野秀敏「縮小する大都市の未来像／ファイバーシティ2050」、『都市農地とまちづくり第59号』財団法人都市農地活用支援センター、2009、pp. 2-6

fc 16J. 大野秀敏、アバンアソシエイツ『シュリンキング・ニッポン—縮小する都市の未来戦略』鹿島出版会、2008

fc 17CJE. 大野秀敏「ファイバーシティ二〇五〇」団紀彦(編著)『東京論 東京の建築とアーバニズム』田園城市、台湾、2008、pp.133-168
Ohno Hidetoshi, "Fiber City 2050" In *Architecture and Urbanism of Tokyo* edited by Dan, Norihiko, pp.133-168. Taipei: Garden City Publishers, 2008.

fc 18F. Ohno Hidetoshi, "La Ville Fibre." In *Mobilité et écologie urbaine* edited by Alain Bourdin, pp.209-232. Paris: Descartes et Cie, 2007.

fc 19J. 大野秀敏「ファイバーシティ 縮小をデザインする」日本建築家協会環境行動委員会編『「2050年」から環境をデザインする—都市・建築・生活の再構築』彰国社、東京、2007、pp.18-49、50-58

fc 20J. 大野秀敏「fibercity/東京2050—縮小をデザインする」、『新都市 61(8)』都市計画協会、2007、pp.68-72 fc 21J. 大野秀敏、鵜飼哲矢、日高仁、山崎由美子、大竹慎、田中義之、井上慎也、田口佳樹、松宮綾子、和田夏子、秋山浩之、大島耕平「2050年の東京圏の都市像―縮小する時代の都市計画はどうあるべきか」、『住宅総合研究財団研究論文集No.33』財団法人住宅総合研究財団、2006、pp.135-146

fc 22J. 大野秀敏「縮小時代の都市ビジョン」、『毎日新聞』夕刊 2007年3月20日(47122号)文化欄、毎日新聞社、2007

fc 23J. 大野秀敏、六鹿正治(対談)「人口減少時代の都市デザイン―ファイバーシティ」、『LIVE ENERGY』vol. 83、東京ガス株式会社・都市エネルギー事業部、2007、pp12-15

fc 24J. 大野秀敏、フィリップ・オズワルト(対談)「shrinking cities x fibercity @ akihabara 縮小する都市に未来はあるか?」、『10＋1(テンプラスワン)』第46号、LIXIL出版、2007、pp.161-170

fc 25JE. 東京大学大野研究室(雑誌特集号責任編集)「ファイバーシティ東京2050」、『JA(The Japan Architect)』63号(全巻特集136頁)、新建築社、2006
Ohno Laboratory. "Fibercity Tokyo 2050." *The Japan Architect 63*, Autumn, 2006, full volume 136 pages.

fc 26J. 東京大学大野研究室「FIBER CITY 東京2050」、『新建築』第81巻7号2006年6月号、新建築社、2006、pp. 44-64

fc 27E. Ohno Hidetoshi, "Fiber City Tokyo." In *Shrinking Cities Volume 2 Interventions* edited by Philipp Oswalt, pp. 204-211. Ostfildern: Hatje Cantz Verlag, 2005.

fc 28G. Ohno Hidetoshi, "Faserstadt Tokio." In *Schrumpfende Städt Band 2* edited by Philipp Oswalt, pp. 204-211. Leipzig: Hndlungskonzepte, 2005.

fc 29J. 大野秀敏(取材記事)「縮小時代における都市ビジョンを探る」、『建築家』第205号、日本建築家協会(JIA)、2005、pp. 2-6

fc 30J. 大野秀敏「環境と人口減少の時代のモビリティ――ハイ・モビリティ、コンパクト、シームレス」、日本デザイン機構編『クルマ社会のリ・デザイン―近未来のモビリティへの提言』鹿島出版会、2004、pp.28-33

fc 31J. 大野秀敏「ひきざんいっかい たしざんいっかい」、伊藤公文・松永ця美編『時間都市―Chronopolis 時間のポリフォニーとしての都市像』東京電機大学出版局、東京、2003、pp.74-103

fc 32J. 大野秀敏(取材記事)「過剰な都市環境をいかに減らすか―減築で都市荒廃防ぐ」、『東京大学新聞』2003、p.3

fc 33E. Ohno Hidetoshi, "Tokyo Ring: Mobility as a Culture." In *Mobility: A Room with a View* edited by Francine Houben, Luisa Maria Calabrese 164-185. Rotterdam: Nai Publishers, 2003.

fc 34J. 大野秀敏「線分都市」、『建築雑誌』2002年11月号(VOL.117 NO.1495)、日本建築学会、2002、pp.24-25

fc 35JE. 東京大学大野研究室(雑誌特集号責任編集)「香港―超級都市 HongKong: Alternative Metropolis」、『SD』330号、鹿島出版会、1992、pp.5-84
Ohno Laboratory as a Guest Editor, *SD Special Issue*, "Hong Kong: Alternative Metropolis."no.330 (1992): *SD*, 5-84

fc 36JE. 大野秀敏「オルタナティヴ・メトロポリス香港」、『SD』330号、鹿島出版会、1992、pp.7-14
"Hong Kong: Alternative Metropolis." In *Hong Kong: Alternative Metropolis* edited by Ohno Lboratory pp. 7-14, Tokyo: Kajima Institute, 1992.

fc 37J. 大野秀敏「〈線分〉概念による現代都市の空間構造の研究」、『東京大学工学部紀要』A-27、東京大学、1989、pp.10-11

fc 38J. 大野秀敏「周縁に力がある―都市・東京の歴史的空間構造」、『建築文化』1985年8月、彰国社、1985、pp.78-82

展覧会/Exhibitions

fc 39GE. Ohno Laboratory, "Fibercity." Exhibition at "EASTERN PROMISES," Contemporary Architecture and Spatial Practices in East Asia, MAK – Museum of Applied Arts, Vienna, 2013.

fc 40EC. Ohno Laboratory, "Fibercity/Tokyo versin 2.0." Exhibition at "TRI-CIPROCAL the Cities the Time, the Place, the People," 2011-12 Bi-City Biennale of Urbanism/Architecture, (Hong Kong), Hong Kong, 2012.

fc 41J. 東京大学大野研究室「地方都市の素晴らしさを実感できる魅力的な住環境の提案」長岡市(主催)、長岡市、2011

fc 42J. 東京大学大野研究室「ファイバーシティ・シュリンキングシティ」、UIA大会関連展示『2050 EARTH CATALOGUE 展 "サスティナブル・シティ―2050年の都市のビジョン』東京、2011

fc 43J. 東京大学大野研究室「移動巡回型公共サービス」、『MINI 'RAIDING PROJECT Crossover architecture'』ビー・エム・ダブリュー・ジャパン株式会社、東京、2011

fc 44J. 東京大学大野研究室「緑の網」、『東京2050//12の都市ヴィジョン展』東京都東京文化発信プロジェクト室(公益財団法人東京都歴史文化財団)・東京2050//12の都市ビジョン展実行委員会、東京、2011
この展覧会は国際建築家連盟(UIA)東京大会の関連展覧会として企画され、展覧会実行委員会委員長を大野秀敏が務めた

fc 45FE. L'AUC and Hidetoshi Ohno et al. "Le Diagnostic Prospectif de L'aggloération Parisieene." Consultancy and Exhibition hosted by French Government, Paris, 2009.

fc 46J. 東京大学大野研究室「FIBERCITY / Tokyo 2050」、東京、2008
(展覧会カタログ)斎藤公男『アーキニアリング・デザイン展2008―テクノロジーと建築デザインの融合・進化』日本建築学会、2008、p.135

fc 47EC. Ohno Laboratory, "Fibercity: Designing for Shrinkage." Exhibition at "Refabricating City: A Reflection," Hong Kong-Shenzhen Bi-City Biennale of Urbanism/Architecture, Hong Kong, 2008.

fc 48JE. 東京大学大野研究室、フィリップ・オズワルト「shrinking cities x fibercites @ Akihabara　縮小する都市に未来はあるか?」SXF@A組織委員会主催、東京、2007
縮小都市の研究を幅広く展開しているドイツの建築家フィリップ・オズワルト氏と共同で開催し、展覧会（2007年1月から2月）、シンポジウム、トーク・インを行った。

Ohno Laboratory and Project Office Philipp Oswalt, "Shrinking Cities x Fibercites @ Akihabara", Exhibition, Symposium, Lectures, A joint venture of the shrinking cities project, Germany, Tokyo, 2007.

fc 49JE. 東京大学大野研究室「ファイバーシティ東京2050」サステーナブル建築世界会議SB05、国土交通省、2005
"Fibercity Tokyo 2050." Exhibition at the 2005 World Sustainable Building Conference hosted by the Ministry of Land, Infrastructure and Transport (MLIT), Tokyo, JP, 2005.

講演／Lectures

fc 50J. 大野秀敏「ファイバーシティ・第四の交通・CMA」日本建築学会大会地球環境部門PD、平塚、2015

fc 51J. 大野秀敏「流れと場所」国際高等研究所研究プロジェクト「ネットワークの科学」京都、2015

fc 52J. 大野秀敏「縮小の時代の住宅地の管理と運営」法政大学都市法・現代総有研究会まちづくり連続講座1「私たちはどのように21世紀のまちをつくっていくべきか―人口減少時代の到来。町がなくなる。安倍政権の「地方創生」の行方は」、東京、2015

fc 53J. 大野秀敏「これからの郊外住宅地―地域の管理・運営」町田市、2015

fc 54J. 大野秀敏「流れと場所」東京大学最終講義、柏、2015

fc 55E. Ohno Hidetoshi, "In a Gardener's Way." Keynote lecture at "Sustainable Society as Our Challenge," The First International Conference of IASUR, Kashiwa, 2014.

fc 56J. 大野秀敏「流れと場所」横浜国立大学講義、横浜、2014
（講義録）北山恒・大野秀敏他『都市のアーキテクチャー』横浜国立大学大学院都市イノベーション研究院Y-GSA、横浜市、2016

fc 57J. 大野秀敏「縮小の時代のための都市デザインのパラダイムシフト」、「ERES公開フォーラム　2020年以降の東京と日本―「大都市vs.地方」の二項対立を超えて」東京大学公共政策大学院、東京、2014

fc 58E. Ohno Hidetoshi, "Urban Change in the Future – Fibercity for the Shrinking Cities." Lecture at FORMEDIL/ANCE, Milan, 2014.

fc 59E. Ohno Hidetoshi, "Urban Change in the Future – Fibercity for the Shrinking Cities." Lecture at FORMEDIL/ANCE, Rome, 2014.

fc 60E. Ohno Hidetoshi, "Fibercity for the Shrinking Cities." Lecture at the Salon Suisse at the 14th International Architecture Exhibition hosted by Venice Biennale, Venice, 2014.

fc 61E. Ohno Hidetoshi, "Mobility with Urban Expansion and Shrinking." Lecture at "Power, Environment, Society: International Seminar on the Architecture of Accelerated Urbanization" hosted by Nanjing University and the University of Cambridge, School of Architecture and Urban Planning, Nanjing, 2014.

fc 62E. Ohno Hidetoshi, "Designing the City for the 21st Century." Keynote lecture at the 18th Forum on Land Use and Planning hosted by National Chung Kung University, Tainan, 2014.

fc 63J. 大野秀敏「都市の未来戦略」、「第23回住生活月間協賛・まちなみシンポジウム in 東京」、2011
（新聞報道）日本経済新聞社、東京　日本経済新聞（夕刊）、2011

fc 64J. 大野秀敏「東京の未来について」HOUSE VISION実行委員会、東京、2011

fc 65E. Ohno Hidetoshi, "Designing a Shrinking Society." Lecture at the Italy in Japan 2011 Program "Aging Society: From Scientific Technological Knowledge to New Market Opportunities," Tokyo, 2011.

fc 66J. 大野秀敏「縮小社会が描く都市モデル」、「JPN2.0 LIVE ROUNDABOUT JOURNAL」vol.14、東京、2011、配布資料、TEAMROUNDABOUT、2011、pp.2-15

fc 67E. Ohno Hidetoshi, "Fibercity Tokyo." Lecture at "Sustainable Cities Today: Inventing a New Urbanity for Tomorrow," international symposium hosted by Ministere de L'Ecologie du Development durable, des Transports et Lodgement, Paris, 2011.

fc 68J. 大野秀敏「縮小期の都市のグレードアップ」第34回行財政研修会東京セミナー「これからのまちづくり―地域再生の処方箋を求めて」社団法人地方行財政調査会、東京、2010
（講演録）「縮小期の都市のグレードアップ」社団法人地方行財政調査会、2010、pp.27-64

fc 69E. Ohno Hidetoshi, "Fibercity for Tokyo and Nagaoka." Lecture at the University of Auckland, School of Architecture and Planning, Auckland, NZ, 2010

fc 70J. 大野秀敏「もっと魅力的な長岡にむけての挑戦　環境都市長岡の実現に向けて」、「地球温暖化対策シンポジウム」長岡、2010

fc 71E. Ohno Hidetoshi, "Fibercity as a Paradigm Shift of Urban Design." Lecture at "The New Urban Question: Urbanism Beyond Neo-Liberalism," 4th Conference of the International Forum on Urbanism hosted by Technical University Delft, Delft, NL, 2009.

fc 72J. 大野秀敏「21世紀の都市ビジョン」第29回住総研シンポジウム「縮小都市における居住」財団法人住宅総合研究財団、東京、2009

fc 73J. 大野秀敏「縮小をデザインする ファイバーシティ／東京2050」財団法人都市農地活用支援センター／定期借地権推進協議会、平成20年度土地月間講演会「新しい都市農地の利用活用を考える」東京、2009

fc 74J. 大野秀敏「21世紀の都市モデルの構築」、「2050年の低炭素社会の理想都市像をさぐるシンポジウム」日本建築学会低炭素社会特別調査委員会、東京、2009

fc 75E. Ohno Hidetoshi, "Fibercity: Designing for Shrinkage." Lecture at "Great Asian Street Symposium: A Public Forum of Asian Urban Design, FUTURE | ASIAN | SPACE," 5th GASS 2008 hosted by the Department of Architecture, National University of Singapore, Singapore, 2008.

fc 76J. 大野秀敏「ファイバーシティの提案」、2007年度日本建築学会連続セミナー「縮退・成熟する都市の建築を考える」日本建築学会能力開発支援事業委員会、東京、2008

fc 77E. Ohno Hidetoshi, "Fibercity: Designing for Shrinkage." Lecture at "Refabricating City: A Reflection," Hong Kong-Shenzhen Bi-City Biennale of Urbanism/Architecture Hong Kong, 2008

fc 78E. Ohno Hidetoshi, "Fibercity: Designing for Shrinkage." Lecture at "Eco-Urbanity Towards Well-Mannered Built Environment," international symposium, Tokyo, JP, 2007.

fc 79J. 大野秀敏「TOKYO 2050 fibercity—縮小する都市のための都市デザイン戦略」ユビキタス建築都市特別研究委員会、日本建築学会大会（九州）特別研究部門（ユビキタス）PD「21世紀の電脳都市論」福岡、2007

fc 80J. 大野秀敏、吉見俊哉（対談）「吉見俊哉×大野秀敏：縮小社会の都市像を語る」、アキバテクノクラブ第5回「ATCコミュニケーションセミナー」株式会社クロスフィールドマネジメント、東京、2007

fc 81E. Ohno Hidetoshi, "Fibercity: Designing for Shrinkage." Lecture at The University of Sydney, Sydney, AU, 2007.

fc 82E. Ohno Hidetoshi, "Fibercity: Designing for Shrinkage." Lecture at Royal Melbourne Institute of Technology University, Melbourne, AU, 2007.

fc 83J. 大野秀敏「人口減少時代のまちづくり戦略」NPO法人日本都市計画家協会第6回通常総会特別講演、NPO法人日本都市計画家協会、東京、2007

fc 84E. Ohno Hidetoshi, "Fibercity: Designing for Shrinkage." Lecture at Sint-Lucas Higher Institute of Architecture, Gent, BE, 2007.

fc 85E. Ohno Hidetoshi, "Fibercity Tokyo 2050." Lecture at the 2006 Biohousing Symposium hosted by Biohousing Research Institute, Chonnam University, Gwangju, KR, 2007

fc 86E. Ohno Hidetoshi, "Fibercity Tokyo." Lecture at IBA Urban Redevelopment 2010 Conference hosted by IBA Büro, Dessau, DE, 2006.

fc 87J. 大野秀敏「fibercity/東京2050—縮小をデザインする」、国土交通省内部の都市・地域整備局勉強会「人口減少等に対応した新たな都市計画制度の展望」に関する講演会、東京、2007

fc 88J. 大野秀敏「縮小時代の都市風景」、日本建築家協会、「建築家大会2005東海：素の力—川が育んだ暮らしと文化」、岐阜、2005

fc 89E. Ohno Hidetoshi, "Mobility is a Culture." Lecture at "Mobility and Metropolis," 3rd International Contemporary Art Experts Forum – ARCO'05 hosted by ARCOmadrid, Madrid, ES, 2005.

fc 90E. Ohno Hidetoshi, "Towards the Fiber City." Lecture at "Common Grounds – Innovation in Architecture and Urban Planning in the Netherlands and Japan," international symposium hosted by The University of Tokyo and The Dutch Embassy, Tokyo, JP, 2004.

fc 91J. 大野秀敏「居住とモビリティ」、「環境と人口減少の時代のモビリティ—ハイ・モビリティ、コンパクト、シームレス」日本デザイン機構、東京、2003

fc 92J. 大野秀敏「変革時代の空間論／縮小社会の都市計画」東京大学先端まちづくり学校、東京、2002

fc 93E. Ohno Hidetoshi, "The Big Cities in Japan: The Model of the High Mobility and the Compact City." Lecture at "Making the City with Flows – Managing Places of Interchange and Architecture of Mobility," international seminar hosted by Institut pour la ville en mouvement, Paris, FR, 2004.

fc 94E. Ohno Hidetoshi, "Intensive and Mixed-Use Projects in Tokyo." Lecture at "Smart Management of Inside and Outside Urban Growth," symposium hosted by the Planning Department of the University of Amsterdam, Amsterdam, NL, 2003.

fc 95E. Ohno Hidetoshi, "From the Age of Addition to the Age of Subtraction – A Paradigm Shift in Architecture Caused by the Decrease of Population in the Asian-Pacific Region." Lecture at the Fifth International

Symposium on Asia Pacific Architecture hosted by the University of Hawaii, Honolulu, HI, 2003.

fc 96J. 大野秀敏「21世紀の都市の再編成のための戦略」日本建築学会大会（関東）研究協議会、日本建築学会、2001

研究室報／Lab Booklets

fc 97JE. 東京大学大野研究室『ファイバーシティー理論』MPF PRESS、2012
Ohno Laboratory, *Fibercity Theory*. Kashiwa: MPF Press, 2012.

fc 98JE. 東京大学大野研究室『都市のダイエット』MPF PRESS、2012
Ohno Laboratory, *Dieting Cities*. Kashiwa: MPF Press, 2012.

fc 99JE. 東京大学大野研究室『暖かい網』MPF PRESS、2012
Ohno Laboratory, *Orange Web*. Kashiwa: MPF Press, 2012.

fc 100JE. 東京大学大野研究室『暖かい食卓』MPF PRESS、2012
Ohno Laboratory, *Orange Tables*. Kashiwa: MPF Press, 2012.

fc 101JE. 東京大学大野研究室『暖かい巡回』MPF PRESS、2012
Ohno Laboratory, *Orange Rounds*. Kashiwa: MPF Press, 2012.

fc 102JE. 東京大学大野研究室『アーバンリンクル都市の皺』MPF PRESS、2012
Ohno Laboratory, *Urban Wrinkle*. Kashiwa: MPF Press, 2012.

fc 103JE. 東京大学大野研究室『地方都市の素晴らしさを実感できる魅力的な住環境の提案　長岡におけるケーススタディ2009』MPF PRESS、2010
Ohno Laboratory, *Proposals for Attractive Living Environments Designed to Experience the Splendor of the Regional Cities 2009: A Case Study in Nagaoka*. Kashiwa: MPF Press, 2010.

fc 104JE. 東京大学大野研究室『地方都市の素晴らしさを実感できる魅力的な住環境の提案　長岡におけるケーススタディ2008』MPF PRESS、2009
Ohno Laboratory, *Proposals for Attractive Living Environments Designed to Experience the Splendor of the Regional Cities 2008: A Case Study in Nagaoka*. Kashiwa: MPF Press, 2009.

fc 105JE. 東京大学大野研究室『ファイバーシティにむけて　持続可能な都市形態の研究』MPF Press、2004
これは、ロッテルダム建築都市ビエンナーレ出展記録である。
Ohno Laboratory, *Towards the Fiber City: An Investigation of Sustainable City Form*. Tokyo: MPF Press, 2004.

ファイバーシティ　縮小の時代の都市像

2016年8月25日　初　版

［検印廃止］

著者　大野秀敏＋MPF

発行所　一般財団法人　東京大学出版会
代表者　古田元夫

153-0041　東京都目黒区駒場4-5-29
http://www.utp.or.jp
電話 03-6407-1069　Fax 03-6407-1991
振替 00160-6-59964
装丁　矢萩喜從郎
印刷所　株式会社精興社
製本所　誠製本株式会社

© 2016 OHNO Hidetoshi and MPF
ISBN978-4-13-066855-2

JCOPY <(社)出版者著作権管理機構 委託出版物>
本書の無断複写は著作権法上での例外を除き禁じられております．複写される場合は，そのつど事前に，(社)出版者著作権管理機構（電話03-3513-6969, FAX 03-3513-6979, e-mail:info@jcopy.or.jp）の許諾を得てください．